Lecture Notes in Networks and Systems

Volume 199

AF806719

The series "Lecture Notes in Networks and Systems" publishes the latest developments in Networks and Systems—quickly, informally and with high quality. Original research reported in proceedings and post-proceedings represents the core of LNNS.

Volumes published in LNNS embrace all aspects and subfields of, as well as new challenges in, Networks and Systems.

The series contains proceedings and edited volumes in systems and networks, spanning the areas of Cyber-Physical Systems, Autonomous Systems, Sensor Networks, Control Systems, Energy Systems, Automotive Systems, Biological Systems, Vehicular Networking and Connected Vehicles, Aerospace Systems, Automation, Manufacturing, Smart Grids, Nonlinear Systems, Power Systems, Robotics, Social Systems, Economic Systems and other. Of particular value to both the contributors and the readership are the short publication timeframe and the world-wide distribution and exposure which enable both a wide and rapid dissemination of research output.

The series covers the theory, applications, and perspectives on the state of the art and future developments relevant to systems and networks, decision making, control, complex processes and related areas, as embedded in the fields of interdisciplinary and applied sciences, engineering, computer science, physics, economics, social, and life sciences, as well as the paradigms and methodologies behind them.

Indexed by SCOPUS, INSPEC, WTI Frankfurt eG, zbMATH, SCImago.

All books published in the series are submitted for consideration in Web of Science.

More information about this series at http://www.springer.com/series/15179

Mustapha Reda Senouci ·
Mohamed El Yazid Boudaren ·
Faouzi Sebbak · M'hamed Mataoui
Editors

Advances in Computing Systems and Applications

Proceedings of the 4th Conference on Computing Systems and Applications

Springer

Editors
Mustapha Reda Senouci
Ecole Militaire Polytechnique
Algiers, Algeria

Mohamed El Yazid Boudaren
Ecole Militaire Polytechnique
Algiers, Algeria

Faouzi Sebbak
Ecole Militaire Polytechnique
Algiers, Algeria

M'hamed Mataoui
Ecole Militaire Polytechnique
Algiers, Algeria

ISSN 2367-3370 ISSN 2367-3389 (electronic)
Lecture Notes in Networks and Systems
ISBN 978-3-030-69417-3 ISBN 978-3-030-69418-0 (eBook)
https://doi.org/10.1007/978-3-030-69418-0

This Springer imprint is published by the registered company Springer Nature Switzerland AG
The registered company address is: Gewerbestrasse 11, 6330 Cham, Switzerland

Preface

We are pleased to present this collection of scholarly articles from the fourth edition of the Computing Systems and Applications (CSA 2020) conference held at the *Ecole Militaire Polytechnique*, Algiers, Algeria, during December 14, 2020. It succeeded the third edition (CSA 2018) that was held at the same place in April 2018. The main aim of this conference is to bring together academics, engineers, and students to exchange and promote excellence across computer science disciplines.

For this edition, the organizing committee has received 127 submissions. After screening, 108 eligible contributions were independently reviewed, typically by 3 to 5 members of the international program committee. Based on the referees' reports, 32 papers have been accepted, out of which 20 were presented as oral talks and 12 as posters, leading to an acceptance rate of 25%. One paper was awarded the "CSA 2020 best paper award" based on the scientific merits and the technical presentation as assessed by the session chairs. We are delighted to notice the good quality of the papers submitted to this edition of the conference, and we hope they will inspire readers in their future works.

On behalf of the organizing committee, we would like to thank the members of the technical program committee and external reviewers. They have worked hard to assess received papers and make valuable suggestions for the authors to improve their work. We would also like to give special thanks to the proceedings contributors, invited speakers, and session chairs for their valuable contributions to make this edition successful in spite of the circumstances related to the COVID-19 pandemic.

This preface cannot be ended without a warm thank you to all of the participants attending this edition; with your continued support and active participation, the conference will continue its success for a long time. We look forward to meeting you again in future editions. Finally, we hope that the readers will find these proceedings interesting and inspiring.

December 2020

Mustapha Reda Senouci
Faouzi Sebbak
M'hamed Mataoui
Mohamed El Yazid Boudaren

Organization

Conference General Chairs

Faouzi Sebbak	EMP, Algeria
M'hamed Mataoui	EMP, Algeria

Publication Chairs

Mustapha Reda Senouci	EMP, Algeria
Mohamed El Yazid Boudaren	EMP, Algeria

Website Chair

Salim Khamadja	EMP, Algeria

Local Organizing Committee

Khoutir Bouchbout	EMP, Algeria
Abdelhamid Boukabou	EMP, Algeria
Kadda Beghdad Bey	EMP, Algeria
Yacine Amara	EMP, Algeria
Zakaria Sahraoui	EMP, Algeria
Abdelhalim Zaidi	EMP, Algeria
Mohamed Maiza	EMP, Algeria
Tayeb Kenaza	EMP, Algeria
Sofiane Bouznad	EMP, Algeria
Redha Benaissa	EMP, Algeria
Salim Khamadja	EMP, Algeria
Abdenebi Rouigueb	EMP, Algeria
Ali Yachir	EMP, Algeria
Malek Nadil	EMP, Algeria

Rafik Mecibah	EMP, Algeria
Lamia Sadeg	EMP, Algeria
Oualid Demigha	EMP, Algeria
Ahmed Habbouchi	EMP, Algeria
Mohamed Boudali	EMP, Algeria
Sid Ali Hasni	EMP, Algeria
Mohamed El-Arbi Djebbar	EMP, Algeria
Billal Merabti	EMP, Algeria
Abdenour Amamra	EMP, Algeria
Badis Djamaa	EMP, Algeria
Walid Cherifi	EMP, Algeria
Akram Benatia	EMP, Algeria
Said Yacine Boulahia	EMP, Algeria
Abdelhamid Boudane	EMP, Algeria

Program Committee

Djekoune A. Oualid	CDTA, Algeria
Ali Abbassene	EADN, Algeria
Mourad Adnane	ENP, Algeria
Rachid Ahmed-Ouamer	University of Tizi-Ouzou, Algeria
Amar Aissani	USTHB, Algeria
Djamil Aissani	University of Bejaia, Algeria
Yamine Ait Ameur	IRIT/INPT-ENSEEIHT, France
Karima Akli-Astouati	USTHB, Algeria
Riad Akrour	TU Darmstadt, Germany
Zaia Alimazighi	USTHB, Algeria
Abdenour Amamra	EMP, Algeria
Yacine Amara	EMP, Algeria
Idir Amine Amarouche	USTHB, Algeria
Abdelmalek Amine	University of Saida, Algeria
Yacine Amirat	UPEC, France
Saliha Aouat	USTHB, Algeria
Ismail Badache	INSPE Aix-Marseille University, France
Riyadh Baghdadi	MIT, USA
Abdelghani Bakhtouchi	EMP, Algeria
Fairouz Beggas	EMP, Algeria
Kadda Beghdad Bey	EMP, Algeria
Mahmoud Belhocine	CDTA, Algeria
Abdelkader Bellarbi	CDTA, Algeria
Adel Belouchrani	ENP, Algeria
Redha Benaissa	EMP, Algeria
Karima Benatchba	ESI, Algeria
Akrem Benatia	EMP, Algeria

Fatima Benbouzid-Si Tayeb	ESI, Algeria
Djamal Benslimane	Lyon 1 University, France
Sidi Mohamed Benslimane	ESI-SBA, Algeria
Fadila Bentayeb	Lyon University, France
Chafika Benzaid	Aalto University, Finland
Sid Ahmed Berrani	ENP, Algeria
Salim Bitam	University of Biskra, Algeria
Abdelmadjid Bouabdallah	UTC, France
Khoutir Bouchbout	EMP, Algeria
Abdelhamid Boudane	EMP, Algeria
Mohamed El Yazid Boudaren	EMP, Algeria
Malika Boukala-Ioualalen	USTHB, Algeria
Kamel Boukhalfa	USTHB, Algeria
Yacine Boulahia	EMP, Algeria
Omar Boussaid	University of Lyon, France
Rachid Chalal	ESI, Algeria
Walid Cherifi	EMP, Algeria
Abdelghani Chibani	UPEC University, France
Azeddine Chikh	University of Tlemcen, Algeria
Samir Chouali	University of Franche-comté, France
Fabio Cuzzolin	Oxford Brookes University, UK
Bernard De Baets	Ghent University, Belgium
Walid Demigha	EMP, Algeria
Badis Djamaa	EMP, Algeria
Yassine Djouadi	USTHB, Algeria
Habiba Drias	USTHB, Algeria
Mohamed Amine Ferrag	Guelma University, Algeria
Abdessamed Réda Ghomari	ESI, Algeria
Emmanuel Grolleau	University of Poitiers, France
Ahmed Guessoum	USTHB, Algeria
Mohamed Guiatni	EMP, Algeria
Mohand-Said Hacid	University of Lyon 1, France
Allel Hadjali	LIAS/ENSMA, France
Abderrahim Halimi	Heriot Watt University, UK
Mahfoud Hamidia	USTHB, Algeria
Walid-Khaled Hidouci	ESI, Algeria
Rabah Imache	UMBB, Algeria
Samir Kechid	USTHB, Algeria
Tayeb Kenaza	EMP, Algeria
Faiza Khellaf	USTHB, Algeria
Selma Khouri	ESI, Algeria
Assia Kourgli	USTHB, Algeria
Nasreddine Lagraa	University of Laghouat, Algeria
Slimane Larabi	USTHB, Algeria
Mark Leeson	University of Warwick, UK

Rabah Louali	EMP, Algeria
Khaled Loukhaoukha	CRD-CGN, Algeria
Ramdane Maamri	University of Constantine 2, Algeria
M'hamed Mataoui	EMP, Algeria
Oussama Mazari Abdessameud	EMP, Algeria
Samir Medjiah	LAAS-CNRS, France
Billal Merabti	EMP, Algeria
Fatiha Merazka	USTHB, Algeria
Mezghiche Mohamed	UMBB, Algeria
Azzaz Mohamed Salah	EMP, Algeria
Boughanem Mohand	IRIT, France
Aïcha Mokhtari	USTHB, Algeria
Samira Moussaoui	USTHB, Algeria
Senouci Mustapha Reda	EMP, Algeria
Hassina Nacer	USTHB, Algeria
Hassiba Nemmour	USTHB, Algeria
Abdelkrim Nemra	EMP, Algeria
Boubakr Nour	Beijing Institute of Technology, China
Wojciech Pieczynski	Telecom Sudparis, France
Mubashir Husain Rehmani	Cork Institute of Technology, Ireland
Abdenebi Rouigueb	EMP, Algeria
Lamia Sadeg	EMP, Algeria
Said Sadoudi	EMP, Algeria
Zakaria Sahraoui	EMP, Algeria
Lakhdar Sais	CRIL, France
Faouzi Sebbak	EMP, Algeria
Benaoumeur Senouci	ECE-Paris, France
Sérgio Silva	ISEL, Portugal
Mohammad Siraj	King Saud University, KSA
Hui-Yin Wu	Inria, France
Ali Yachir	EMP, Algeria
Abdelhalim Zaidi	EMP, Algeria
Khalil Zebbiche	CRD-CGN, Algeria
Nadia Zenati	CDTA, Algeria
Irki Zohir	EMP, Algeria

Contents

Operational Research and Optimization

IoT, Networks, and Web Technologies

Machine Learning and Data Science

3D Face Identification Using HOG Features and Collaborative Representation

Ahmed Yassine Boumedine[1(✉)], Samia Bentaieb[1,2], Abdelaziz Ouamri[1], and Abderrahmane Mallek[1]

[1] Laboratory Signals and Images LSI, USTO-MB, Oran, Algeria
ahmdiassine@gmail.com, ouamri@yahoo.com, abderrahmane.mallek@outlook.com
[2] The University Center Belhadj Bouchaib, Ain Témouchent, Algeria
bentaiebsamia01@yahoo.fr

Abstract. Scientists have attempted to enhance face recognition accuracy by using 3D images to solve pose variability, lighting issues, and occlusion; but the applicability of these approaches is insufficient in realistic applications due to the slowness and high cost of 3D sensors. In this paper, we developed a face recognition process for low-cost and low-quality Microsoft Kinect sensor based on the Histogram of Oriented Gradient (HOG) descriptor. After locating the nose tip, the face is extracted and filtered to reduce the noise, the feature vectors are extracted by first computing the shape index map and then applying the HOG descriptor. For identification, the Collaborative Representation Classifier (CRC) is used. We compare our approach results with previous ones obtained on the publicly available CurtinFaces database, achieving rank one recognition rates of 99.87% and 82.53% when testing against expression and pose variations respectively.

Keywords: Face recognition · Kinect · HOG · Shape index · Collaborative representation

1 Introduction

Face recognition (FR) is an important technology in computer vision with its non-intrusive properties, particularly with the need for accurate identification /authentication systems. In recent years, the biometrics research community has been very diligent in recognizing facial expressions obtained in an uncontrolled environment. Most of the systems use 2D intensity images, but it has some drawbacks, including facial expressions and its high sensitivity to posture and light changes. 3D FR can solve the shortcomings of 2D structures by introducing three-dimensional (3D) imaging technology. Using the shape of the face and depth information instead of color details has proved to be efficient and reliable Under different conditions of lighting and posture. The disadvantages of current 3D scanners capable of capturing high-resolution 3D data are typically

M. R. Senouci et al. (Eds.): CSA 2020, LNNS 199, pp. 3–13, 2021.
https://doi.org/10.1007/978-3-030-69418-0_1

their slowness and high cost. Today, with the success of high-speed and low-cost 3D scanners such as Kinect, information on depth has become easily acquired and useful for face recognition. Figure 1 offers a comparison of face images with Minolta VIVID 910 scanner (top) versus Microsoft Kinect Sensor (low).

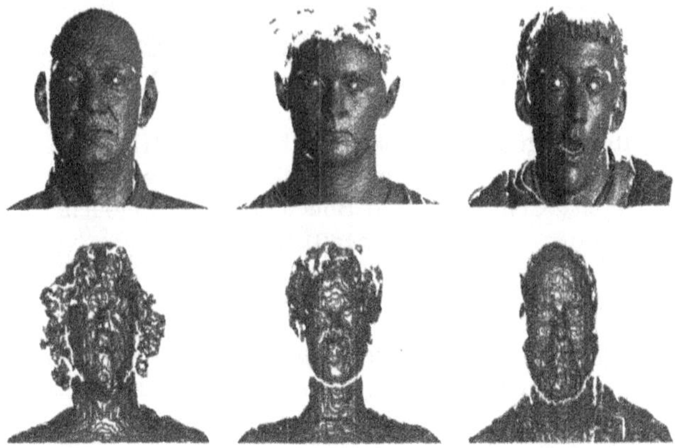

Fig. 1. 3D scans acquired with Minolta from FRGC dataset [1] (top row) and Kinect sensor from CurtinFaces database [2] (low row).

Several approaches have been developed for low-cost and low-quality 3D sensors. Ciaccio et al. [3] proposed a face representation that combines the Local Binary Patterns (LBP) features with covariance descriptor. Goswami et al. proposed in [4] a face recognition algorithm based on the HOG features extracted from the entropy and saliency maps of the RGB-D faces followed by a Random Decision Forest (RDF) classifier. In [2], Li et al. exploited the facial symmetry at the 3D point cloud and the usefulness of Sparse Representation Classifier (SRC) for face recognition. In [5], the Histogram of Oriented Gradient (HOG) features are extracted from three different Regions of Interest (ROI) of facial image. In [6], the authors fused two approaches, Attributes based on Depth Map (ADM) and a descriptor based on Saliency and Entropy (RISE) for face recognition. In [7], the authors explored the usefulness of the Histogram of Averaged Oriented Gradients (HAOG) and 3DLBP fusion. Boutella et al. [8] evaluated the performance of four descriptors: Binarized Statistical Image Features (BSIF), HOG, local phase quantization (LPQ) and LBP when used for learning ethnicity, gender, and identity. Azakhnini et al. [9] studied effectiveness of each facial part when used for gender, ethnicity and emotional state recognition tasks. Nourbakhsh et al. in [10], evaluated the performance of LBP, 3DLBP and HOG descriptors when combined with the Support Vector Machine (SVM) classifier.

Several remarks can be made from the approaches mentioned above. The first is the use of RGB and depth images as inputs. Exploring the complementary information of both modalities can yield more accurate recognition, but increases computational and time costs. The second concerns the extraction and the fusion of different kinds of descriptors to benefit of their strengths and achieve better results.

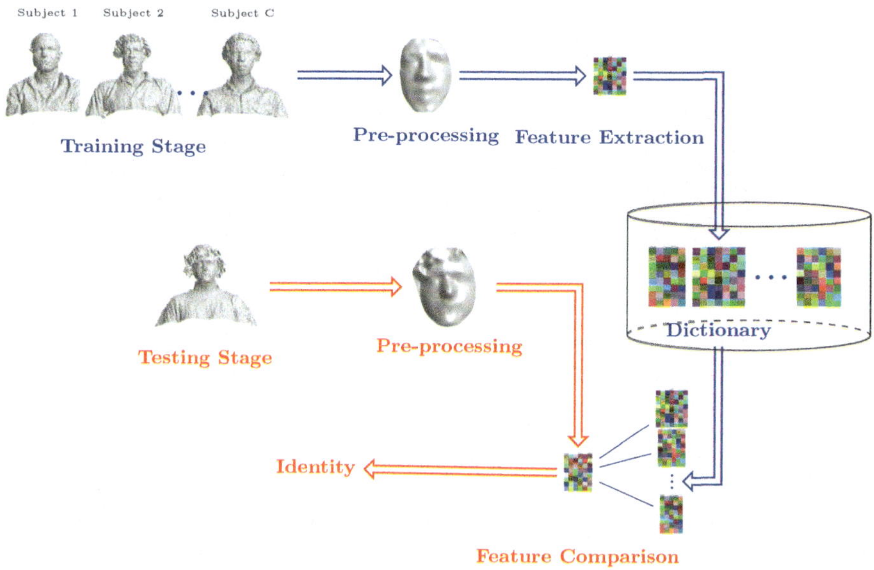

Fig. 2. The proposed approach diagram

2 Collaborative Representation for Face Recognition

In this paper, we investigate the potential use of the depth information provided by low-cost device using one kind of feature. To achieve this aim, we propose and evaluate a three-dimensional (3D) method for face identification, as shown in Fig. 2, which extracts HOG descriptors from the shape index representation of the preprocessed depth maps, and a Collaborative Representation Classifier (CRC) is used to find the best match between a probe face and the dictionary built from all the gallery descriptors. These stages are detailed in the next subsections.

2.1 Preprocessing

The depth maps of the Kinect sensor are known to have a high level of noise that affects negatively the recognition process. Also, the scans cover not only the face but also unnecessary parts, such as the chest, shoulders, neck, and clothing, as shown in Fig. 3a.

At the face detection step, the nose tip is located then the useful part is extracted using a sphere (see Fig. 3b) centered at the tip of the nose as shown in Fig. 3c. Pose normalization is important to improve the performance of the recognition. At this step, a face with a frontal pose is chosen as reference and the Iterative Closet Point (ICP) algorithm [11] is used to keep all the faces at the same pose.

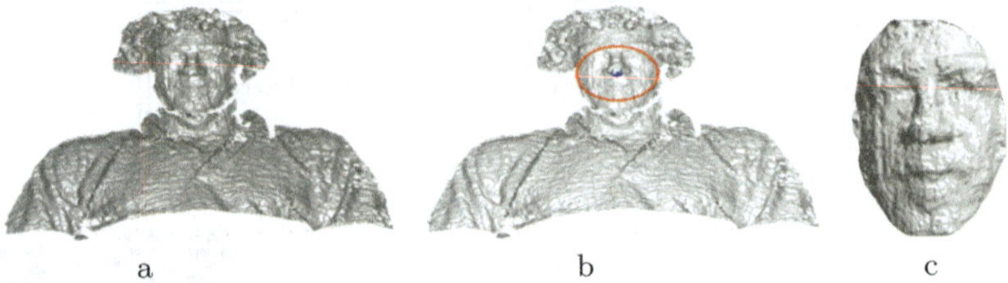

Fig. 3. (a) sample from CurtinFaces dataset, (b) face cropping using a sphere of radius 90 mm around the nose tip, (c) the cropped face.

Eliminating noise is an inevitable step, particularly while processing the Kinect sensor data. In our approach, we use a simple, popular but powerful tool for smoothing noisy 3D shapes. It consists of applying the mean filter by investigating the neighborhood of each point and replacing its depth value with the average depth values of the points in that neighborhood. Finally, the holes are filled using a 2D interpolation algorithm. Figure 4 shows an example of a noisy facial surface and its smoothed version.

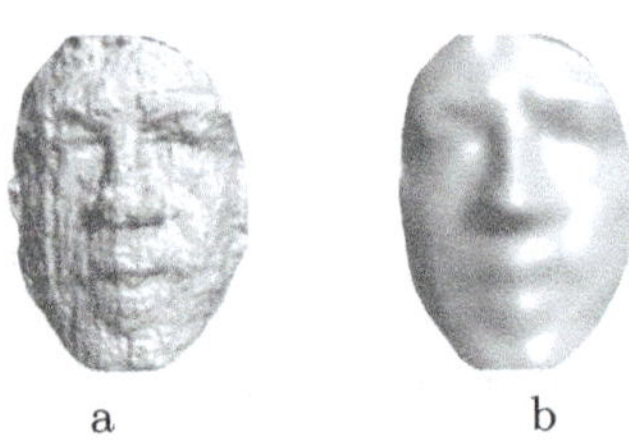

Fig. 4. Face denoising: (a) facial surface, (b) denoised facial surface.

2.2 Shape Index Computation

Curvature is a property that calculates how the surface changes. It has been frequently used for facial recognition [12–15] and facial feature detection [16,17].

One of the most common methods for characterizing a 3D surface is the extraction of the principal curvatures.

Given a 3D facial surface S, a curve is formed by the intersection of the surface and the normal plane in a tangent direction t at a point P.

The curvature of this planar curve is the normal curvature κ_n. Rotating the normal plane will yield to different curvatures; the maximum value and the minimum value of these curvatures form the principal curvatures κ_1 and κ_2, respectively.

Different approaches were proposed to calculate curvatures. In this paper, we used a quadratic patch [18]. First, the local neighborhood points in radius r around a point P so that the normal of the current point is aligned with the Z-axis, then, a least-square problem

$$f(x, y) = ax^2 + by^2 + cxy + dx + ey + f, \tag{1}$$

is solved. The eigenvectors and eigenvalues of the Hessian matrix are used to calculate the principal curvatures.

The Shape Index (SI), as proposed by Koenderink et al. [19], describes surface attributes of shapes. The shape index value at point P is a function of the principal curvatures:

$$SI(P) = \frac{1}{2} - \frac{1}{\pi} \arctan \frac{\kappa_1(P) + \kappa_2(P)}{\kappa_1(P) - \kappa_2(P)}, \tag{2}$$

As the principal curvatures are invariant to translation, rotation and scale so is the shape index.

The parameter r is quite important for the process. If its value is too small, the curvature value will be local, and the map will not be representative enough for the shape analysis. If its value is too big, a great part of the details will be lost. In this paper, this parameter is set to 15 mm [15].

Once the shape index is computed for every point of the facial surface, it is mapped into a range image as shown in Fig. 5.

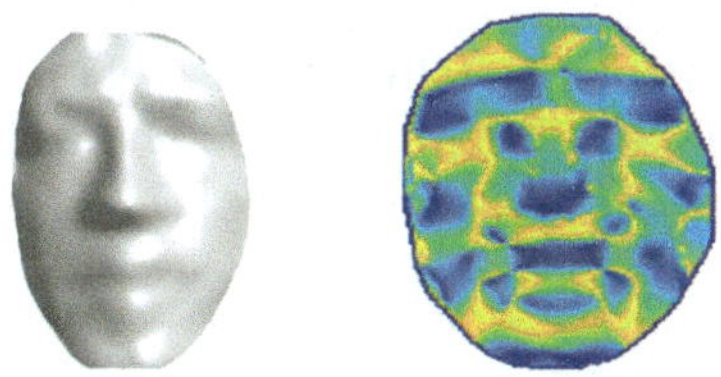

a b

Fig. 5. An illustration of the shape index map of 3D facial surface.

a b c d e

Fig. 6. Examples of samples from the CurtinFaces dataset used for training: (a) $-60°$ yaw, (b) $-30°$ yaw, (c) frontal, (d) $+30°$ yaw, (e) $+60°$ yaw.

2.3 The Histogram of Oriented Gradient (HOG) Descriptors

The basic idea is that shape index range image can be described by the selection of localized intensity gradients or edge orientation. In fact, this is achieved by splitting the image into smaller regions ("8×8 cells") [20], on each cell the gradient orientation and the regional contrast is determined using only a standard filter $[-1, 0, 1]$. If dx and dy are the derivatives of each pixel, then the gradient orientation is $arctan(dx, dy)$ and the contrast is $\sqrt{(dx)^2 + (dy)^2}$.
For each cell, a histogram vector of gradient directions is accumulated. These histogram vectors are concatenated to form the feature vector.

2.4 Face Identification

Gallery Dictionary Construction. Suppose the gallery consists of C persons, and for each person, k samples are used for training. Their corresponding N-dimensional HOG vector features, denoted $V_{c1}, V_{c2}, ...V_{ck}$, are computed and concatenated to build the gallery sub-dictionary of the person c as:

$$D_c = [V_{c1}, V_{c2}, ...V_{ck}], \tag{3}$$

The sub-dictionaries of all the C persons are built in the same manner and gathered to construct the gallery dictionary:

$$D = [D_1, D_2, ...D_C] \in \mathbb{R}^{N \times K}, \tag{4}$$

where $K = k \times C$ represents the total number of HOG feature vectors in the gallery dictionary.

Collaborative Representation Classifier. Let $y \in \mathbb{R}^N$ the HOG feature vector of a probe face, y can be expressed as a linear combination of all the HOG feature vectors in the gallery dictionary [21], as follows:

$$y = Dx_0 \tag{5}$$

Since the solution of Eq. 5 is not unique, to make the vector coefficient x_0 useful for identification, an L_2-norm sparsity term is introduced to x_0 to obtain:

$$x_0 = \underset{x}{argmin} \left\{ \|y - Dx\|_2^2 + \lambda \|x\|_2^2 \right\} \tag{6}$$

where λ is a scalar weight, the Eq. 6 has a closed-form solution as:

$$x_0 = \left(D^T D + \lambda I\right)^{-1} D^T y. \tag{7}$$

where I is an identity matrix, let $A = \left(D^T D + \lambda I\right)^{-1} D^T$, since A is independent of y , it can be pre-computed entirely from the gallery dictionary.
Then the residuals are computed:

$$Res_c(y) = \|y - D\delta_c(x_0)\|_2, \tag{8}$$

where $\delta_c(x_0)$ entries are almost zero except the ones affiliated with person c. To determine the identity of the probe face, we use:

$$identity(Y) = \underset{c}{argmin}\{Res_c(y)\} \tag{9}$$

3 Experimental Results and Discussion

The proposed approach is implemented using a MATLAB code on Intel (R) core(TM) i5-2430M CP@ 2.40 GHz Processor using CurtinFaces database [2]. It contains over 5000 images of 52 subjects in both RGB and depth maps obtained by Kinect sensor with variations in pose, illumination, facial expression, and sunglasses disguise.

To evaluate the proposed system, the identification (one-to-many) mode is performed. This mode determines the identity of a probe face and Rank-1 Recognition Rate (R1RR) and Cumulative Match Characteristic (CMC) curves are used to report its accuracy. Under this setting, we conducted two experiments. In the first one, five neutral with pose variations scans (frontal, ±30°, ±60°) (k = 5) for each subject are used for training as shown in Fig. 6, and a two testing sets are used for testing.

- Set 1: contains 1560 samples with variation of expression and illumination
- Set 2: contains 1924 samples with the variation of pose and expression.

Detailed results of this experiment, reported in Table 1, show that the proposed approach achieves the best performance (98.85%) on the Sadness expression, because it has the least face shape deformation when compared to a neutral expression which is used for training. On the other hand, the Surprise expression has the most face shape deformation resulting in the least performance (88.46%) in Set 1. In Set 2, we clearly see that the face pose affects negatively the system performance, especially the pitch variation which has the least RRIR (46.31%).

The CMC curves for the proposed approach shown in Fig. 7 clearly show that our process is robust against illumination expression and pose variation.

The effect of the number of trained samples (k) on the performance of the proposed face recognition approach is studied in the second experiment. Under the same scheme, the number of trained samples k is changed to k= 18 and the RRIRs are computed and presented in Table 2. We can clearly see that the proposed approach performance has increased for all expressions and pose variations. Regarding Set 1, the proposed method gives a RRIR of 100% on all expressions except for Fear expression, where a recognition rate of 99.23% is reached. On Set 2, the proposed approach achieves the best performance (except the frontal pose) with a RRIR of 91.51% for ±30° yaw rotation compared to 87.33% when using five samples. A significant improvement (almost 40%) is reached with a rate of 85.58% for ±60° pitch rotation.

In Table 3, the recognition rates obtained with our method are compared to those of other methods using the CurtinFaces database. It should be noted that these approaches use only depth information.

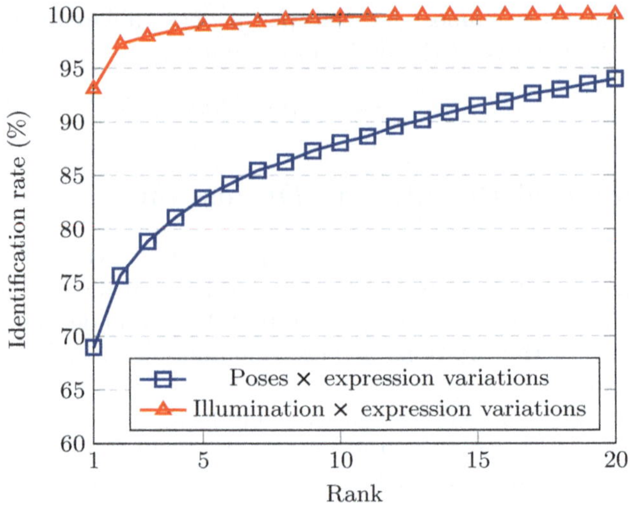

Fig. 7. CMC curves on CurtinFaces database.

Table 1. Expression, illumination variations (Set 1) and poses and expression variations (Set 2) R1RRs tested on the CurtinFaces database.

Testing set	Identification rates (%)
Happiness (260)	90.77
Disgust (260)	93.08
Anger (260)	95.38%
Sadness (260)	98.85
Surprise (260)	88.46
Fear (260)	92.31
Set 1 (1560)	**93.14**
Frontal (52)	98.08
±30°yaw (624)	87.33%
±60°yaw (624)	70.99
±60°pitch (624)	46.31
Set 2(1924)	**69.02**

Li et al. [2] achieved an accuracy of 54.74% and 88.8% when testing their approach against pose and expression variations respectively using 18 samples of the depth maps for training. Kaashki et al. [10] reported an accuracy of 69.7% and 96.1% when using the same evaluation method as Li et al. [2]. Our approach reaches 99.87% on Set 1 and 83.0% on Set 2, outperforming approaches presented in Table 3.

Table 2. Expression, illumination variations (Set 1) and poses and expression variations (Set 2) R1RRs obtained using 18 samples per person for training.

Testing set	Identification rates (%)
Happiness (260)	100
Disgust (260)	100
Anger (260)	100
Sadness (260)	100
Surprise (260)	100
Fear (260)	99.23
Set 1 (1560)	**99.87**
Frontal (52)	100
±30°yaw (624)	91.51
±60°yaw (624)	70.51
±60°pitch (624)	85.58
Set 2(1924)	**83.00**

Table 3. RRIRs of the proposed approach on the CurtinFaces database compared to those of the other approaches.

Approach	Set 1	Set 2
Li et al. Without Symmetric Filling [2]	88.8%	54.74%
Kaashki et al. [10]	96.1%	69.7%
Our approach	**99.87%**	**83.00%**

4 Conclusion

In this paper, we presented an automatic 3D face identification approach for low-quality Kinect 3D scans. The proposed approach used HOG on shape index maps to find feature descriptors. In the classification process, the Collaborative Representation is used to find the best match between the HOG descriptors of a probe face and the dictionary. The proposed approach is based only on the depth maps but can integrate RGB modality. It has been evaluated on the publicly available CurtinFaces database, reaching an accuracy of 99.87% and 83.0% when testing against expression and pose variations respectively. The proposed approach has several limitations. CRC is relying on the inversion of a matrix whose computational cost can be constraining in a large gallery set.

For future works, more aspects are to be explored. The effects of the extreme yaw rotation (±90°) and occlusions on the process will be investigated. Information provided from multiple descriptors can contribute to improve face recognition performance. The effectiveness and the robustness of the proposed approach have to be validated on additional public databases.

References

1. Phillips, P.J., Flynn, P.J., Scruggs, T., et al.: Overview of the face recognition grand challenge. In: Computer Vision and Pattern Recognition, CVPR 2005. IEEE Computer Society Conference, pp. 947–954. IEEE (2005)
2. Li, B.Y.L., et al.: Using kinect for face recognition under varying poses, expressions, illumination and disguise. In: 2013 IEEE Workshop on Applications of Computer Vision (WACV). IEEE (2013)
3. Ciaccio, C., Wen, L., Guo, G.: Face recognition robust to head pose changes based on the RGB-D sensor. In: 2013 IEEE Sixth International Conference on Biometrics: Theory, Applications and Systems (BTAS). IEEE (2013)
4. Goswami, G., Bharadwaj, S.,, Vatsa, M., et al.: RGB-D face recognition using Kinect. In: 2013 IEEE Sixth International Conference on Biometrics: Theory, Applications and Systems (BTAS), pp. 1–6. IEEE (2013)
5. Segundo, M.P., Sarkar, S., Goldgoe, D., et al.: Continuous 3D face authentication using RGB-D cameras. In: Proceedings of the IEEE Conference on Computer Vision and Pattern Recognition Workshops, pp. 64–69 (2013)
6. Goswami, G., Vatsa, M., Singh, R.: RGB-D face recognition with texture and attribute features. IEEE Trans. Inf. Forensics Secur. **9**(10), 1629–1640 (2014)
7. Neto, C., Baptista et MARANA, J., Nilceu, A.: 3DLBP and HAOG fusion for face recognition utilizing Kinect as a 3D scanner. In: Proceedings of the 30th Annual ACM Symposium on Applied Computing, pp. 66–73. ACM (2015)
8. Boutellaa, E.: Contribution to face analysis from rgb images and depth maps. In: Diss. Ecole nationale Supérieure en Informatique Alger (2017)
9. Azzakhnini, S., Ballihi, L., Aboutajdine, D.: Combining facial parts for learning gender, ethnicity, and emotional state based on RGB-D Information. ACM Trans. Multimed. Comput. Commun. Appl. (TOMM) **14**(1s), 19 (2018)
10. Kaashki, N.N., Safabakhsh, R.: RGB-D face recognition under various conditions via 3D constrained local model. J. Visual Commun. Image Represent. **52**, 66–85 (2018)
11. Li, B., Mian, A., Liu, W., Krishna, A.: Using kinect for face recognition under varying poses, expressions, illumination and disguise. In: 2013 IEEE Workshop on Applications of Computer Vision (WACV), pp. 186–192 (2013)
12. Mahoor, M.H., Abdel-Mottaleb, M.: 3D face recognition based on 3D ridge lines in range data. In: IEEE International Conference on Image Processing, ICIP 2007, vol. 1. IEEE (2007)
13. Tang, Y., et al.: 3D face recognition with asymptotic cones based principal curvatures. In: 2015 International Conference on Biometrics (ICB). IEEE (2015)
14. Tang, Y., et al.: Principal curvature measures estimation and application to 3D face recognition. J. Math. Imaging Vis. **59**(2), 211–233 (2017)
15. Bentaieb, S., et al.: Face recognition from unconstrained three-dimensional face images using multitask sparse representation. J. Electron. Imaging **27**(1), 013008 (2018)
16. Colombo, A., Cusano, C., Schettini, R.: 3D face detection using curvature analysis. Pattern Recogn. **39**(3), 444–455 (2006)
17. Szeptycki, P., Ardabilian, M., Chen, L.: A coarse-to-fine curvature analysis-based rotation invariant 3D face landmarking. In: IEEE 3rd International Conference on Biometrics: Theory, Applications, and Systems, 2009. BTAS'09. IEEE (2009)
18. Berretti, S., et al.: Matching 3D face scans using interest points and local histogram descriptors. Comput. Graph. **37**(5), 509–525 (2013)

19. Koenderink, J.J., Van Doorn, A.J.: Surface shape and curvature scales. Image Vision Comput. **10**(8), 557–564 (1992)
20. Dalal, N., Triggs, B.: Histograms of oriented gradients for human detection (2005)
21. Zhang, L., et al.: 3D palmprint identification using block-wise features and collaborative representation. IEEE Trans. Pattern Anal. Mach. Intell. **37**(8), 1730–1736 (2014)

A Study of Dimensionality Reduction for Facial Expression Recognition

Yacine Yaddaden[1(✉)], Mehdi Adda[1,2], and Abdenour Bouzouane[2]

[1] Département de mathématiques, d'informatique et de génie, Université du Québec à Rimouski (UQAR), Rimouski, Canada
{yacine_yaddaden,mehdi_adda}@uqar.ca

[2] Laboratoire d'Intelligence Ambiante pour la Reconnaissance d'Activités (LIARA), Département d'informatique et de mathématique, Université du Québec à Chicoutimi (UQAC), Saguenay, Canada
abdenour_bouzouane@uqac.ca

Abstract. With the widespread use of smart devices, finding effective and intuitive ways of interaction becomes a real challenge. One of the most relevant sources of information remains the facial expressions. Indeed, they provide various information related to the user such as his emotional state. In the last years, several methods of automatic facial expression recognition based on *deep learning* algorithms have been proposed. Even if they achieve high accuracy in terms of recognition, they require high-performance hardware and huge training datasets. In this paper, we propose an automatic facial expression recognition approach based on the Histogram of Oriented Gradients along with two distinct *dimensionality reduction* techniques namely Principal Component Analysis and Autoencoder. The proposed approach allows recognizing the *six basic emotions* using a *multi-class* Support Vector Machine. We evaluated its performance with three facial expression datasets namely **JAFFE**, **RaFD** and **KDEF**. The obtained results attest to its efficiency with 93.75%, 87.26%, and 96.27%, respectively.

Keywords: Facial expression recognition · Histogram of oriented gradients · Principal component analysis · Autoencoder · Multi-class support vector machine

1 Introduction

The scientific community has always been interested in how humans interact with each other. Charles Darwin was one of the first and most famous scientists to be interested in human emotions. Indeed, they represent a universal way of interaction. Charles Darwin conducted various studies and experimentation in order to understand how to recognize the expressed emotions and their impact on the human mind state. Other scientists in psychology have also been interested in this field such as Mehrabian [13] who conducted an important study to measure

M. R. Senouci et al. (Eds.): CSA 2020, LNNS 199, pp. 14–24, 2021.
https://doi.org/10.1007/978-3-030-69418-0_2

and estimate the contribution of three different emotion *modalities* during a common human interaction. The obtained results state that facial expressions have the highest contribution with 55% while the verbal and vocal parts contribute by 7% and 38%, respectively. Expressing emotions is a complex process and the reaction might be different depending on several factors such as the ethnicity, the stimulus and the gender. However, Ekman & Friesen [3] have made a breakthrough in the field by introducing a cross-cultural categorization of emotions. Indeed, they stated about the presence of *six basic emotions* namely happiness (**HA**), fear(**FE**), anger (**AN**), surprise (**SU**), disgust (**DI**) and sadness (**SA**). Moreover, the neutral (*ne*) state is considered as a reference.

Based on these works and findings in *affective science*, several applications have been proposed in various fields such as human-computer interaction, security, entertainment, healthcare, and e-learning. Moreover, with the recent advances in Information and Communication Technology, a new field of study emerged and raised the interest of the research community. Indeed, *affective computing* provides to the smart devices the ability to recognize *autonomously* the human emotional state. Even if various *modalities* (speech, body gesture, facial expressions, etc.) might be used for emotion recognition, the most relevant one remains the facial expressions according to Mehrabian [13] works. Therefore, several Automatic facial Expression Recognition (AFER) methods have been proposed. Most of them allow the recognition of the *six basic emotions* according to Ekman & Friesen works. In the last previous years, most of the proposed methods focuses on the use of *deep learning* algorithms [20] for two main reasons: 1) the *high accuracy* they achieve, 2) the use of *non-handcrafted* features. However, *deep learning* algorithms are not used without cost. Indeed, two conditions have to be satisfied: 1) high-performance hardware is required to perform the training phase, 2) huge datasets are needed in order to achieve a good generalization without *overfitting*.

In this paper, we propose an AFER approach based on an efficient and widely used descriptor. Indeed, after applying various *pre-processing* operations on the input image, the Histogram of Oriented Gradients (HOG) [2] is extracted. Next, in order to reduce the size of the generated representation, two distinct *dimensionality reduction* techniques are employed namely Principal Component Analysis (PCA) and Autoencoder (AE). One of the main aims of this work consists of defining the most efficient and suitable dimensionality reduction technique in the context of AFER. The last component of the proposed approach performs the recognition and it is based on one of the most effective classical machine learning techniques namely *multi-class* Support Vector Machine (SVM). Finally, in order to measure the effectiveness of the proposed AFER approach, we used three different benchmark facial expression datasets. Each one includes the *six basic emotions* in addition to the *neutral* state.

The remainder of the paper is organized as follows. Section 2 is dedicated to presenting the basic knowledge about AFER. Moreover, some existing methods are also discussed. In Sect. 3, we present and explain each one of the various components of the proposed AFER approach. All the details about the evaluation process and experimental protocol are explained in Sect. 4. The obtained results are presented and discussed in Sect. 5. Finally, concluding remarks about this work are presented in Sect. 6.

2 Background

The proposed approach in this paper has been designed following a specific architecture commonly adopted for AFER methods. Furthermore, we present some exiting *state-of-the-art* AFER methods using different types of descriptors and classification techniques.

2.1 Fundamentals

An AFER method might be defined as a *pattern recognition* system. Indeed, it performs recognition based on the *input* signal. In the case of an AFER system, we distinguish two categories depending on the input type: 1) *static* [17–19] systems employ a single image in order to perform recognition while 2) *dynamic* [16] ones use image sequences including temporal variations. The first type of system is faster since it takes into account a relatively small amount of information while the second one is usually more precise.

Both categories of AFER systems consist of *five* different components. The first one is the *input* which might be either *static* or *dynamic*. The second component performs an important operation namely *feature extraction*. It consists in generating a relevant *representation* of the input in order to ease and optimize the recognition process. In the field of AFER, three different types of descriptors might be employed in order to generate a representation: 1) *appearance-based* [17] features focus in emphasizing the textural information generated from the whole image, 2) *geometric-based* [19] ones are based on *models* generated using extracted facial landmarks or Facial Fiducial Points (FFPs), 3) *hybrid-based* [18] features combine both previous types of descriptors in order to increase the relevancy of the generated representation. Usually, the generated representation might be large and includes noisy and redundant information. Therefore, *feature reduction* is useful, but not mandatory, to keep only the relevant information. The component that performs the recognition in a common AFER system consists in *classification*. It is based on a *supervised* machine learning technique. It might be either classical techniques or *deep learning* based ones [20]. The results or *output* of this last operation consists of the *six basic emotions* [3].

2.2 Related Works

Due to the important role of emotions in interaction, especially from facial expressions, several AFER methods have been introduced over the last past years. Among these methods, several ones have employed the HOG based descriptor for its efficiency. Indeed, Liu et al. [8] used this descriptor along with PCA for dimensionality reduction and *multi-class* SVM for recognition. Nurzynska [14] used the same type of descriptor and classification technique, but without applying any feature reduction technique. Moreover, Meena et al. [12] also used the same type of representation, but with a different feature selection technique which consists in Graph Signal Processing (GSP). For the recognition part, they used a different type of classifier namely k-Nearest Neighbors (k-NN). In Ali et al. [1] work, they proposed the use of another type of classification technique namely the Sparse Representation Classifier (SRC). Other authors have employed different types of descriptors such as González-Hernández et al. [4] who proposed Local Binary Patterns (LBP) along with an Artificial Neural Network (ANN) classifier. The LBP is another common and widely used *appearance-based* feature. Therefore, Yaddaden et al. [17] have employed it in their proposed AFER approach with Independent Component Analysis (ICA) for dimensionality reduction and a *multi-class* SVM for the recognition part. Using the same type of classifier, Olivares-Mercado et al. [15] employed 2D Gabor based features to generate a relevant representation. In order to gather a larger amount of information, other authors proposed to combine different descriptors. Thus, Happy & Routray [5] proposed to combine the Pyramid Histogram of Oriented Gradients (PHOG) with LBP based features. They reduced the size of the resultant descriptor using the PCA and Linear Discriminant Analysis (LDA) before performing classification using *multi-class* SVM. Finally, Liu et al. [9] introduced an AFER method that combines LBP and 2D Gabor based features. Similar to previous methods, they applied the PCA for dimensionality reduction and the Extreme Learning Machine (ELM) classifier for recognition.

3 Proposed Approach

In this section, we detail each one of the different components of the proposed AFER approach.

3.1 Pre-processing

Before generating a relevant representation, various *pre-processing* operations have to be applied to the input image in order to optimize the *feature extraction* process. As shown in Fig. 1, the first operations consist in performing *face alignment* and *extraction*. Both of them are based on the result of another operation that consists in extracting 68 *facial landmarks* or FCPs. In order to achieve that, the method proposed by Kazemi & Sullivan [6] has been employed. It is based on the application of a *regression function* $r_t()$ aiming to adjust in an *iterative* way a predefined face shape $\hat{S}^t$ using the *input* image I. From the Eq. 1, the obtained result consists of $|S| = N = 68$ *facial landmarks* defined by $S = \{P_1, P_2, \ldots, P_N\} \in \mathbb{R}^{2N}$ where $P_i(x_i, y_i)$ represents the *Cartesian coordinates*.

$$\hat{S}^{t+1} = \hat{S}^t + r_t(I, \hat{S}^t) \tag{1}$$

Based on the $|S| = N = 68$ *facial landmarks*, the *alignment* of the *input* face image is corrected using the position of the eyes. Then, the face region is extracted before being resized to 128×128 pixels. Finally, the last *pre-processing* operation consists in applying *histogram equalization* that aims to reduce the illumination and contrast issues.

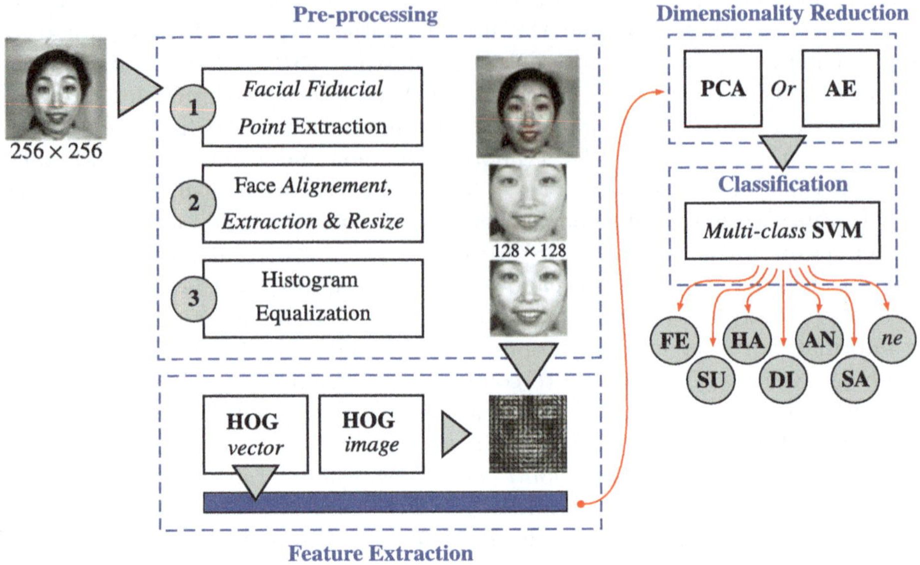

Fig. 1. Overview of the proposed *static* AFER system.

3.2 Feature Extraction

HOG is a well-known and widely used descriptor that has been used in various fields. Moreover, it belongs to *appearance-based* feature category and has been used for the first time by Dalal & Triggs [2] in the context of pedestrian detection. The generation of such representation is described by the following steps:

1. *Compute the gradient* by applying a convolution to the *input* image in order to form horizontal and vertical gradient maps,
2. *Magnitude & orientation* are computed based on the previously obtained horizontal and vertical gradient maps,
3. *Cell division* which means dividing the image in smaller and equal sub-parts,
4. *Cell quantization*, the orientation value of each distinct cell is quantized within a histogram with a specific number of bins,
5. *Block normalization*, a group of cells constitutes a block and each one has 50% overlapping with the adjacent block. Finally, the orientation histograms of each block are normalized and merged to form a single feature vector.

The HOG feature extraction is illustrated in Fig. 2(a). Morevoer, the used parameters (*block size* and the *number of orientations*) have been defined *empirically* for each dataset.

3.3 Dimensionality Reduction

In the context of this work, we use a dimensionality reduction technique in order to reduce the size of the feature vector. The main objective lies in optimizing the classification step. One of the main objectives of this work consists of comparing the performances of two distinct *dimensionality reduction* techniques. Depending on the obtained results, we define the most suitable to use.

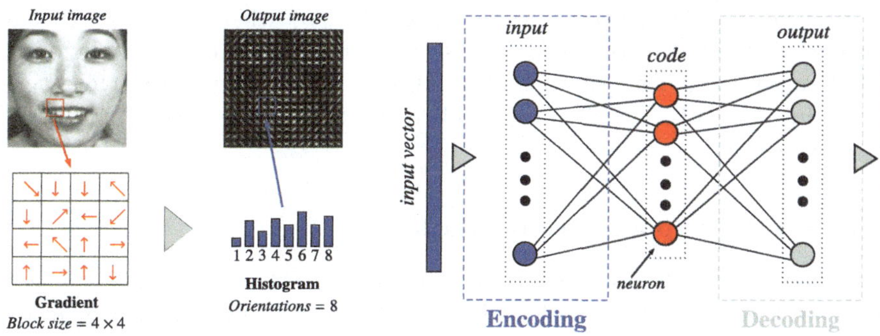

Fig. 2. (a) **HOG** extraction process, (b) Overview of the **AE** architecture.

Principal Component Analysis: The first technique is based on *classical* machine learning while the second one is a *deep learning* based technique. The motivation behind using such techniques lies in reducing the large dimension of the generated feature vectors. Indeed, after extracting the HOG based features, the obtained representation has between $|V| = 2048$ and 8192 different attributes. Due to its large size, the application of a dimensionality reduction technique is necessary in order to ease and optimize the recognition process.

The first technique consists in the PCA that is an *unsupervised* machine learning technique. It is based on the extraction of the important and *relevant* information as new *orthogonal* variables called *principal components* from a set of *observations* defined by $x = \{x_1, x_2, \ldots, x_n\}$. Each component might be defined by $z = \{z_1, z_2, \ldots, z_m\}$ and generated by the Eq. 2.

$$z_j = a_j^T x = \sum_{i=1}^{n} a_{i,j} \times x_i \tag{2}$$

Where $a = \{a_1, a_2, \ldots, a_n\}$ and for the first component z_1, the *variance* $var(z_1)$ is *maximized*. From the jth component (with $j > 1$) several conditions have to be satisfied:

1. the $var(z_j)$ is *maximized*,
2. subject to the *co-variance* $cov(z_j, z_k) = 0$ $(j > k \geqslant 1)$,
3. $a_j^T a_j = 1$.

Autoencoder: The second dimensionality reduction technique used in the context of this work consists in the AE. It is an *unsupervised* neural network that usually consists of three different layers namely: *input*, *code* and *output* (see Fig. 2(b)). This technique has various applications such as compression, information retrieval and dimensionality reduction.

Basically, it applies the *backpropagation* algorithm on the *input* layer which is represented by a set of *unlabeled* data defined by $x = \{x_1, x_2, \ldots, x_n\}$ where $x_i \in \mathbb{R}^n$. The main objective consists in trying to learn a function $h_{\mathbf{w},b}(x) \approx x$. The *output* layer is identified by $\hat{x} = \{\hat{x}_1, \hat{x}_2, \ldots, \hat{x}_n\}$ where $\hat{x}_i \in \mathbb{R}^n$. In our case, we are interested in the *code* layer that contains a compressed version of the *input* emphasizing the *relevant* information. The proposed AE architecture has a *code* layer with 128 *neurons* defined by $y = \{y_1, y_2, \ldots, y_m\}$ where $y_i \in \mathbb{R}^m$ and $m < n$.

3.4 Multi-class Support Vector Machine

In the proposed AFER approach, the recognition part is performed using a widely used *supervised* machine learning. The SVM technique has been introduced by Cortes & Vapnik as a *binary* classifier. Thus, it consists in finding a *hyperplane* that ensures an optimal *separation* between two distinct classes. It might be defined by $y_i = sign(\langle \mathbf{w}, x_i \rangle + b)$ where the *maximum-margin hyperplane* is represented by $(\mathbf{w}, b)$, the feature vectors by $x_i \in \mathbb{R}^d$ and labels by $y_i \in \{\pm 1\}$. However, the proposed AFER approach has to distinguish between *six basic emotions*. Therefore, we proposed to use the *One-Against-All* architecture that consists of a combination of six different *linear binary*-SVM classifiers.

The choice of the SVM technique for classification is motivated by its efficiency and the yield of high performances in our previously proposed AFER methods [17–19].

4 Experimentation and Evaluation

In order to evaluate and estimate the performances of the proposed AFER approach, we employed three different facial expression datasets namely JApanese Female Facial Expression (**JAFFE**) [11], Radboud Faces Database (**RaFD**) [11] and Karolinska Directed Emotional Faces (**KDEF**) [10]. Each one of the benchmark datasets contains a specific number of samples where the participants express one of the *six basic emotions* in addition to the *neutral* state (see in Table 1).

Furthermore, the evaluation is achieved using the *ten-folds cross-validation* strategy. Basically, the dataset is divided into ten *stratified* groups. During each one of the ten iterations, nine groups are used for training and the last one for evaluation. The final accuracy is obtained by averaging all the obtained accuracies.

Due to the presence of a *random aspect* in the initialization of the neural network (AE). The training process has been repeated *five times* before extracting

Table 1. Image facial expression datasets.

Dataset		JAFFE [11]	KDEF [10]	RaFD [7]
Emotions	FE	30	140	67
	SU	30	140	67
	HA	29	140	67
	DI	28	140	67
	AN	30	140	67
	SA	30	140	67
	ne	30	140	67
	Total	**207**	**980**	**469**
Resolution		256 × 256	562 × 762	681 × 1024

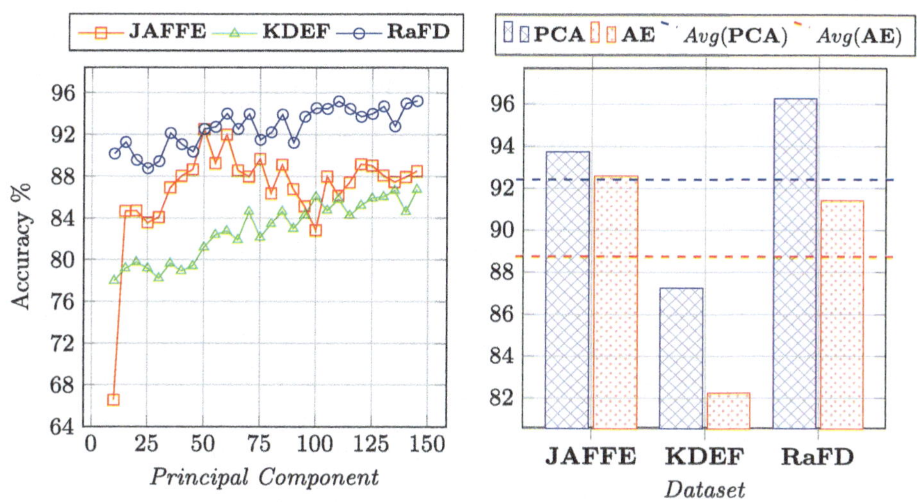

Fig. 3. (a) Accuracy's convergence with **PCA**, (b) Accuracy's comparison.

the top and *average* accuracies. During the training, we configured the AE in order to continue the learning until reaching 300 *epoch*.

5 Results and Discussion

In this section, we present the obtained results after evaluation. Then, we discuss the impact of each one of the two dimensionality reduction on the proposed AFER approach. We also compare its accuracy with *state-of-the-art* methods.

5.1 Comparison Between PCA and AE

In the Fig. 3(a) is illustrated the variation in terms of accuracy depending on the number of used *principal components*. We notice that the PCA allows to reach *convergence* in a relatively fast way and achieve the top accuracy with a relatively low number of *principal components*. For example, when evaluating using the **JAFFE** [11] dataset, the top accuracy is reached with only 50 *principal components*.

We also compared the obtained results when using the PCA and AE. Indeed, in Fig. 3(b) is illustrated the impact of using the PCA and AE on the AFER approach accuracy. We notice that the highest accuracy is achieved by the use of the PCA for the whole three datasets. Moreover, the average accuracy using the PCA is estimated at 92.50% while with the AE, we achieve 87.50%. We might conclude from this first part of our evaluation that the PCA performs better compared to the AE with a difference of **5.00%** in terms of accuracy. Moreover, training the AE takes more time than the PCA.

5.2 Performance Evaluation of the AFER approach

The second part of the evaluation is dedicated to measuring the performance of the proposed AFER approach in terms of accuracy.

Table 2. Accuracy for each emotion on the three benchmark datasets.

Dataset	FE	SU	HA	DI	AN	SA	*Overall*
JAFFE [11]	93.33%	96.67%	93.10%	92.86%	96.67%	90.00%	93.77%
	90.00%	**93.33%**	**96.55%**	**85.71%**	**96.67%**	**93.33%**	**92.60%**
KDEF [10]	75.00%	88.57%	97.14%	87.14%	90.00%	85.71%	87.26%
	65.00%	**87.86%**	**92.86%**	**79.29%**	**87.86%**	**80.71%**	**82.26%**
RaFD [7]	89.55%	95.52%	100.00%	100.00%	98.51%	94.03%	96.27%
	86.57%	**91.04%**	**98.51%**	**100.00%**	**88.06%**	**83.58%**	**91.29%**

In Table 2 is presented the recognition rate for each emotion when evaluating with the three benchmark facial expression datasets (namely **JAFFE** [11], **RaFD** [11] and **KDEF** [10]). For each dataset, the first row refers to the use of the PCA while the second one is for the AE.

The final evaluation is dedicated to comparing the performance of the proposed AFER approach with existing *state-of-the-art* methods. As shown in Table 3, the performance of the proposed approach outperforms the existing methods even those using the same type of descriptor [1,8,12,14]. The reason might be the efficiency of the *pre-processing* operations and the choice of the *classification* technique. Even the methods that combine two different descriptors do not outperform the performance of the proposed AFER approach [5] and [9]. They obtained an accuracy lower than 90.00% while ours is higher.

Table 3. Comparison with *state-of-the-art* methods in terms of accuracy.

Dataset	*Method*	*Feature type*	*Classification*	*Classes*	*Accuracy*
JAFFE [11]	[8]	HOG	PCA + SVM	Six	89.60%
	[5]	PHOG + LBP	PCA-LDA + SVM		83.86%
	Ours	HOG	PCA + SVM		**93.75%**
	[9]	LBP + 2D Gabor	PCA + ELM	Seven	85.71%
	[12]	HOG	GSP + k-NN		88.57%
	Ours		PCA + SVM		**92.69%**
KDEF [10]	[1]	HOG	SRC	Six	82.17%
	Ours		PCA + SVM		**87.26%**
	[15]	2D Gabor	SVM	Seven	79.29%
	[14]	HOG			84.08%
	Ours		PCA + SVM		**88.47%**
RaFD [7]	[4]	LBP	ANN	Six	88.00%
	[1]	HOG	SRC		90.67%
	Ours		PCA + SVM		**96.27%**
	[17]	LBP	ICA + SVM	Seven	93.16%
	Ours	HOG	PCA + SVM		**93.06%**

6 Concluding Remarks

In this work, we have been able to design an AFER system using an *appearance-based* feature namely HOG. The obtained results are relatively good compared to the existing *state-of-the-art* AFER methods. We have also been able to compare the performance of two distinct dimensionality reduction techniques in order to define which is the most suitable to use. Based on the results, we conclude that the PCA which is a *classical* machine learning technique performs better than the AE which is a *deep learning* technique.

Acknowledgment. We would like to acknowledge the providers of the used benchmark facial expression datasets namely **JAFFE** [11], **KDEF** [10] and **RaFD** [7].

References

1. Ali, A.M., Zhuang, H., Ibrahim, A.K.: An approach for facial expression classification. IJBM **9**(2), 96–112 (2017)
2. Dalal, N., Triggs, B.: Histograms of oriented gradients for human detection. In: Schmid, C., Soatto, S., Tomasi, C. (eds.) CVPR vol. 1, pp. 886–893. IEEE Computer Society (Jun 2005)
3. Ekman, P., Friesen, W.V.: Constants across cultures in the face and emotion. J. Pers. Soc. Psychol. **17**(2), 124–129 (1971)
4. González-Hernández, F., Zatarain-Cabada, R., Barrón-Estrada, M.L., Rodríguez-Rangel, H.: Recognition of learning-centered emotions using a convolutional neural network. JIFS **34**(5), 3325–3336 (2018)

5. Happy, S., Routray, A.: Robust facial expression classification using shape and appearance features. In: ICAPR, pp. 1–5. IEEE (2015)
6. Kazemi, V., Sullivan, J.: One millisecond face alignment with an ensemble of regression trees. In: CVPR, pp. 1867–1874 (2014)
7. Langner, O., Dotsch, R., Bijlstra, G., Wigboldus, D.H., Hawk, S.T., Van Knippenberg, A.: Presentation and validation of the radboud faces database. Cogn. Emot. **24**(8), 1377–1388 (2010)
8. Liu, Y., Li, Y., Ma, X., Song, R.: Facial expression recognition with fusion features extracted from salient facial areas. Sensors **17**(4), 712 (2017)
9. Liu, Z.T., Li, S.H., Cao, W.H., Li, D.Y., Hao, M., Zhang, R.: Combining 2d gabor and local binary pattern for facial expression recognition using extreme learning machine. JACIII **23**(3), 444–455 (2019)
10. Lundqvist, D., Flykt, A., Öhman, A.: The karolinska directed emotional faces - kdef, cd rom from department of clinical neuroscience, psychology section, karolinska institutet (1998)
11. Lyons, M., Akamatsu, S., Kamachi, M., Gyoba, J.: Coding facial expressions with gabor wavelets. In: AFGR, pp. 200–205. IEEE (1998)
12. Meena, H.K., Joshi, S.D., Sharma, K.K.: Facial expression recognition using graph signal processing on hog. IETE J. Res., 1–7 (2019)
13. Mehrabian, A.: Communication without words, pp. 51–52. Transaction Publishers, 2 edn. (1968)
14. Nurzynska, K.: Emotion recognition: the influence of texture's descriptors on classification accuracy. In: BDAS, pp. 427–438. Springer (2017)
15. Olivares-Mercado, J., Toscano-Medina, K., Sanchez-Perez, G., Portillo-Portillo, J., Perez-Meana, H., Benitez-Garcia, G.: Analysis of hand-crafted and learned feature extraction methods for real-time facial expression recognition. In: IWBF, pp. 1–6. IEEE (2019)
16. Yaddaden, Y., Adda, M., Bouzouane, A., Gaboury, S., Bouchard, B.: Facial expression recognition from video using geometric features. In: ICPRS, pp. 1–6. IET (2017)
17. Yaddaden, Y., Adda, M., Bouzouane, A., Gaboury, S., Bouchard, B.: Facial subregions for automatic emotion recognition using local binary patterns. In: SIVA, pp. 1–6. IEEE (2018)
18. Yaddaden, Y., Adda, M., Bouzouane, A., Gaboury, S., Bouchard, B.: Hybrid-based facial expression recognition approach for human-computer interaction. In: MMSP, pp. 1–6. IEEE (2018)
19. Yaddaden, Y., Adda, M., Bouzouane, A., Gaboury, S., Bouchard, B.: One-class and bi-class SVM classifier comparison for automatic facial expression recognition. In: ICASS, pp. 1–6. IEEE (2018)
20. Yaddaden, Y., Adda, M., Bouzouane, A., Gaboury, S., Bouchard, B.: User action and facial expression recognition for error detection system in an ambient assisted environment. ESWA **112**, 173–189 (2018)

How Would Image Down-Sampling and Compression Impact Object Detection in the Context of Self-driving Vehicles ?

Imene Bouderbal(✉), Abdenour Amamra, and Mohamed Akrem Benatia

Ecole Militaire Polytechnique, Bordj El -Bahri BP 17, Algiers, Algeria
imene.bouderbal@yahoo.com

Abstract. Accurate and real-time detection of road objects is a necessary component that self-driving cars must be equipped with to drive as safely as humans do. This task is critical since it is essential for more crucial tasks such as tracking, free slot identification, road-risk assessment and path planning. In this work, we provide an analysis of some image preprocessing techniques for real-time object detection in the context of autonomous vehicles. Image resolution and quality are often overlooked in the design of deep neural networks. To this end, we first examine the impact of image resolution on road object detection accuracy. Then, we examine the impact of image compression on object detection. To this end, several experiments have been carried out on the state-of-the-art YOLOv3 detector. The experimental results show that the network is resilient to compression given that its level is sufficient. The inference time is reduced without affecting considerably the other metrics. However, downsampling affects adversely the overall accuracy.

Keywords: Autonomous driving · Real-time object detection · Convolutional Neural Network (CNN) · YOLOv3 · Image compression · Image resolution

1 Introduction

To achieve complete image understanding, the focus shouldn't only be about classifying different images but also on precisely estimating the concepts and locations of objects contained in images. This task is referred to as object detection [1], which is one of the fundamental computer vision challenges. Object detection can provide valuable information for semantic understanding of images and videos and is related to many applications such as video surveillance, human behavior analysis, face recognition and autonomous driving [1], which is our subject of interest.

Successful autonomous vehicles should be able to ensure the safety of its passengers and pedestrians while navigating smoothly without disrupting the

M. R. Senouci et al. (Eds.): CSA 2020, LNNS 199, pp. 25–37, 2021.
https://doi.org/10.1007/978-3-030-69418-0_3

traffic or causing discomfort to their passengers. To achieve this goal, real-time and accurate on-road object detection for a rapid response of the vehicle is primordial. Common approaches usually employ multiple sensors, including cameras, LASERs, LIDARs, and millimeter-wave RADARs (all-weather). However, these sensors have some drawbacks, including high cost, low spatial resolution, interference among the different sensing modalities and preprocessing time [1]. The best compromise of cost and quality at a reasonable preprocessing load is offered by cameras. Therefore, most of the object detection deep neural models rely on *RGB* color images to locate and identify objects in the image [1].

In this work, a study of the impact of input images resolution and compression for real-time object detection was conducted. To this end, the remainder of the paper is structured as follows: In Sect. 2 we describe the main related works that were contributed in: first, the study of the impact of image quality factors on CNN performance; second, an overview of object detection using convolutional neural networks. Section 3 presents the experimental setup and protocols. Section 4 presents and discusses the obtained results. Finally, we conclude the paper, and we trace the potential perspectives that can extend the current work.

2 Related Works

2.1 Impact of Image Quality on CNN Performance

Nowadays, a large amount of image data is generated and processed in a real-world application. The resolution and quality of the image vary due to different conditions (input sources, imaging devices, weather, illumination, etc.). Variation in images resolution alters the visual information of images [2]. Simple visual information does not vary significantly, but complex visual information varies drastically with the reduction of image resolution. On the other hand, image compression pursues to encode the original raw image with lesser bits [2]. It is used to diminish image redundancy such to save or transfer image data in a more effective way.

Commonly, machine vision systems are trained and tested on high-quality image datasets, yet in practical applications, the input images can not be assumed to be of high quality. Downsampling (or downsizing) and compression are two commonly used preprocessing methods to handle large images. The main difference between downsampling and compression is that downsizing reduces both the input image's file size and image size (resolution), but compression-only reduces the input image's file size at the expense of some compression artifacts.

In the literature, some works investigated how image quality factors (such as resolution, compression, noise, contrast, blur) affect the CNN performance. The authors in [3] explore the effect of image quality distortions (blur, noise, contrast, JPEG and JPEG2000 compression) on the performance of deep neural networks (VGG-16, VGG-CNN-S, GoogleNet) but not explore the effect of image resolution. While, in [4], a modified MNIST dataset is generated using motion blur, noise, and contrast variation. These image distortions are successfully handled by the probabilistic quadtree DBN framework. In [5], the authors show the impact of compression on CNN classification performance. They also suggest that an image can be compressed by a factor 7, 16, 40 for JPEG, JPEG2000 compression while still maintaining a correct classification. In [6], an analysis of the impact of contrast in large-scale recognition by estimating different illumination quality. The effect of image resolution on classification is also considered. In [7], an LR-CNN model is proposed and the authors study the effect of varying resolution on fine-grained classification. The authors in [8] explore the effects of image resolution on the classification task, realized by both humans and CNNs. While in [2], an experimental study of the effects of varying resolution on performance of CNN based image classification.

In the current work, we propose to study the impact of both image resolution and compression on object detection task. Concerning the resolution, this latter will be degraded from higher to lower. Concerning compression, the JPEG standard [9] will be used with different rates. Their impact on both accuracy and inference time will be investigated.

2.2 Object Detection

The problem of object detection is to determine where objects are located in a given image (object localization) and to which category each object belongs (object classification) [1]. Recent advances in object detection have been deeply and strongly driven by deep convolutional neural networks (CNN). In this section, we'll only focus on object detectors based on CNN.

Deep-learning based object detection algorithms can be divided into two categories: two-stage and one-stage detectors. The two-stage detectors, e.g., Fast R-CNN [10], Faster R-CNN [11], and R-FCN [12] conduct a first stage of region proposal generation, followed by a second stage of object classification and bbox regression. These methods generally show a high accuracy but have a disadvantage of slow detection speed and lower efficiency. On the other hand, one-stage detectors, e.g., SSD [13] and YOLO [14], conduct object classification and bbox regression concurrently without a region proposal stage. These methods generally have a fast detection speed and high efficiency but low accuracy.

Recent Object Detectors. Recently, to take advantage of both categories and to compensate for their respective disadvantages, object detectors combining various schemes have been widely investigated [15–19]. MS-CNN [15], a two-stage detector improves the detection speed by conducting detection on various intermediate network layers. SINet [16], also a two-stage detector, enables a fast detection using a scale-insensitive network. CFENet [17], a one-stage detector, uses a comprehensive feature enhancement module based on SSD to improve the detection accuracy. RefineDet [18], also a one-stage detector, improves the detection accuracy by applying an anchor refinement module and an object detection module. Another one-stage detector, RFBNet [19], applies a receptive field block to improve the accuracy.

Nevertheless, the above-mentioned studies [15–18] couldn't reach a real-time detection speed of above 30 fps because of the use of an input resolution of

512 × 512 or higher, which is widely applied in object detection algorithms for achieving high accuracy. Even if real-time detection is possible in [19], it is difficult to apply to autonomous driving due to low accuracy.

We conclude that even the previous recent schemes are incomplete in terms of a trade-off between accuracy and detection speed, and consequently, have a limitation in their application to self-driving systems. In this work, YOLOv3 was chosen to conduct our study. This choice was motivated by the fact that YOLOv3 is a state of the art object detector belonging to the one-stage category, providing relatively good accuracy and inference time [1].

3 Experimental Setup and Protocols

3.1 Object Detector: YOLOv3

To evaluate the influence of image resolution and file size on object detection, we used YOLOv3 [20] as our CNN detector. In our experiment, YOLOv3 used darknet53 as its backbone and was pre-trained on COCO dataset [21]. The architecture of the object detector is illustrated in Fig. 1.

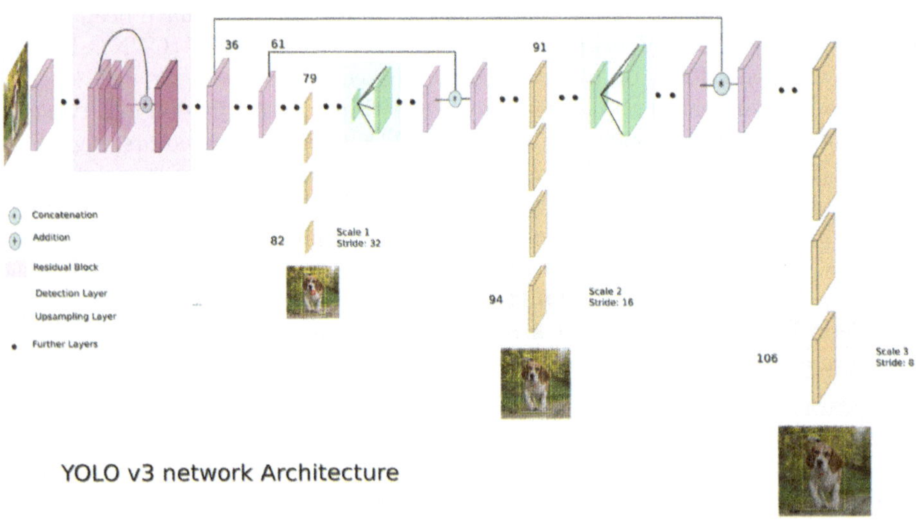

Fig. 1. YOLOv3 architecture.

3.2 Dataset: LISA-A

The experiments were conducted on LISA-A dataset [22], one of the latest published autonomous driving datasets. The testbed incorporates LIDARs, RADARs, cameras, and GPS/IMU sensors to holistically observe and understand the scene. The testbed was chosen because it offers high-resolution images

(1200 × 1600) compared to another autonomous driving dataset. Concerning the output of the camera, LISA-A dataset consists only of one class: vehicle, and consists of 4 full-surround sequences recorded for a very diverse set of weather, traffic and daylight conditions at 30 FPS. Each sequence contains 8 folders, each one of them relative to one of the views: front, front-right, right, rear-right, rear, rear-left, left, front-left.

3.3 Hardware

We carry out experiments by using Google Colab environment which provides a Tesla K80 GPU.

3.4 Evaluation Metrics

For assessing experimental results, we use:

1. Structural similarity index measures $SSIM$ and peak-signal-to-noise ratio $PSNR$ to evaluate image quality after compression.
2. Mean average precision (mAP) and mean inference time as the main evaluation metrics of the detector.

$SSIM$ and $PSNR$ are image similarity tests [23]. They are used to determine, how much quality is lost with the encoding process. $SSIM$ assesses the visual impact of three characteristics of an image: luminance, contrast, and structure. It was developed to calculate the visual quality of an image. $PSNR$ is the ratio between a signal's maximum power and the power of the signal's noise. It measures the quality of the reconstructed image compared to the original image.

The mAP calculates the area under the precision/recall curve. Precision is the ratio of the number of correctly classified positive instances to the number of instances labeled by the classifier as positive. Whereas recall is the ratio of the number of correctly positive instances to the number of instances labeled that are relevant. Therefore, precision measures how accurate are the predictions and recall refers to the percentage of total relevant results that were correctly classified.

The inference is where capabilities learned during deep learning training are put to work, so inference time is the time that takes the model to make its prediction. Also, we further drill down into what the preprocessing (downsizing and compression) affects – precision or recall.

3.5 Experiments

In this work, we want to evaluate to what extent the original uncompressed (raw) images can be resized whilst maintaining the detection accuracy. To study this, we downsize the original images of resolution 1024 × 1024 to 128 × 128, 256 × 256, 416 × 416, and 512 × 512 resolution. Then for each of the resolutions obtained plus the original one, we compressed the images using the following compression rates: 20%, 40%, 60% and 80%.

4 Results and Discussion

Figure 2 shows the average file size per image at each resolution and compression rate. It's important to note that the chosen resolutions are multiples of 32 since YOLOv3 predicts at three scales, which are precisely given by downsampling the dimensions of the input image by 32, 16 and 8 respectively [20]. As shown in the figure, it can be seen that after applying the first compression of 20%, the file size drops considerably for the three first resolutions (1024×1024, 416×416 and 512×512), then continue declining gradually for the next compression rates (40%, 60% and 80%). For resolutions 256×256 and 128×128, the compression affects, in the same way, the file size, but the decline isn't as important as in the first resolutions since the file size is already smaller. On the other hand, downsampling also reduces the file size. However, its effect is less important than that of compression.

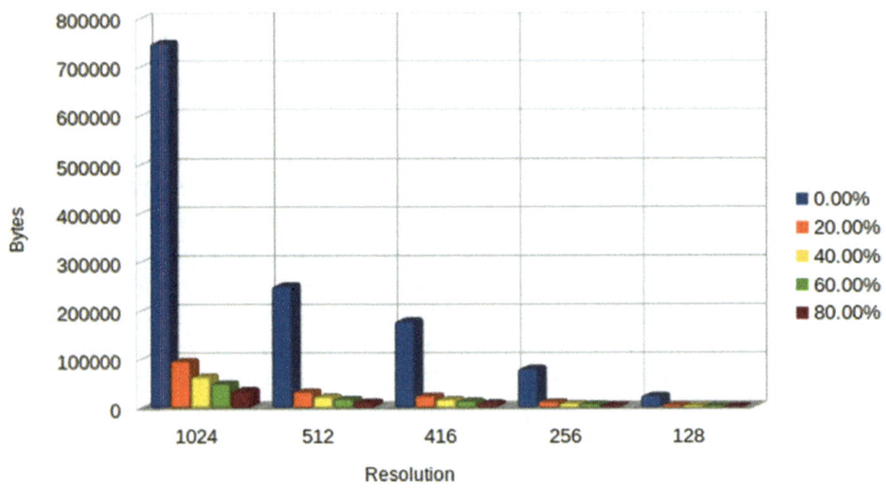

Fig. 2. Average file size per image against resolution and compression rate.

As shown in Fig. 3, both $SSIM$ and $PSNR$ values decrease while increasing the compression rate. This means the degradation of the quality of the image. $PSNR$ values indicate that after applying the first compression rate the quality of the compressed images drops by approximately 15 dB compared to the original ones, then decreases slightly for the rest of the compression rates. On the other hand, $SSIM$ values indicate that after each compression, the visual quality of the compressed images is degrading with practically the same rate.

Detection speed for the different resolutions and compression rates was analyzed. Table 1 shows the average detection speed for a batch of 128 images at the defined resolutions and compression rates. As the table shows, for the three first compression rates (0%, 20% and 40%), downsampling the images by a factor of

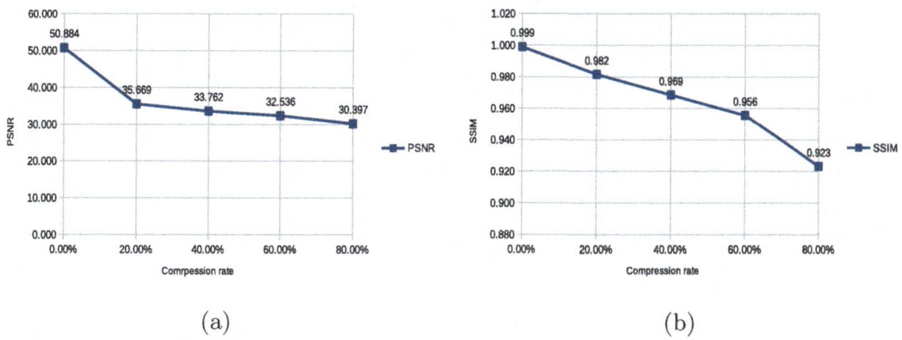

Fig. 3. Values of *PSNR* and *SSIM* for each resolution and compression rate.

2 reduces considerably the inference time. This is due to the gain in computation and memory transfer times with smaller file sizes. The inference time stays practically at the same level for the other downsampling rates.

The only exception to this is at resolutions 416 and 128 with no compression. We noticed that for these cases, the inference time was the highest

We can explain this by the fact that in small resolutions, objects would appear smaller and distinguishing them from the background would take more time. This also could be due to the fact that in the training dataset, there were not enough small objects.

For the remaining compression rates (60% and 80%), the inference time decreases quietly while increasing the downsampling rate. For the starting resolution (1024×1024), the gain in inference time is only obtained after a compression rate of 60%, since the gain in the file size is only considered from this compression rate. For the other resolutions, 20% is sufficient.

Table 1. Average detection speed for a batch of 128 images at different resolutions and compression rates.

Resolution	Compression rate				
	0%	20%	40%	60%	80%
1024	37.56	38.86	38.70	24.22	25.46
512	29.00	24.98	24.53	24.85	25.27
416	39.73	24.49	25.31	24.22	24.46
256	28.92	23.93	24.46	23.95	24.21
128	40.42	24.06	23.59	23.57	23.87

Figure 4 shows the mAP performance of the detector against the different compression rate for each resolution. Overall, there was a drop in the mAP while downsizing or compression the images. For resolution 1024, the mAP started dropping only at a compression of 40%, to reach a value of 66.67 at a compression rate of 80%. However, for the other resolutions, the drop in the mAP starts at the first compression rate. This drop is gradual, at the exception of:

- Resolution 416 × 416 at a compression rate of 40%: there is a slight increase in the mAP, followed by a steady decrease for the rest of the compression rates.
- Resolution 256 × 256: after a compression rate of 40%, the mAP reaches its lowest point with a null value.

Therefore, we can say that the detector is more resilient to compression than to lower resolutions.

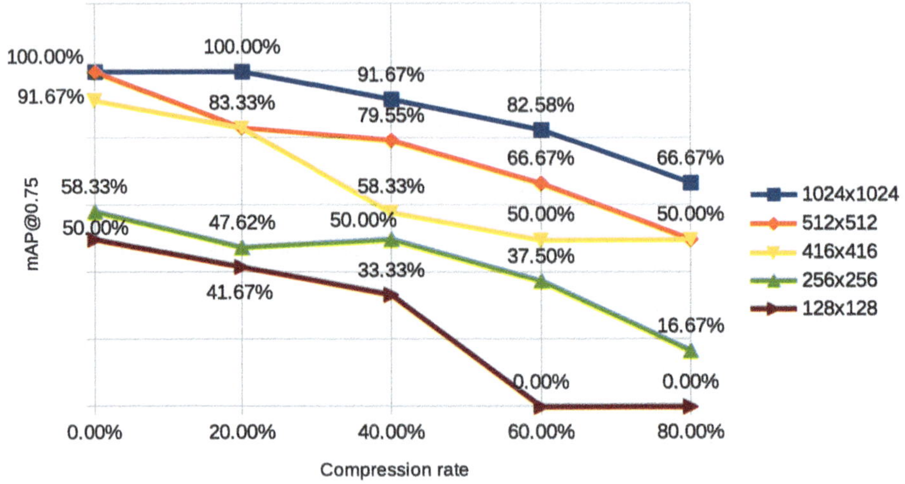

Fig. 4. The mAP performance of the detector against compression rate for each resolution.

Figures 5 and 6 shows precision and recall performance of the detector against the different resolutions and compression rate. Overall, we can see that downsizing the images does not affect much the precision, except at high compression rates (60% and 80%), where it decreases and becomes null at resolution 128×128. On the other hand, recall decreases while increasing the compression rate and downsampling factor, which is due to the rise of the false-negative rate (Figs. 7 and 8).

4.1 Concluding Remarks

From the above observations and results, we conclude that YOLOv3 is more resilient to compression than too small resolutions. So, compression is more interesting to investigate to improve the real-time aspect of object detection for

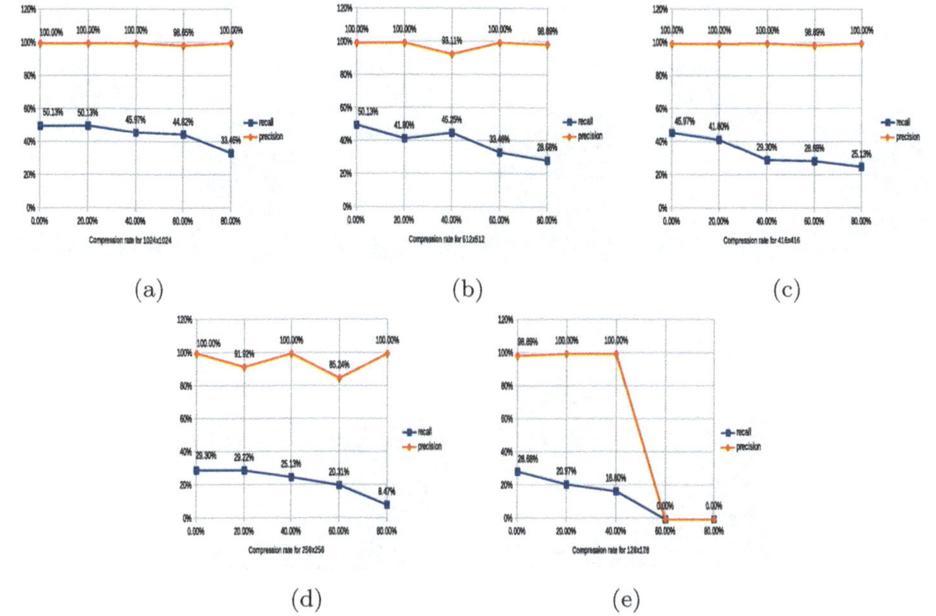

Fig. 5. Precision and recall performance of the detector against compression rate for resolutions: (a) 1024, (b) 512, (c) 416, (d) 256 and (e) 128.

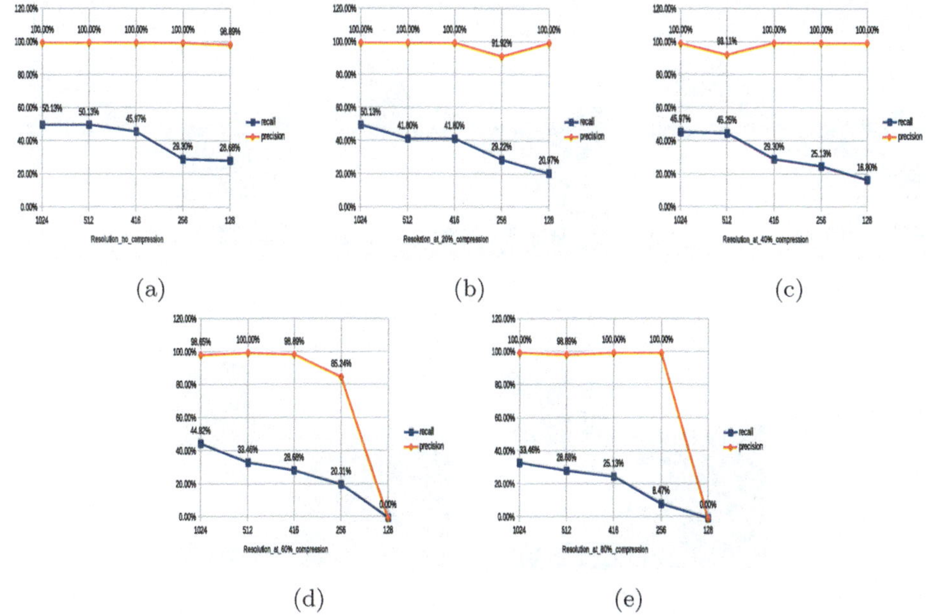

Fig. 6. Precision and recall performance of the detector against resolution for each compression rate: (a) 0%, (b) 20%, (c) 40%, (d) 60% and (e) 80%.

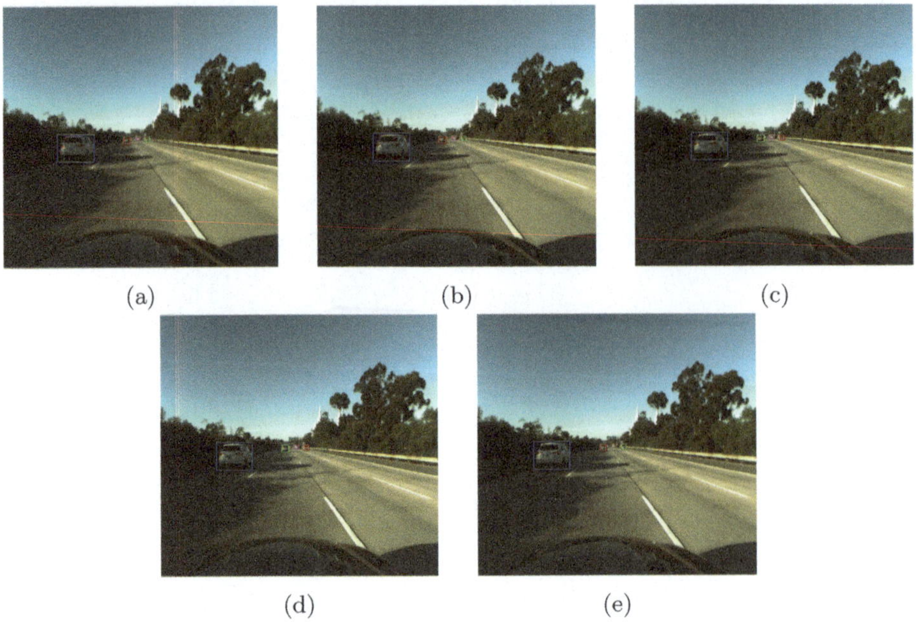

Fig. 7. Example of object detection results for the same image over the different compression rates: (a) 0%, (b) 20%, (c) 40%, (d) 60% and (e) 80%.

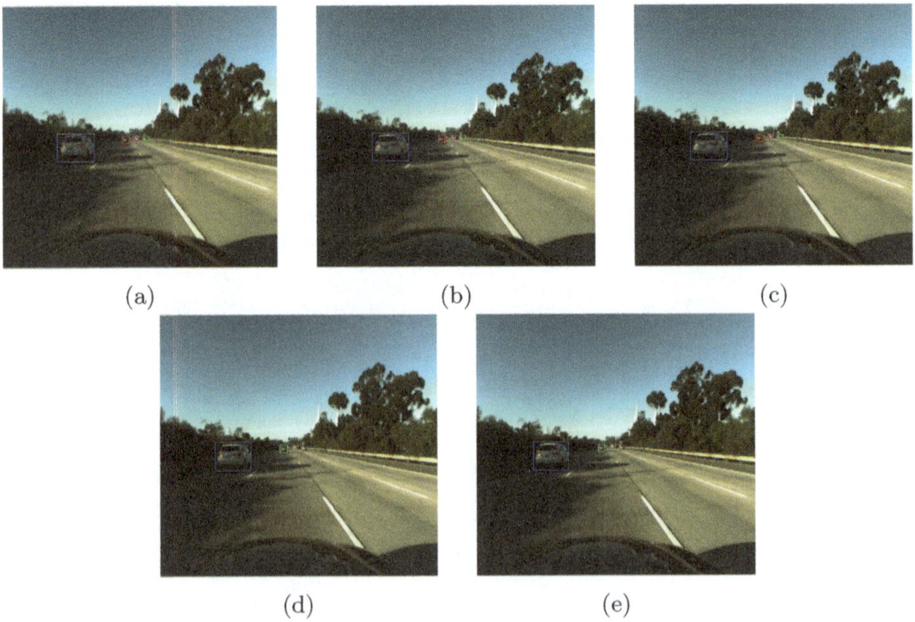

Fig. 8. Example of object detection results for the same image over the different resolutions: (a) 1024, (b) 512, (c) 416, (d) 256 and (e) 128.

autonomous driving. Our results indicate that applying compression to the input images of the detector reduces the inference time while maintaining relatively good accuracy. This mainly to the gain in computation and memory transfer times with images of smaller file sizes. Also, using high-resolution images for detection provides better accuracy. So, data images delivered by the camera should be first compressed before being feed to the detector, since small image file size reduces the computation time and memory consumption. However, the adequate compression level should be determined according to the resiliency of the used network to compression distortions.

5 Conclusion and Perspectives

In this paper, we investigated the impact of image resolution and compression on the object detection task in the context of autonomous driving scenes. We conclude that:

- high-resolution images provide better accuracy.
- downsampling the images reduces the inference time at the expense of loss in precision and recall, and so on the accuracy.
- compression of the images reduces the inference time without affecting considerably the other metrics (precision and recall).

In our future work, we think some topics need to be deeply investigated to improve real-time object detection for self-driving cars. First, the exploitation of sophisticated lossless compression methods in the preprocessing of images in the case of object detection. Second, investigate patch detection in the case of autonomous driving. Third, the incorporation of the output of other sensors and the development of fusion-based models. Third, the re-annotation of the used dataset to extend the existing class to classes usually encounter in the self-driving cars field.

References

1. Zhao, Z.-Q., Zheng, P., Shou-tao, X.: and Xindong Wu. A review. IEEE transactions on neural networks and learning systems, Object detection with deep learning (2019)
2. Kannojia, S.P., Jaiswal, G.: Effects of varying resolution on performance of CNN based image classification: an experimental study. Int. J. Comput. Sci. Eng. **6**(9), 451–56p (2018)
3. Dodge, S., Karam, L.: Understanding how image quality affects deep neural networks. In: 2016 Eighth International Conference on Quality of Multimedia Experience (QoMEX), pp. 1–6. IEEE (2016)
4. Basu, S., Karki, M., Ganguly, S., DiBiano, R., Mukhopadhyay, S., Gayaka, S., Kannan, R., Nemani, R.: Learning sparse feature representations using probabilistic quadtrees and deep belief nets. Neural Process. Lett. **45**(3), 855–867 (2017)

5. Dejean-Servières, M., Desnos, K., Abdelouahab, K., Hamidouche, W., Morin, L., Pelcat, M.: Study of the impact of standard image compression techniques on performance of image classification with a convolutional neural network (2017)
6. Sanchez, A., Moreno, A.B., Velez, D., Vélez, J.F.: Analyzing the influence of contrast in large-scale recognition of natural images. Integr. Comput.-Aided Eng. **23**(3), 221–235 (2016)
7. Chevalier, M., Thome, N., Cord, M., Fournier, J., Henaff, G., Dusch, E.: LR-CNN for fine-grained classification with varying resolution. In: 2015 IEEE International Conference on Image Processing (ICIP), pp. 3101–3105. IEEE (2015)
8. Ullman, S., Assif, L., Fetaya, E., Harari, D.: Atoms of recognition in human and computer vision. Proc. Natl. Acad. Sci. **113**(10), 2744–2749 (2016)
9. Leger, A.M., Omachi, T., Wallace, G.K.: Jpeg still picture compression algorithm. Optical Eng. **30**(7), 947–955 (1991)
10. Girshick, R.: Fast R-CNN. In: Proceedings of the IEEE International Conference on Computer Vision, pp. 1440–1448 (2015)
11. Ren, S., He, K., Girshick, R., Sun, J.: Faster R-CNN: towards real-time object detection with region proposal networks. In: Advances in Neural Information Processing Systems, pp. 91–99 (2015)
12. Dai, J., Li, Y., He, K., Sun, J.: R-FCN: object detection via region-based fully convolutional networks. In: Advances in Neural Information Processing Systems, pp. 379–387 (2016)
13. Liu, W., Anguelov, D., Erhan, D., Szegedy, C., Reed, S., Fu, C.-Y., Berg, A.C.: SSD: single shot multibox detector. In: European Conference on Computer Vision, pp. 21–37. Springer (2016)
14. Redmon, J., Divvala, S., Girshick, R., Farhadi, A.: You only look once: Unified, real-time object detection. In: Proceedings of the IEEE Conference on Computer Vision and Pattern Recognition, pp. 779–788 (2016)
15. Cai, Z., Fan, Q., Feris, R.S., Vasconcelos, N.: A unified multi-scale deep convolutional neural network for fast object detection. In: European Conference on Computer Vision, pp. 354–370. Springer (2016)
16. Xiaowei, H., Xuemiao, X., Xiao, Y., Chen, H., He, S., Qin, J., Heng, P.-A.: Sinet: a scale-insensitive convolutional neural network for fast vehicle detection. IEEE Trans. Intell. Transp. Syst. **20**(3), 1010–1019 (2018)
17. Zhao, Q., Wang, Y., Sheng, T., Tang. Z.: Comprehensive feature enhancement module for single-shot object detector. In: Asian Conference on Computer Vision, pp. 325–340. Springer (2018)
18. Zhang, S., Wen, L., Bian, X., Lei, Z., Li, S.Z.: Single-shot refinement neural network for object detection. In: Proceedings of the IEEE Conference on Computer Vision and Pattern Recognition, pp. 4203–4212 (2018)
19. Liu, S., Huang, D., et al.: Receptive field block net for accurate and fast object detection. In: Proceedings of the European Conference on Computer Vision (ECCV), pp. 385–400 (2018)
20. Redmon, J., Farhadi, A.: Yolov3: an incremental improvement. arXiv preprint arXiv:1804.02767 (2018)
21. Lin, T.-Y., et al.: Microsoft coco: common objects in context. In: European Cnference on Computer Vision, pp. 740–755. Springer (2014)

22. Rangesh, A., Yuen, K., Satzoda, R.K., Rajaram, R.N., Gunaratne, P., Trivedi, M.M.: A multimodal, full-surround vehicular testbed for naturalistic studies and benchmarking: design, calibration and deployment. arXiv preprint arXiv:1709.07502 (2017)
23. Hore, A., Ziou, D.: Image quality metrics: Psnr vs. ssim. In: 2010 20th International Conference on Pattern Recognition, pp. 2366–2369. IEEE (2010)

Extreme Subject Targeting for Deep Learning Based Action Recognition

Said Yacine Boulahia(✉), Abdenour Amamra, Said Daikh, Mohamed Ridha Madi, Mohamed Akrem Benatia, and Yacine Amara

Ecole Militaire Polytechnique, Chahid Abderrahmane Taleb, Bordj El-Bahri BP 17, Algiers, Algeria
boulahia.yacinesaid@gmail.com
http://www.emp.mdn.dz/

Abstract. Human action recognition occupies an important place in the computer vision community. This is particularly true as deep learning-based approaches have emerged. Recent attempts to solve this problem suggested developing attention driven architectures that look for the most discriminant features of the RGB frames to model and recognize an action. In this paper, we push this concept to its extreme by intentionally providing the model with the only information it needs. In particular, we propose to investigate the contribution of focusing the learning of actions onto the targeted subjects in each frame of an RGB video instead of considering the whole frame. The proposed learning procedure has been evaluated using a state-of-the-art recognition model on a subset of the well known NTU RGB+D dataset. Results reported in the experimental section demonstrate the interest of the proposed approach.

Keywords: Action recognition · RGB videos · Subject targeting · Extreme attention

1 Introduction

Computer vision is an exciting research theme aimed at providing computer systems with analysis and interpretation capabilities of the visual content of a scene close to that of human vision. One of the major objectives of computer vision is to recognize and understand human movement and in particular to allow the classification of human actions.

In recent decades, several approaches that can recognize 3D actions have been proposed with the purpose of improving the interpretation of actions in different contexts. These approaches do not all deal with the same problem. In fact, it depends on whether the actions are pre-segmented or not and if these actions have been fully performed or not. In this paper, we particularly focus on recognising fully performed actions, which we refer to as pre-segmented action recognition.

Among the proposed approaches, most performing ones are based on recent advances in deep learning. These approaches can be categorized according to

M. R. Senouci et al. (Eds.): CSA 2020, LNNS 199, pp. 38–49, 2021.
https://doi.org/10.1007/978-3-030-69418-0_4

several taxonomies. One of the most adopted taxonomies lists approaches in three families namely, spatial based, multi-streams based and hybrid approaches. Each of those approaches aims at modeling the most discriminant information of an action in order to ease the distinction with other actions. The observation we did is that such discriminant information lies mainly on the subjects performing the actions and not too much on the surrounding background. In other words, we hypothesise that the less information about the surrounding background is provided to our recognition model during the training process the more likely this model will focus on the subject performing the actions which would ease the identification of the most discriminant information in the processed videos.

Therefore, in this paper, we propose to investigate the contribution of focusing the learning of actions onto the targeted subjects in each frame of an RGB video instead of considering the whole frame. Our aim is to reduce the searching space by intentionally pushing the model to focus on most discriminant part of the frame. By doing so, our approach is a an extreme way to focus on the most discriminant information during the learning. That is why we refer to our study as an extreme learning based on subjects targeting.

The remainder of the paper is organized as follows: Sect. 2 reviews recent related work regarding action recognition. In this section, we mainly focus on recent deep learning-based approaches. Section 3 contains a description of the proposed study. In this section, we explain the main steps to process the data and the adopted deep learning architecture. Various results obtained using different testing protocols are presented and discussed in Sect. 4. Finally, in Sect. 5 conclusions and future research directions are drawn.

2 Related Work

The action recognition process has two key stages: the representation of the action and then its classification. Representing actions amounts to synthesizing the data of successive frames into a representation in the form of descriptors. This allows us to reduce the action from its initial representation space (very large) to a more synthetic representation (smaller).

The first proposed approaches in the literature used to manually design the representation of the actions by selecting a set of features that are likely to model the considered action. These features are then passed to a classifier to be trained. There are plenty of such approaches in the literature and covering all of them is beyond the scope of this paper. These approaches can be divided into three distinct families, including those based on a model of the human body, holistic approaches, and the so-called local approaches.

Approaches based on a model of the human body [3,14] are based on the reconstruction of the human model, which gives an accurate estimate of the posture to recognize human actions. Holistic approaches [2,13] use the overall structure and dynamics of the body to represent human actions. Last, local based approaches [5,9] describe the observations as a series of local descriptors or patches.

Although the global and local features based approaches have been very successful, these handcrafted features require human labor and in-depth knowledge in order to develop efficient feature extraction methods. Also, they do not generalize very well over large datasets.

In recent years, learning features using deep learning techniques has been gaining increasing attention due to their ability to design powerful and very generalizable features. The success of deep networks in action recognition can also be attributed to the extension of networks to tens of millions of parameters and huge labeled datasets. Recent deep networks have achieved surprisingly high recognition performance on various action datasets.

Existing deep learning recognition approaches can roughly be categorized into three main family of approaches namely spatial based, multi-streams based and hybrid approaches.

Historically, the spatial based recognition approaches are the first ones that have been proposed in literature. They are based on the convolutional neurons which are a deep adaptation of multilayer neurons to process data from the real world. One such seminal architecture is the convolutional architecture introduced by LeCun et al. [6,7] named ConvNet. This architecture is a succession of layers alternating 2D convolutions (to capture salient information). More recent architectures that are only based on spatial information have been proposed. For instance, Baccouche et al. [1] proposed an architecture using 3D convolutions instead of the original 2D CNNs.

The most important drawback of those first approaches is the fact that they only rely on spatial information. Discarding the temporal sequencing of an action is likely to confuse several action classes. To tackle that drawback, authors in the literature proposed multi streams architectures. For instance, Simmoyan and Zisserman [11] proposed an architecture which has two distinct networks: one for the spatial context and the other for the context of movement.

In particular, the space network is supplied with a single image of the video. The authors experimented with the inputs into the temporal network and concluded that the bidirectional optical flow stacked during 10 successive frames gave better results. The two streams were trained separately and their results were then combined using an SVM. For the final decision, multiple clips were sampled from the entire video and the prediction scores from those videos were averaged. Although this type of approach markedly improves the performance of the single-flow method by explicitly capturing local time motion, there were still some drawbacks. On the one hand, it does not allow interactions between the two flows. On the other hand, since the video level forecasts were obtained from average forecasts on sampled clips, the long-term time information was always missing in the learned functions.

Therefore, authors have recently proposed authentically hybrid architectures to better aggregate spatial and temporal information. These architectures mainly rely on the use of the well known Long Short Term Memory (LTSM) neurons. These later allow to better capture long relations among a video without having to explicitly split it. That is why, building most of such architectures involves adding a repeating layer on top of CNNs, such as LSTM to take advantage of the benefits of CNN networks and LSTM modules. For instance, Donahue et al. [4] proposed the Long-term Recurrent Convolutional Networks architecture named as LRCN. Authors extended the coder-decoder architecture for video representations by relying on the idea of using LSTM blocks (decoder) after the convolution blocks (encoder). Finally, their architecture has the advantage of being trained end-to-end for action recognition which implies a better interaction between spatial and temporal information.

Based on our review of the state of the art in action recognition, we notice that the existing architectures proposed to model an action using the entire RGB frame regardless of the relevance of the information provided by the different parts of this RGB frame. Therefore, in this work, we propose to investigate the contribution of focusing the learning of actions onto the targeted subjects in each frame of an RGB video instead of considering the whole frame. To the best of our knowledge, our approach is the first attempt to push the attention to its extreme degree by focusing the learning on the main part of an RGB frame during acting, namely the subject itself. This approach is described in the next section.

3 Proposed Study

As introduced above, our study aims at investigating the interest of focusing the learning of actions onto the targeted subjects in each frame of an RGB video instead of considering the whole frame. This is an extreme way of applying attention to the input data. Different from previous works that conceived architectures that look for the image parts containing the most discriminant information, we propose to provide a recognition architecture with that information as an input. This would likely reduce the searching space by intentionally pushing the model to focus on the most discriminant part of the frame, and thus would both improve recognition performance and reduce convergence time.

To this end, we based our model on a performing deep architecture namely VGG16. It is a convolutional neural network model proposed by K. Simonyan and A. Zisserman in their paper "Very Deep Convolutional Networks for Large-Scale Image Recognition" [12]. The model achieves 92.7% top-5 test accuracy in ImageNet, which is a dataset of over 14 million images belonging to 1000 classes. It was one of the famous model submitted to the ILSVRC-2014 challenge. It makes the improvement over AlexNet by replacing large kernel-sized filters (11 and 5 in the first and second convolutional layer, respectively) with multiple 3×3 kernel-sized filters one after another. VGG16 was trained for weeks and was using NVIDIA Titan Black GPU's.

We provide in Fig. 1 an overview of the VGG16 architecture. As shown, each RGB frame is passed through a stack of convolutional (conv.) layers, where the filters were used with a very small receptive field: 3×3 (which is the smallest size to capture the notion of left/right, up/down, center). In the last layer of the architecture, the original model classifies each frame in one among 1000 classes.

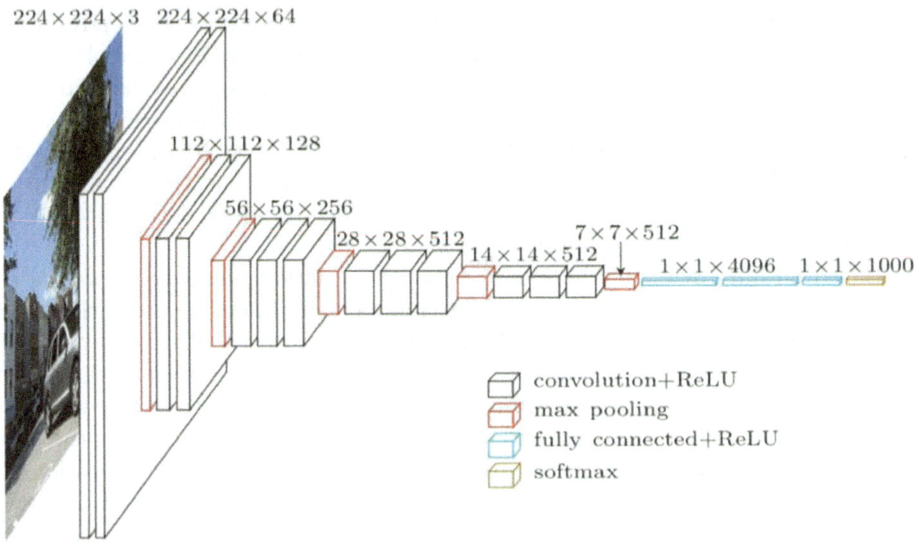

Fig. 1. Illustration of the overall architecture of VGG16 [12].

In our case, we adapted accordingly the architecture to a fewer number of classes. To get the final class label, we average the score provided by the model for each frame. Thus, the class getting the highest score is the label of the input sequence.

We trained the VGG16 based architecture according to two different protocols. According to the first protocol, the model is trained as done in previous works. That is to say, we used the whole RGB images and passed them to the network, in train/valid and test split. The results obtained according to this first protocol are used as baselines.

The proposed (second) protocol is composed of several steps. First, we need to crop the videos by limiting them to the region in which the person doing the action is located and eliminate all the other components of the scene that do not intervene in the action. To locate the subject in the frame we rely on the very efficient object detection model called YOLO [8]. Besides being very efficient, YOLO is a very fast object detection architecture that processes images in real-time at 45 frames per second.

After cropping each frame, a size-related problem should be fixed. The cropped frames issued from different videos are likely to have different sizes. To fix this problem, we thought of two possible solutions. The first one consists in training the model with a batch containing a single video at a time. The disadvantages of this solution are the long learning time and also the low learning rate because the network aggregate weakly information extracted from different videos. The second solution is to process videos per packet, by grouping the data in subsets of the same size.

In the present work, we propose to enforce all videos to have the same size. This is obtained by normalizing the frames concerning the size estimated as being the most frequent in our data. For this, we went through the entire collection of cropped videos and have drawn the two distributions (width and length) to have the most frequent size in the new set of videos. The width and height size distribution obtained after this process on the NTU RGB+D dataset [10] are shown in Fig. 2a and Fig. 2b, respectively.

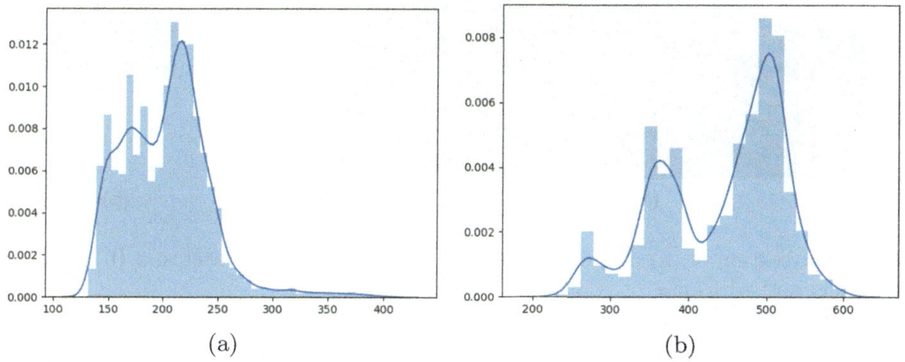

Fig. 2. (a) Width and (b) height distribution of the cropped videos from NTU RGB+D dataset.

According to the two distributions, we noticed that for the NTU RGB+D dataset on which we conducted tests, the two quantities (width and length) follow the normal law with averages of 220 pixels for the width and 500 pixels for the height. From these two sizes, we went through the collection a second time to crop the videos. Using YOLO we locate the region in which the person doing the action is located in the video, we fall into one of the following cases: the size of the cropped frame is larger than the fixed sizes, or it's smaller than these. If the size is larger, we have cropped the video to the values we found by deleting part of the frames. Otherwise, if the size is smaller, we have cropped the video to the values we found and then we reproduce the pixels of the last rows and columns of each frame, i.e. duplicate the pixels at the cut boundaries. To further accelerate the recognition process, we converted original RGB videos to grayscale videos. An example of the new frames obtained after this processing is shown into Fig. 3.

4 Experimental Results

The proposed approach is evaluated on a public dataset, namely NTU RGB+D dataset [10]. This dataset is one of the largest depth-based action recognition dataset. It is collected by Kinect v2 and contains more than 56 thousand sequences and 4 million frames. A total of 60 different action classes including daily actions, pair actions, and medical conditions are performed by 40 subjects

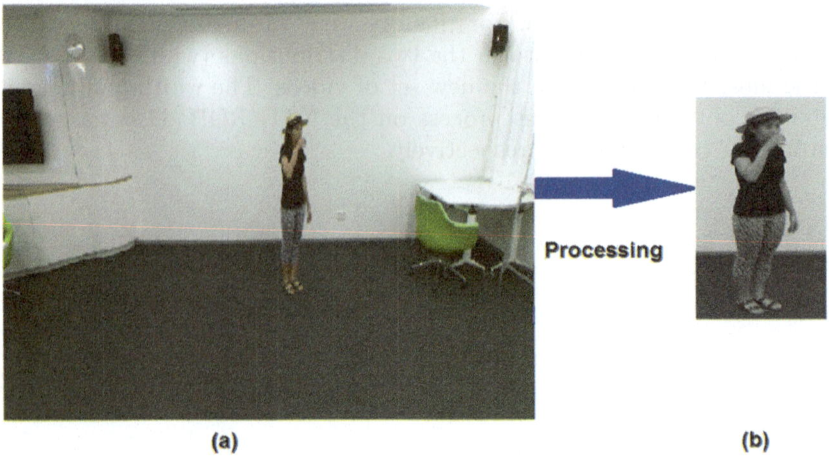

Fig. 3. Illustration of (a) original frame from NTU RGB+D dataset and (b) its corresponding cropped frame.

aged between 10 and 35. The 3D coordinates of 25 joints are provided in this dataset. The large intra-class and viewpoint variations make this data set very challenging. Due to a large number of samples, this dataset is highly suitable for deep learning-based action recognition.

Given that our aim is to evaluate the interest of the proposed processing, in our experiment we selected three classes namely drink, clapping and hopping. Then we split these three class data into training/validation and testing sets. The parameters of the retained architectures were set to default. For both settings, we trained the model for 100 epochs.

We first present and discuss results obtained with the raw data and then those obtained with processed data. Figure 4 provides the evolution of the loss and accuracy during the training and validation of the raw data.

We mainly notice that, even after several epochs, the accuracy stills very low, i.e. around 50%. Besides, by comparing the training and the validation accuracy evolution, it looks like even after 100 epochs, the model still can learn as the validation accuracy is continuously growing. That is to say that the model is unable to locate the most discriminant information in the different frames. We then recorded results on the testing set. For measuring the performance we compute for each experiment the different metrics namely the precision, the recall, the F1-score, the accuracy, the macro and the weighted averages.

The precision is the ratio $tp/(tp+fp)$ where tp is the number of true positives and fp the number of false positives. The precision is intuitively the ability of the classifier not to label as positive a sample that is negative. The recall is the ratio $tp/(tp+fn)$ where tp is the number of true positives and fn the number of false negatives. The recall is intuitively the ability of the classifier to find all the positive samples. The F1-score can be interpreted as a weighted average of the precision and recall, where an F1-score reaches its best value at 1 and worst

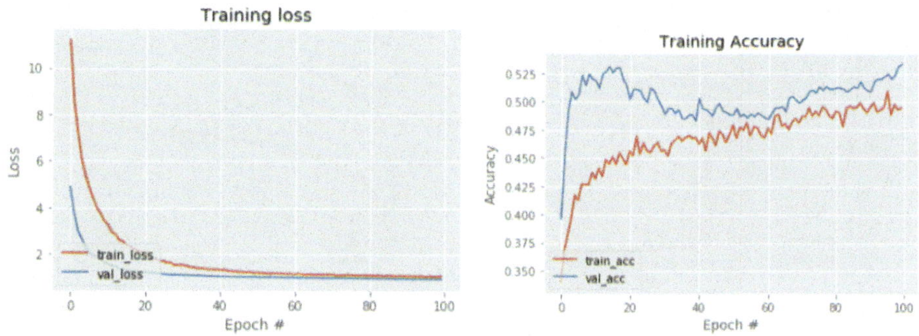

(a) (b)

Fig. 4. (a) Evolution of the training loss and (b) accuracy on raw videos from NTU RGB+D dataset.

score at 0. The relative contribution of precision and recall to the F1-score are equal. The support is the number of ground truth occurrences of each class.

We also provide two extra metrics namely macro and weighted averages. On the one hand, the macro average calculates metrics for each label, and find their unweighted mean. This does not take label imbalance into account. On the other hand, the weighted average calculates metrics for each label, and find their average weighted by support (the number of true instances for each label). This alters the macro average to account for label imbalance. Table 1 gives a summary of those results for the three classes.

Table 1. Results on raw data (without preprocessing)

	Precision	Recall	F1-score	Support
Class 1: Drink	0.60	0.77	0.68	1108
Class 2: Clapping	0.54	0.54	0.54	842
Class 3: Hopping	0.88	0.61	0.72	1015
Accuracy	–	–	**0.65**	2965
Macro average	0.67	0.64	0.64	–
Weighted average	0.68	0.65	0.65	–

As shown in Table 1, the model overall performance equals 65%. Giving that the three classes are quite distinct, this score represents a low performance. Also, as detailed in this table, the model is particularly weak for recognizing the clapping action with an F1-score value of 54%.

Similarly, we provide in Fig. 5 the evolution of the loss and accuracy while training on processed data. We clearly notice that there is a great improvement on the validation set. In particular, the score on the validation set surpasses

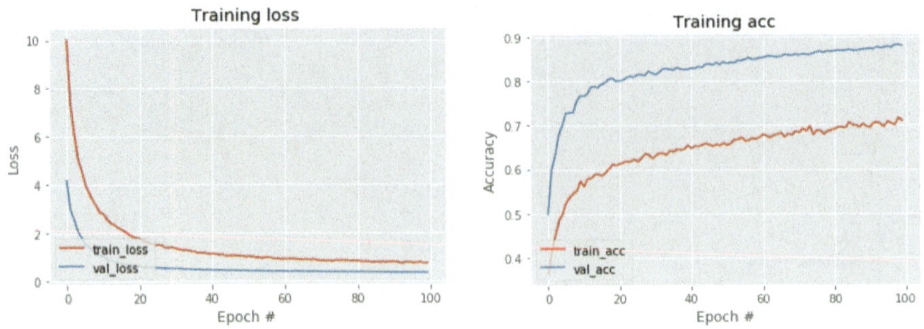

(a) (b)

Fig. 5. (a) Evolution of the training loss and (b) accuracy on processed videos from NTU RGB+D dataset.

70%, which is a far greater value than the one obtained using raw data, namely 50%. Besides, it appears that the convergence of the stationary state during the training process was reached quicker than when using raw data. This suggests that providing the model with the processed data helps to find and modeling the most important and discriminant information.

We also provide the detailed scores on the test set in Table 2.

Table 2. Results on processed data

	Precision	Recall	F1-score	Support
Class 1: Drink	0.88	0.90	0.89	1108
Class 2: Clapping	0.82	0.83	0.82	842
Class 3: Hopping	0.90	0.87	0.89	1015
Accuracy	–	–	**0.87**	2965
Macro average	0.867	0.867	0.867	–
Weighted average	0.869	0.869	0.870	–

Compared to results on raw data, it is clear that processing the data the way we did improves drastically the recognition performance. In particular, the overall performance achieved using the processed data equals 87%, which corresponds to the improvement of more than 20%. Besides, we notice that the recognition improvement concerns all action classes which suggests that the model has properly identified the most discriminant information for each class.

In Fig. 6, we provide some recognition results on raw samples from NTU RGB+D dataset. While the model manages to properly classify some frames, like in Fig. 6b, it appears also that the model misclassified samples at a frame that clearly shows which action is performed. For instance, in Fig. 6d, the subject clearly performs a drink action while the model predicted it as a clapping action.

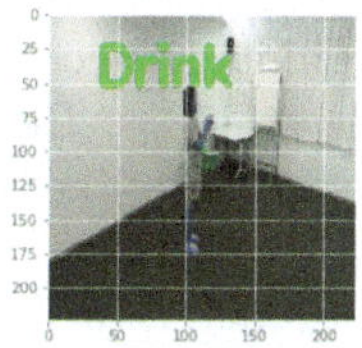

(a) GT = Clapping

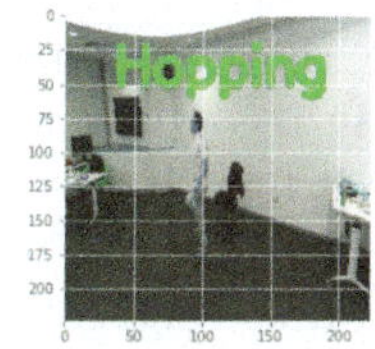

(b) GT = Hopping

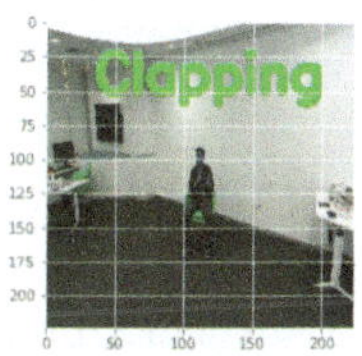

(c) GT = Clapping

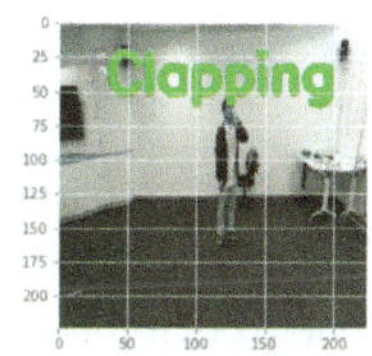

(d) GT = Drink

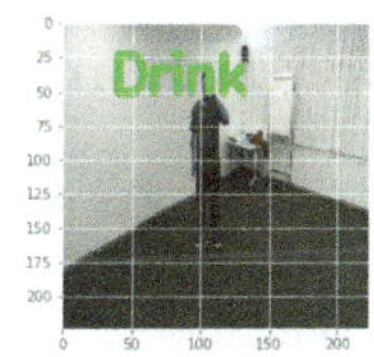

(e) GT = Drink

Fig. 6. Recognition results on raw samples from NTU RGB+D dataset. The predicted label is reported on the frame. GT stands for Ground Truth label.

In Fig. 7, we provide some recognition results on processed samples from NTU RGB+D dataset.

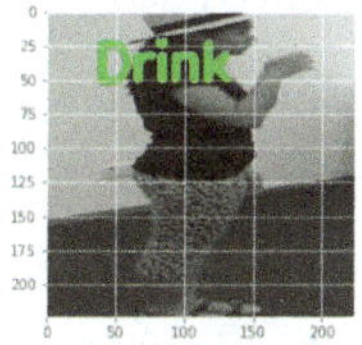

(a) GT = Drink

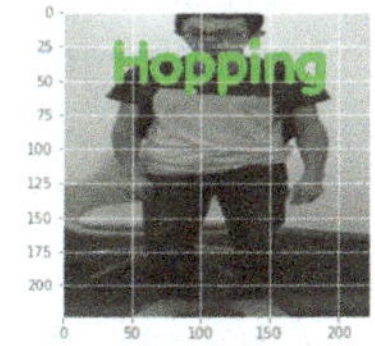

(b) GT = Hopping

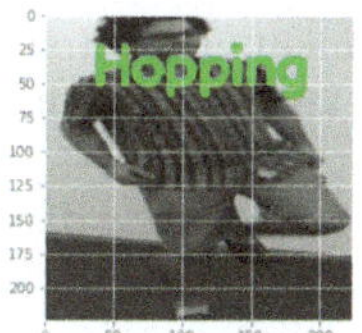

(c) GT = Hopping

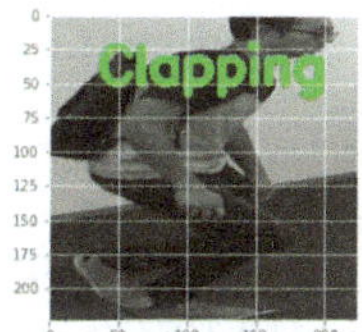

(d) GT = Hopping

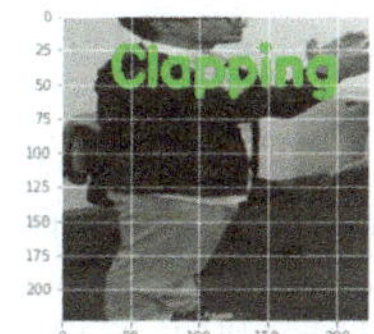

(e) GT = Clapping

Fig. 7. Recognition results on processed samples from NTU RGB+D dataset. The predicted label is reported on the frame. GT stands for Ground Truth label.

As shown in these frames, the information related to the subject is dominating the frame. In these examples, we can see that the model succeeds to properly recognize actions despite the difference in perceiving angles. Besides, in case it fails, it is most likely because the action has not yet started like in the illustration of Fig. 7b.

As far as time consumption is concerned, the proposed approach is at least as fast as existing state-of-the-art methods. In particular, it allows the processing and the classification of a single frame in about 0.016 s.

5 Conclusion

In this paper, we propose a learning procedure to improve the action recognition process. In particular, we intend to push the attention while training a recognition model to an extreme degree by focusing on the subject. The proposed learning procedure is composed of several steps. First, we apply an object detection model, here YOLO, to locate the subject in each frame. We then statistically identify the most common width and height of the box surrounding the subjects and process the frames according to these sizes. Last, the selected recognition model is trained on the processed data to improve its recognition capacity. Our approach is thus an extreme way of guiding the model during the training process. This is a very light and more efficient alternative compared to the complex mechanisms used to identify the interesting parts of each frame. As reported in the paper, we noticed a significant improvement after the data are processed enhancing the recognition score by almost 30%. In future work, we intend to extend this process to recognizing actions in unsegmented streams and evaluate its performance for real usage.

References

1. Baccouche, M., Mamalet, F., Wolf, C., Garcia, C., Baskurt, A.: Sequential deep learning for human action recognition. In: International Workshop on Human Behavior Understanding, pp. 29–39. Springer (2011). https://doi.org/10.1007/978-3-642-25446-8_4
2. Bobick, A.F., Davis, J.W.: The recognition of human movement using temporal templates. IEEE Trans. Pattern Anal. Mach. Intell. **3**, 257–267 (2001). https://doi.org/10.1109/34.910878
3. Boulahia, S.Y., Anquetil, E., Multon, F., Kulpa, R.: Cudi3d: Curvilinear displacement based approach for online 3d action detection. Comput. Vis. Image Underst. **174**, 57–69 (2018). https://doi.org/10.1016/j.cviu.2018.07.003
4. Donahue, J., et al.: Long-term recurrent convolutional networks for visual recognition and description. In: Proceedings of the IEEE Conference on Computer Vision and Pattern Recognition pp. 2625–2634 (2015). https://doi.org/10.1109/TPAMI.2016.2599174
5. Laptev, I.: On space-time interest points. Int. J. Comput. Vision **64**(2–3), 107–123 (2005). https://doi.org/10.1007/s11263-005-1838-7

6. LeCun, Y., Bottou, L., Bengio, Y., Haffner, P., et al.: Gradient-based learning applied to document recognition. Proc. IEEE **86**(11), 2278–2324 (1998). https://doi.org/10.1109/5.726791
7. LeCun, Y., Kavukcuoglu, K., Farabet, C.: Convolutional networks and applications in vision. In: Proceedings of 2010 IEEE International Symposium on Circuits and Systems, pp. 253–256. IEEE (2010). https://doi.org/10.1109/ISCAS.2010.5537907
8. Redmon, J., Divvala, S., Girshick, R., Farhadi, A.: You only look once: unified, real-time object detection. In: Proceedings of the IEEE Conference on Computer Vision and Pattern Recognition, pp. 779–788 (2016). https://doi.org/10.1109/CVPR.2016.91
9. Schuldt, C., Laptev, I., Caputo, B.: Recognizing human actions: a local SVM approach. In: Proceedings of the 17th International Conference on Pattern Recognition, 2004. ICPR 2004, vol. 3, pp. 32–36. IEEE (2004). https://doi.org/10.1109/ICPR.2004.1334462
10. Shahroudy, A., Liu, J., Ng, T.T., Wang, G.: Ntu rgb+ d: a large scale dataset for 3d human activity analysis. In: Proceedings of the IEEE Conference on Computer Vision and Pattern Recognition, pp. 1010–1019 (2016). https://doi.org/10.1109/CVPR.2016.115
11. Simonyan, K., Zisserman, A.: Two-stream convolutional networks for action recognition in videos. In: Advances in Neural Information Processing Systems, pp. 568–576 (2014)
12. Simonyan, K., Zisserman, A.: Very deep convolutional networks for large-scale image recognition. arXiv preprint arXiv:1409.1556 (2014)
13. Yamato, J., Ohya, J., Ishii, K.: Recognizing human action in time-sequential images using hidden Markov model. In: Proceedings 1992 IEEE Computer Society Conference on Computer Vision and Pattern Recognition, pp. 379–385. IEEE (1992). https://doi.org/10.1109/CVPR.1992.223161
14. Ziaeefard, M., Ebrahimnezhad, H.: Hierarchical human action recognition by normalized-polar histogram. In: 2010 20th International Conference on Pattern Recognition, pp. 3720–3723. IEEE (2010). https://doi.org/10.1109/ICPR.2010.906

Medium Term Electricity Load Forecasting Using Machine Learning Techniques

Bakir Abderrahim(✉) and Gherbi Fatima Zohra

Intelligent Control and Electrical Power System Laboratory, University Djillali Liabes, Sidi-Bel-Abbes, Algeria
rmbakir8@gmail.com, fzgherbi@gmail.com

Abstract. Accurate load forecasting plays a key role in the economic and reliable operation of electrical utilities. Therefore, a rigid forecasting approach with strong general nonlinear mapping abilities is important. Among the existing forecasting models, support vector regression (SVR) has received significant attention. As the heart of SVR, the kernel function is used to map the input vectors into a higher dimensional feature space to build the non-linear algorithm for original space. However, the selection of kernel function and corresponding parameters directly affects the SVR's ability to learn and generalize. This paper compares the performances of six SVR models based on different kernel functions and different kernel parameters values, the kernel functions used are as follows: linear, polynomial and Gaussian radial bias (RBF). The data used in this study were historical half hourly load and weather observations from the Australian energy market operator (AEMO). Several error indicators were used to compare and evaluate the performance of the predictive models. The simulation results show that the SVR-RBF model with medium gamma parameter value is a superior method for medium term load forecasting due to its ability to process load data and it has fewer errors which is a much better result than the other proposed forecasting models.

Keyword: Load forecasting · Medium term · Support vector regression · Kernel functions

1 Introduction

With the implementation of deregulation in power industry, many challenges have been faced by the participants of the electricity market. Forecasting electricity parameters such as load and energy prices has become a major concern in power systems. The main objective of electric power industry deregulation is to maximize efficient generation and consumption of electricity and reduce energy prices. To achieve these goals, accurate and efficient electricity load forecasting is becoming more and more important. An accurate load forecasting is very useful for reducing cost and maintaining security and stability of power grid. However, because the electric load is affected by various factors, such as social factors, economic situations, irregular behaviors, climate factors (temperature

M. R. Senouci et al. (Eds.): CSA 2020, LNNS 199, pp. 50–60, 2021.
https://doi.org/10.1007/978-3-030-69418-0_5

and humidity) and other non-linear factors, it is very difficult to forecast the electricity load accurately [1, 2].

Depending on the length of the time period, load forecasting can be classified into three categories: short-term load forecasting (STLF) which is typically from one hour to one week, medium-term load forecasting (MTLF), typically from one week to one year, and long-term load forecasting (LTLF) which is longer than one year. MTLF plays a major role in power systems, Accurate MTLF provides information to utilities to better plan (or purchase) power generation expansion, schedule maintenance activities, improve system performance, trade forward contracts and develop cost efficient fuel purchasing strategies [3].

Various techniques have been developed for electricity demand forecasting during the past few years. Lingyi Han et al. [4] proposed two deep learning methods, time-dependency convolutional neural network (TD-CNN) and cycle-based long short-term memory (C-LSTM) network to improve the STLF and MTLF accuracy. In [5] a single and multi- sequence LSTM and gated recurrent unit (GRU) models were developed to achieve better performances for STLF and MTLF. In addition, the application of support vector machine (SVM) technology for load forecasting in power system has received much attention in recent years. SVM was first time applied to load forecasting in [6] where it was observed that support vector regression (SVR) performs well for time series analysis. SVR is an extension of SVM which is used for regression analysis, the main reason of SVR becoming so popular lies in its ability to learn complex and nonlinear relationships that are difficult to model with conventional techniques and its ability to solve complex problems with higher accuracy under uncertain conditions.

Ajmera et al. [7] presented an SVR model for MTLF using PJM market data. The results show that the SVR model have great potential. Arghavan et al. [8] proposed a hybrid model based on SVR which was optimized by organism search optimization (SOSO) method, and had good forecasting results for MTLF on the EUNITE competition dataset. Zhongyi Hu et al. [9] used the firefly algorithm (FA) to select proper input features for a multi output SVR model which can give more accurate forecasting results in North America and Australia markets.

This paper compares the accuracy of six SVR models for MTLF based on three types of kernel functions with different kernel parameters values.

2 Support Vector Machine SVM

Support vector machine (SVM) analysis is a popular machine learning tool for classification and regression, first identified in 1992 by Vladimir Vapnik et al. [10]. Support vector machines (SVMs) are supervised learning methods which create input-output mapping functions from a set of labeled training data. The mapping function can be either a classification function, used to categorize the input data; or a regression function, used for the estimation of the desired output. In classification, nonlinear kernel functions are often used to transform the input data (inherently representing highly complex nonlinear relationships) to a high dimensional feature space in which the input data becomes more separable, i.e., linearly separable, compared to the original input space. Then, the maximum-margin hyperplanes are constructed to optimally separate the classes in the training data. Two parallel hyperplanes are constructed on each side of the hyperplane that separates the data by maximizing the distance between the two parallel hyperplanes. An assumption is made that the larger the margin or distance between these parallel hyperplanes the better the generalization error of the classifier will be.

A version of SVM for regression has been proposed in 1997 by Vapnik, Steven Golowich, and Alex Smola [11]. The study of MTLF belongs to the latter that is called support vector machine regression (SVR).

3 SVR Methodology

3.1 The Concept of Support Vector Regression

The SVR is a linear regression in the feature space. This space is implemented by a mapping function which maps the input space to the feature space, i.e. $\phi : R^n \rightarrow R^v$. In general $v > n$ so that the nonlinear function under approximation in the input space becomes a linear function in the feature space. Given input output data $\{x_i, y_i\}_{i=1}^N$ where each $x_i \in R^n$ reflects the input space of the sample and has a relevant target value $y_i \in R$ for $i = 1, \ldots, N$, where N corresponds to the size of the training data. The estimating function takes the form as follows [12]:

$$f(x) = w, x + b \tag{1}$$

Where, w, x represents the dot product, w contains the coefficients that have to be estimated from the data and b is a real constant. Support vector regression can be used to solve the regression estimation problem by introducing Vapink's ν insensitive loss function [13].

$$|\xi|_\nu = \begin{cases} 0 & \mathit{if}\ |\xi| \leq \nu \\ |\xi| - \nu & \mathit{otherwise} \end{cases} \tag{2}$$

The overall optimization is formulated as:

$$\min_{w,b,\xi,\xi^*} \frac{1}{2} w^T w + C \sum_{i=1}^{N} (\xi + \xi_i^*) \tag{3}$$

$$s.t. \begin{cases} y_i - \left(w^T \phi(x_i) + b\right) \leq \nu + \xi_i^* \\ \left(w^T \phi(x_i) + b\right) - y_i \leq \nu + \xi_i^* \\ \xi, \xi_i^* \geq 0, \qquad i = 1, \ldots, N \end{cases} \tag{4}$$

where x_i is mapped to higher dimensional space by the function ϕ, ν is a real constant, ξ_i (the lower training error) and ξ_i^* (the upper training error) are slack variables subject to ν-insensitive loss function, the regularization parameter (C) determines the trade-off between the flatness of f and training errors.

Through introducing Lagrange multipliers α_i and α_i^*, and according to the Karush-Kuhn Tucker optimality conditions [12], the SVR training procedure amounts to solve the convex quadratic problem:

$$\min_{\alpha,\alpha^*} \frac{1}{2} \sum_{i,j=1}^{N} \left(\alpha_i - \alpha_i^*\right)\left(\alpha_j - \alpha_j^*\right) K\left(x_i + x_j\right) + \nu \sum_{i=1}^{N} \left(\alpha_i - \alpha_i^*\right) - \sum_{i=1}^{N} y_i\left(\alpha_i - \alpha_i^*\right) \tag{5}$$

With the constraints:

$$\sum_{i=1}^{N}(\alpha_i - \alpha_i^*) = 0, i = 1, \ldots, N\, for\, 0 \le \alpha_i, \alpha_i^* \le C \quad (6)$$

The regression output takes the following form:

$$\hat{y} = \hat{f}(x) = \sum\nolimits_{i=1}^{N}(\alpha_i - \alpha_i^*)Q(x_i - x) + b \quad (7)$$

Where, $Q(x_i - x)$ is a kernel function. By employing kernels in SVR, all necessary computations can be carried out directly in the input space. For MTLF in this study, SVR is performed based on three forms of kernel function Eqs. (8, 9 and 10) where each kernel function has a particular parameter that must be optimized to obtain the best performance results [14].

Linear

$$k(x_n, x_i) = (x_n.x_i) \quad (8)$$

Polynomial. With d is the kernel parameter to be optimized.

$$k(x_n, x_i) = ((x_n.x_i) + 1)^d \quad (9)$$

Gaussian Radial Basic Function (RBF). With σ is the kernel parameter to be optimized.

$$k(x_n, x_i) = \exp\left(-\frac{x_n - x_i^2}{2\sigma^2} = \exp(-\gamma x_n - x_i^2\right) \quad (10)$$

Where d is the degree of the polynomial kernel, $\left(\gamma = \frac{1}{2\sigma^2}\right)$ is an adjustable hyper parameter or gamma parameter and σ is the width of the RBF kernel.

3.2 SVR Models for MTLF

In this study, we propose a medium-term load forecasting method based on six SVR models. In order to develop the SVR models, the initial dataset was divided into training and testing as, 80% training data with 70080 observations and 20% test data with 17520 observations. The training set is used for building the prediction models, and the test set is adopted for evaluating the effectiveness of the SVR models.

Support Vector Regression with Linear Kernel Function (SVR-L). The linear kernel implementing the simplest of all kernel functions. In this paper no parameters of kernel functions are used to optimize the SVR-L model.

Support Vector Regression with Polynomial Kernel Function (SVR-P). In support vector regression with polynomial kernel functions, two models were developed based on the polynomial degree parameter. The models are the quadratic SVR-P with (d = 2) and the cubic SVR-P with (d = 3).

Support Vector Regression with Radial Basis Function (SVR-RBF). In support vector regression with radial basis function (SVR-RBF), three models were selected based on three values of gamma (γ) parameter, a lower value of ($\gamma = 0.006$), a medium value of ($\gamma = 0.098$) and a higher value of ($\gamma = 1.602$).

For increasing the accuracy of the models and avoid overfitting, k-fold cross-validation technique with (k = 5) was used in the development of the SVR models.

4 Dataset Description

The dataset used in this paper is a table of historical half-hourly loads and temperature observations from the AEMO and BOM for Sydney, Australia [15]. Weather information includes dry bulb, wet bulb temperatures, dew point and humidity. The models are trained and validated on half-hourly data from January 01, 2006 to December 31, 2009 and tested on out-of-sample data from January 01, 2010 to December 31, 2010. The temperature variable is included because it has a nonlinear relationship with electricity load.

For the load forecast, the input data for the models include the following features:

- Dry bulb temperature, Wet bulb temperature, Humidity, Dew point temperature.
- Hour of day, Day of the week, Holiday/weekend indicator (0 or 1).
- Previous 24-h average load, 24-h lagged load, 168-h (previous week) lagged load.

5 Simulation and Results

In this section, results of electricity load forecasting are presented. All the developed models are implemented in MATLAB 2019a and the simulations are performed using a server of 2.50 GHz Intel Core i5 CPU and 8 GB RAM.

5.1 Training Results

Measurement of Training Accuracies. The forecasting performance of our six models for load are evaluated using four statistics as, coefficient of determination (R^2), root-mean-square error (RMSE), mean square error (MSE) and mean absolute error (MAE) which are calculated using Eqs. (11–14) [16], respectively:

$$R^2 = 1 - \frac{\sum_{tm=1}^{TM} (A_v - F_v)^2}{\sum_{tm=1}^{TM} (A_v - \overline{A})^2} \tag{11}$$

$$RMSE = \sqrt{\frac{1}{T} \sum_{tm=1}^{TM} (A_v - F_v)^2} \tag{12}$$

$$MSE = \frac{1}{T} \sum_{tm=1}^{TM} (A_v - F_v)^2 \tag{13}$$

$$MAE = \frac{\sum_{tm=1}^{TM} |F_v - A_v|}{T} \tag{14}$$

where A_v is the actual value at time TM and F_v is the forecasted value at time TM.

The best model has a small RMSE, MSE and MAE values and the coefficient of determination R^2 approaches the value of one.

Table 1 provides a summary of all the error metric results obtained for the three kernels based SVR models.

Table 1. Training accuracy of the proposed MTLF models.

Kernel	Error metric			
	R^2	RMSE (MWh)	MSE (MWh)	MAE (MWh)
Linear	0.90	454.11	206220	314.6
Polynomial ($d = 2$)	0.94	353.22	124770	255.27
Polynomial ($d = 3$)	0.96	284.05	80683	204.7
RBF ($\gamma = 0.006$)	0.94	344.22	118490	247.07
RBF ($\gamma = 0.098$)	0.97	226.06	51103	161.51
RBF ($\gamma = 1.602$)	**0.99**	**154.57**	**23892**	**115.74**

Table 1 describes the results obtained from training the SVR models using 5-fold cross validation. As shown in the Table, the SVR-RBF model with ($\gamma = 1.602$) gave the best result out of the six models with $R^2 = 0.99$ the closest value to one, and the lowest RMSE, MSE and MAE values of 154.57 MWh, 23892 MWh and 115.74 MWh respectively, while the SVR-L model gave the worst.

The SVR-RBF with ($\gamma = 0.098$) gave the second best performance and the cubic SVR-P outperforms the quadratic SVR-P and SVR-RBF ($\gamma = 0.006$) models, but it should be mentioned here that this model took the longest training time compared to all other models, which makes it a less important option when dealing with large datasets such as our case study.

5.2 Forecasting Results

Measurement of Forecast Accuracies. In this section, four accuracies are selected to assess the prediction performance of our models, they are mean absolute percentage error MAPE Eq. (15) [16], MAE, RMSE and the daily peak MAPE Eq. (16).

MAPE is one often used metric, which measure the percentage error between the actual and predicted values. The smaller the values of MAPE, the closer the predicted values to the actual values.

$$MAPE = \frac{1}{T}\sum_{tm=1}^{TM} 100\left|\frac{A_v - F_v}{A_v}\right| \tag{15}$$

$$daily\,peak\,MAPE = MAPE\,for\,load\,spike \tag{16}$$

In order to verify the performance of the proposed models and to validate their robustness, two well-known deep learning: artificial neural network trained by Levenberg-Marquardt backpropagation algorithm (BP-ANN) and machine learning: random forest (RF) models were also developed and applied to the dataset described in Sect. 4.

The following Table contains a summary of all the error metric results obtained for the three kernels based SVR models as well as the BP-ANN and RF models.

Table 2. Forecasting accuracy of the proposed MTLF models.

Model	Error metric			
	MAPE (%)	MAE (MWh)	RMSE (MWh)	Daily peak MAPE (%)
Linear	3.52	319.27	452.71	4.26
Polynomial ($d = 2$)	2.89	254.27	343.36	3.22
Polynomial ($d = 3$)	2.37	210.29	284.19	2.75
RBF ($\gamma = 0.006$)	2.77	245.22	331.43	3.17
RBF ($\gamma = 0.098$)	**2.02**	**179.60**	**244.76**	**2.26**
RBF ($\gamma = 1.602$)	2.93	255.97	404.75	3.01
BP-ANN	2.46	216.75	289.94	2.52
RF	2.16	192.34	268.22	2.47

The results shown in Table 2 suggest that the electricity demand forecasts based on the six SVR models are relatively acceptable. It is also clear that the BP-ANN and RF models have performed well. However, the findings of Table 2 highly confirm that the SVR-RBF model with ($\gamma = 0.098$) is more efficient in forecasting electric load, where the lower MAPE and daily peak MAPE values of 2.02% and 2.26% respectively, demonstrate the positive ability of the model to process load data, and also the lower MAE and RMSE values of 179.60 MWh and 244.76 MWh respectively confirm that the SVR-RBF model with medium gamma value is capable of efficiently and accurately forecasting the medium term electricity load in the Australian market.

In comparison to the training results, the SVR-RBF model with ($\gamma = 1.602$) shows very poor performance in forecasting electricity load, the higher error measurements indicate that the model has an over-fitting problem which makes it less relevant for load forecasting.

The following section shows the forecasting performance of our superior method, where the results are generated in graphic form.

Figure 1 shows the forecasted load of the year 2010 using SVR-RBF model with ($\gamma = 0.098$). Where, the blue line shows the actual load and the predicted load is represented by the orange line. SVR-RBF prediction depicts that the predicted load is nearly close to the actual load.

Multiple series plots between the actual load in the blue line and the forecasted load in the red line on the first week of the months of January, April, July and October 2010 have been shown in Fig. 2. A visual comparison of the two curves in the figure indicate that

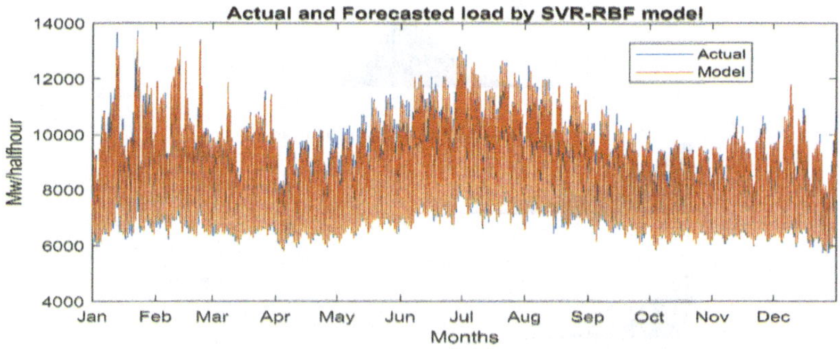

Fig. 1. Load forecasting in year 2010 by SVR-RBF model.

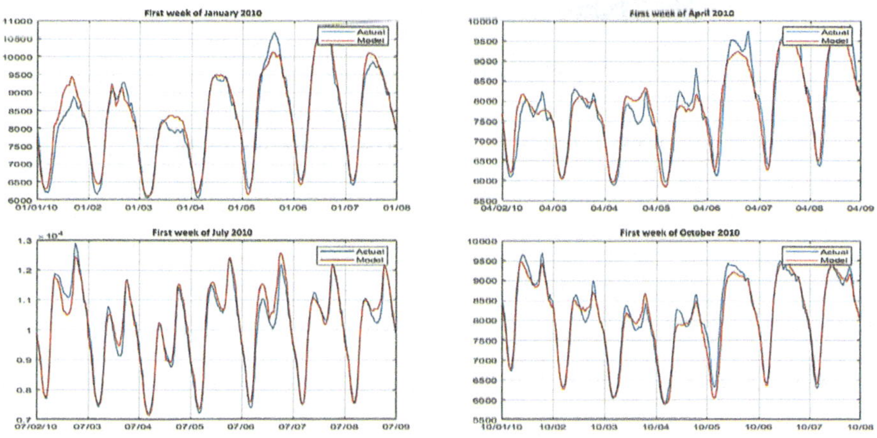

Fig. 2. Load forecasting of the first week of January, April, July and October 2010.

they are almost of the similar patterns presenting excellent forecasting performance of the proposed model.

The upper part of Fig. 3 shows how the forecasting error is normally distributed around zero, and the mean value of the error is very close to zero, which makes the model appropriate for forecasting purposes. Additionally, the lower parts of Fig. 3 reveal the distributions of MAE, MAPE and RMSE on the tails of error distribution. We can, moreover, notice that MAE and RMSE are less than 500 and MAPE is less than 5 which are clarified numerically in Table 2.

The box-plot of the error distribution of the forecasted load as a function of the hour of the day is presented in Fig. 4. It shows the percentage error statistics of hour of the day in the year of 2010. It is also evident that the maximum error is for the 8th hour and the minimum error is for the 10th hour of the day in the year of 2010.

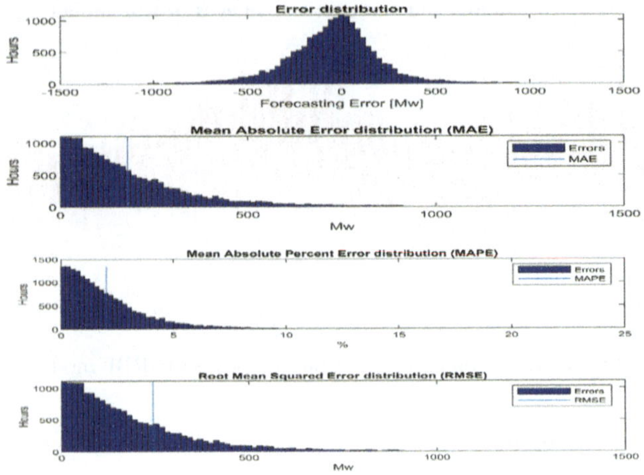

Fig. 3. Error distribution.

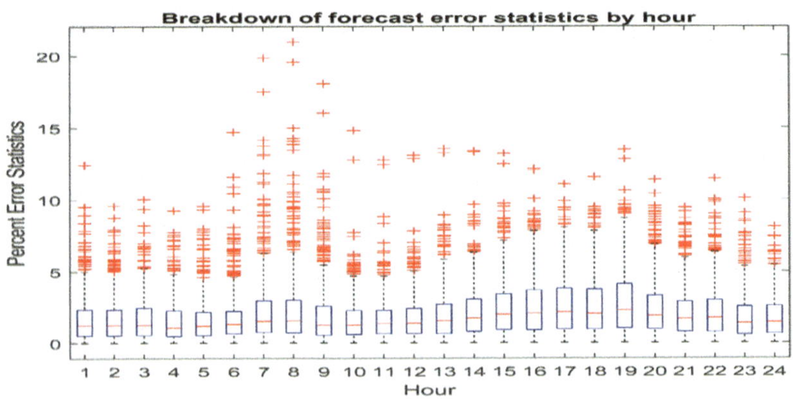

Fig. 4. Box-plot of the error distribution of forecasted load as a function of hour of the day.

Regarding the breakdown of forecasting error for days of the week, Fig. 5 shows that Tuesdays and Mondays have relatively higher forecasting error, and Wednesdays have the minimum error in the year of 2010. It is worth mentioning here that higher error does not necessarily mean higher electricity demand.

In the same manner, the last breakdown in Fig. 6 shows the forecasting error by month in the year of 2010, and it is clearly to see that the month of December exhibits the highest forecasting error and the month of November has the lowest.

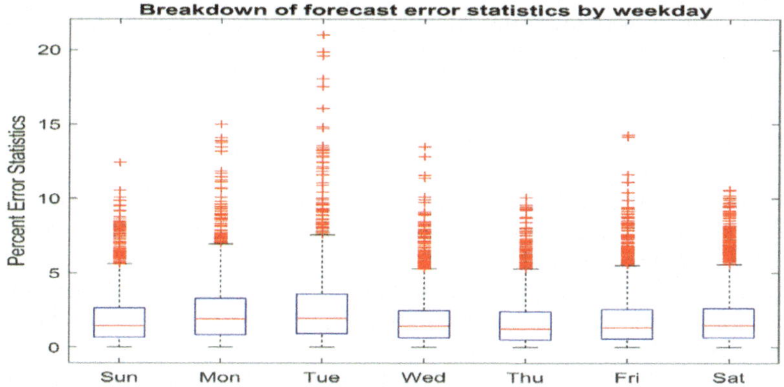

Fig. 5. Box-plot of the error distribution for the forecasted load as a function of day of the week.

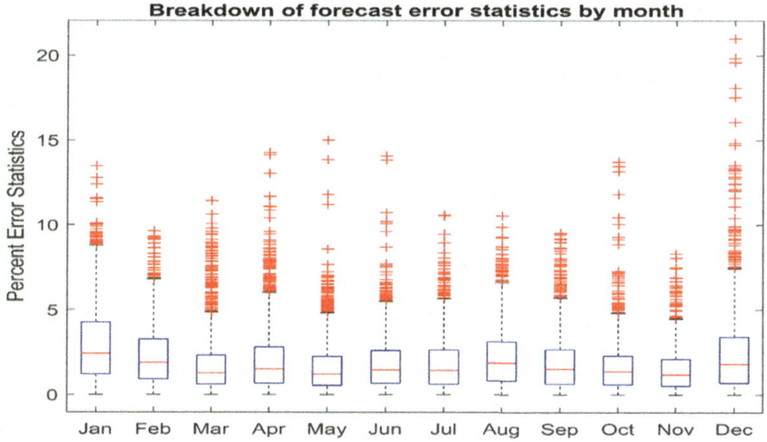

Fig. 6. Box-plot of the error distribution for the forecasted load as a function of month of the year.

6 Conclusion

In this paper, the prediction performances of support vector regression for load forecasting in electric power systems has been presented. Six SVR models with three types of kernel functions and kernel parameters values were developed and tested and the simulation results showed that the SVR model with RBF kernel seems to be more appropriate than linear and polynomial functions to solve medium term load forecasting problem, with high accuracy, in addition the SVR-RBF model with medium gamma parameter value outperforms the other proposed forecasting models in terms of MAPE = 2.02%, MAE = 179.60 MWh, RMSE = 244.76 MWh and daily peak MAPE = 2.26%.

Future work will incorporate SVR set tuned with three parameters, i.e., RBF kernel parameter (γ), cost penalty (C), and incentive loss function parameter (ε), where a metaheuristic algorithm will be used to determine the best parameters of SVR with high forecasting accuracy. Moreover, investigating the effectiveness of our approach for price forecasting in competitive electricity markets.

References

1. Negnevitsky, M., Mandal, P.: An overview of forecasting problems and techniques in power systems. In: Power & Energy Society General Meeting, Canada, pp. 1–4. IEEE (2009). https://doi.org/10.1109/PES.2009.5275480
2. Wang, J., Li, L.: An annual load forecasting model based on support vector regression with differential evolution algorithm. Appl. Energy **94**, 65–70 (2012). https://doi.org/10.1016/j.apenergy.2012.01.010
3. Ghiassi, M., Zimbra, D.K., Saidane, H.: Medium term system load forecasting with a dynamic artificial neural network model. Electr. Power Syst. Res. **76**(5), 302–316 (2006). https://doi.org/10.1016/j.epsr.2005.06.010
4. Han, L., Peng, Y., Li, Y., Yong, B., Zhou, Q., Shu, L.: Enhanced deep networks for short-term and medium-term load forecasting. IEEE Access **7**, 4045–4055 (2019). https://doi.org/10.1109/ACCESS.2018.2888978
5. Bouktif, S., Fiaz, A., Ouni, A., Serhani, M.A.: Single and multi-sequence deep learning models for short and medium term electric load forecasting. Energies **12**(1), (2019). https://doi.org/10.3390/en12010149
6. Chen, B.J., Chang, M.W., lin, C.J.: Load forecasting using support vector Machines: a study on EUNITE competition 2001. IEEE Trans. Power Syst. **19**(4), 1821–1830 (2004). https://doi.org/10.1109/TPWRS.2004.835679.
7. Ajmera, S., Singh, A.K., Chauhan, V.: An approach towards medium term forecasting based on support vector regression. In: IEEE 7th Power India International Conference 2016, PIICON, Institute of Electrical and Electronics Engineers Inc. (2017). https://doi.org/10.1109/POWERI.2016.8077232
8. Zare-Noghabi, A., Shabanzadeh, M., Sangrody, H.: Medium-term load forecasting using support vector regression, feature selection, and symbiotic organism search optimization. In: IEEE Power & Energy Society General Meeting (PESGM), Atlanta, GA, USA, pp. 1–5. IEEE (2019). https://doi.org/10.1109/PESGM40551.2019.8973726
9. Hu, Z., Bao, Y., Chiong, R., Xiong, T.: Mid-term interval load forecasting using multi-output support vector regression with a memetic algorithm for feature selection. Energy **84**(1), 419–431 (2015). https://doi.org/10.1016/j.energy.2015.03.054
10. Vapnik, V.: The Nature of Statistical Learning Theory. 2nd edn. Springer, New York (1995). https://doi.org/10.1007/978-1-4757-3264-1
11. Vapnik, V., Golowich, S., Smola, A.: Support vector method for function approximation, regression estimation, and signal processing. In: Mozer, M.C., Jordan, M.I. (eds.) Neural Information Processing Systems 1996, NIPS, pp. 281–287 (1996)
12. Smola, A.J., Scholkopf, B.: A tutorial on support vector regression. Stat. Comput. **14**(3), 199–222 (2004). https://doi.org/10.1023/B:STCO.0000035301.49549.88
13. Vapnik, V.N.: Statistical Learning Theory. Wiley, Hoboken (1998)
14. Karatzoglou, A., Meyer, D., Hornik, K.: Support Vector Machines in R. J. Stat. Softw. **15**(9) (2006). https://doi.org/10.18637/jss.v015.i09
15. MathWorks Homepage, https://www.mathworks.com/matlabcentral. Accessed 29 Nov 2019
16. Zahid, M., Ahmed, F.: Electricity price and load forecasting using enhanced convolutional neural network and enhanced support vector regression in smart grids. Electronics **8**(1) (2019). https://doi.org/10.3390/electronics8020122

Can Judges Trust the I-Vectors Scores?: A Comparative Study of Voices Comparison in the Forensic Domain

Kawthar Yasmine Zergat(✉), Yazid Kahil, and Abderrahmane Amrouche

USTHB, FEI, LCPTS, Speech Com. and Signal Proc. Lab,
P.O. Box 32, El Alia, 16111 Bab Ezzouar, Algeria
{kzergat,namrouche}@usthb.dz, yazid.kahil@myport.ac.uk

Abstract. Phonetic forensics is now widely used. Indeed, in some cases, the voice can be the only potential proof for investigations. In order to evaluate the performance of Forensic Voice Comparison domain (FVC), we will study two factors limiting the robustness of verification task. One depends on the type of transmission channel (telephone, microphone, etc.) and the other is related to the physiological difference between the speakers' voices. Our work consist in adapting an open source platform for Automatic Speaker Recognition (ASR), "ALIZE", for use in the forensic domain to estimate and represent the voice as an exhibit. For this, we will study and compare the two models GMM-UBM and i-vectors to evaluate performance.

Keywords: I-vectors · MFCC · ASR · UBM · Forensic · FVC · ALIZE/LIA_RAL toolkit

1 Introduction

Kidnapping, death threat, malicious call or terrorist claim, the crime rate continues to increase. The judge decision is sometimes very delicate for lack of evidence. The world of justice is thus a world of uncertainty where every detail can help to make the right decision. In fact, it is important to rely on biometrics to make the decisions. The highly developed and fully automated existence of telephone networks and the omnipresence of speech in communication technology have resulted in the use of the human voice as a crime commissioner. Therefore, Automatic Speaker Verification (ASV) has become a very important tool in forensic science and one of the technical methods applied to criminal investigations to demonstrate the existence of a crime and help the authorities to expose the perpetrators.

Forensic Voice Comparison (FVC) was emerged as the use of scientific methods and techniques, for legal and law enforcement purposes, to provide information if a questioned speech recording (evidence) originates from a suspected

M. R. Senouci et al. (Eds.): CSA 2020, LNNS 199, pp. 61–69, 2021.
https://doi.org/10.1007/978-3-030-69418-0_6

speaker or not [1,2]. There are two different approaches in FVC. The acoustic-phonetic statistical approach and the automatic approach [3]. The common statistical models used in the automatic approach are the Gaussian mixture model –universal background model (GMM-UBM) [4] and i-vectors with probabilistic linear discrimination analysis (PLDA) [5].

In forensic context, the quality of speech is strongly interfered by noise, variable channel, and different duration. This is very important for forensics to find a speaker recognition model that is unsusceptible to those factors. In a previous work [6] , authors build a new speaker recognition model based on Wavelet Cepstral Coefficient (WCC), i-vector, and cosine distance scoring (CDS). For evaluation the TIMIT database has been used. The experimental results show that the proposed model can effectively solve the troubles of forensic scenario, but the time cost of the method is high.

Another work presented in [7], authors implemented a Text-dependent Forensic Voice Comparison using Hidden Markov Model (HMM) and Gaussian Mixture Model – Universal Background Model (GMM-UBM) Approaches. Obtained results show that, the HMM system constantly outperformed the GMM system in terms of average Cllr values. The study also demonstrates that the outcomes (scores) of the GMM and HMM systems are not well-calibrated; thus calibration is an essential part of the FVC if they are to be used as models in the system.

This work investigates and explores the performances of both GMM/UBM and i-vectors on the open source ALIZE/LIA_RAL [8] toolkit used in Forensic Voice Comparison. The rest of the paper is laid out as follows: Sect. 2 describes the i-vectors framework. Section 3 describes the Likelihood-Ratio framework adopted in forensic comparison. In Sect. 4, the design of the dataset used and the experimental analysis is presented. Obtained results and discussion are shown in Sect. 5. Finally, conclusion and remarks are summarized in Sect. 6.

2 I-Vectors Paradigm

The Joint Factor Analysis (JFA) [9] model is based on the assumption that speech recording can be divided into two static supervectors, one dependent on the speaker and the other on the channel. The latter has largely prepared the ground for the arrival of small sized vectors called i-vectors. The most basic definition of an i-vector is to consider it as the projection of a supervector into a reduced size space. This space, unlike the two subspaces of JFA, makes no distinction between the different types of variability (the interlocutor variability and that due to the channel), it is thus named the space of the total variability [10,11]. Mathematically, a S-supervectors of a speaker and channel dependent audio segment can be written in the following form:

$$S = m + Tx \tag{1}$$

Where:

- m: is the super average vector of the UBM world model.

- T: is the matrix of total variability (interlocutor and channel variability).
- x: is a hidden vector that follows the normal distribution.

Thus, the super vector S follows a normal distribution of the mean vector m and the covariant matrix. At this stage, the i-vectors contain the information related to the speakers but also the useless information (channel, environment...). To remedy this two models are used.

2.1 Probabilistic Linear Discriminant Analysis (PLDA)

Discriminant Analysis is a tool for predicting group membership from a linear combination of variables. Its main purpose consists of maximizing separation between groups and discarding variables which are little related to group distinctions. For a speaker I having a set $X = \{x_r\}_{r=1\cdots R}$ of R records represented by D-dimensional i-vectors. In a manner similar to JFA model, a speaker and channel dependent i-vector can be divided into two components as:

$$x_r = s + c_r \tag{2}$$

Where $S = m + U_1Y_1$ is the component depending only on the identity of the speaker and $C_r = U_2y_{2r} + \varepsilon_r$ is the component dependent only on the channel, often referred to as the noise component. Thus, the equation can be rewritten in the following form:

$$x_r = m + U_1y_1 + U_2y_2r + \varepsilon_r \tag{3}$$

Where:

- m is the mean dimension D vector estimated from a large population of speakers
- U_1 of dimension $D \times d1$ and U_2 of dimension $D \times d2$ are projection matrices often named respectively, the matrix of clean voices and the matrix of clean channels.
- The hidden vectors, y_1 of dimension $d1$ and y_2 of dimension $d2$ are respectively denominated the vector of the factors of the speaker and the vector factors of the channel.
- ε_r is a vector of dimension D representing the residual noise, it follows a Gaussian distribution of mean vector 0 and diagonal precision matrix [12].

2.2 Angular Similarity

The Cosine-based Angular Similarity (CAS) is based on the calculation of the angle between two vectors in order to measure their similarity while ignoring their amplitudes. In the i-vector space, this metric has been successfully applied to the speaker verification problem, as well as to other problems such as language recognition [13]. Given a verification test $x = \{e, t\}$ where x is composed of two voice records represented by i-vectors respectively for enrollment and for testing, the equation of the angular similarity of the cosine S is written as follows:

$$S(e,t) = \frac{e.t}{\|e\|\,\|t\|} \tag{4}$$

Where $\|e\|$ and $\|t\|$ are the Euclidean norms.

3 Scoring Method

The judicial voice comparison is based on the comparison of a record of the unknown criminal (the trace) and the recording of a known suspect, a comparison piece. The latter, aims to indicate whether the evidence supports the prosecution (the two extracts of speech are pronounced by the same speaker) or defense (both speeches are delivered by two different speakers). The Likelihood-Ratio (LR) framework has been advocated by many as the logically and legally correct framework for assessing forensic evidence and reporting the outcome in court [7]. The LR calculation process is illustrated in Fig. 1.

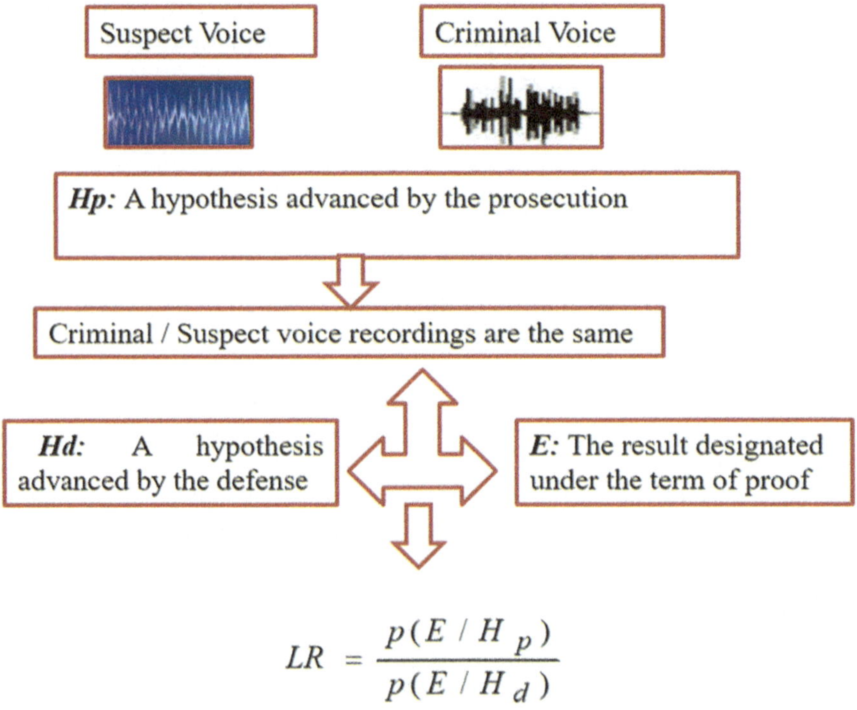

Fig. 1. Likelihood-Ratio Scoring method.

The degree of similarity of the evidence with respect to the suspect's samples is calculated by $P(E/Hp)$, where $P(E/Hd)$ stands for the degree of typicality in relation to the population concerned [7]. If the evidence occurs under the origin hypothesis than under the different origin hypothesis, in that case the value of the true likelihood ratio will be greater than one. Otherwise, it will be less than the threshold [7].

In the end, forensic science has the responsibility to provide an LR, and it is up to the judges to make a final decision using other sources of information about the case in question. This decision is in itself a probabilistic statement known as "posterior chances" and can be expressed mathematically as:

$$\underbrace{\frac{P(^{Hp}/_{E})}{P(^{Hd}/_{E})}}_{posteriorside} = \underbrace{\frac{P(^{E}/_{Hp})}{P(^{E}/_{Hd})}}_{LR} \times \underbrace{\frac{P(Hp)}{P(Hd)}}_{frontside} \tag{5}$$

In fact, the likelihood ratio tells us how to connect these two coasts so that the judge can take a logical decision after obtaining the evidence [7].

4 Corpora and Experimental Method

4.1 Corpora

The experiments are carried out using AISHELL [14] corpus, which is defined as a free-source Chinese Mandarin Word Corpus by Beijing Shell Technology, where 400 people from different accent areas in China are recorded. Information on age and gender of speakers are given in Table 1.

Table 1. Information on age and gender of speakers.

Age range	Speakers	Male	Female
16–25 years	316	140	176
26–40 years	71	36	35
>40 years	13	10	3

The recording conditions were made in a quiet indoor environment, using three different devices at the same time: a high fidelity microphone (44.1 kHz, 16bit), a mobile phone with Android system (16 kHz, 16bit), and a mobile phone with iOS system (16 kHz, 16bits). 178 h of Mandarin speech are published with high quality human transcriptions. The audio was resampled at 16 kHz to build AISHELL-ASR0009-OS1. The corpus includes a set of training, development and tests as follows:

- Train: train set contains 120,098 statements of 340 speakers.
- Development: The development set contains 14,326 statements of 40 speakers.
- Test: the test set contains 7 176 statements of 20 speakers.

4.2 Features Extraction and Speaker Modeling

The first step implemented on ALIZE consists of features extraction. For this, the Speech Signal Processing toolkit (SPro 5) implemented on LIA_RAL toolkit is used. The audio files were converted into 13 Mel Frequency Cepstral Coefficients (MFCCs) excluding the zeroth one. The delta and acceleration coefficients were appended to form a 39-dimensional feature vector.

In fact, a preemphasized with a first order high pass filter with a coefficient of 0.97 is applied, then, short-term frames of 20 ms with an overlap of 50% was used.

Once the acoustic parameters have been extracted, low energy frames, corresponding to silence and noise, can be removed using the function Energy-Detector. After that, the NormFeat function allows the feature normalization which changes the acoustic vectors to fit the zero mean and the unit variance distribution [8].

The GMM-UBM framework in the ALIZE/LIA RAL platform is based on a single GMM (namely UBM), trained offline using the function TrainWorld and a substantial amount of collected data from various speakers across multiple channels and environments to cover various acoustic conditions.

Indeed, for the i-vector paradigm, the entire database (400 speakers) is used to train the UBM and the total variability matrix, while 100 speakers are used to train the UBM gender dependant model for the GMM-UBM classifier. A detailed diagram of the FVC task is given in Fig. 2.

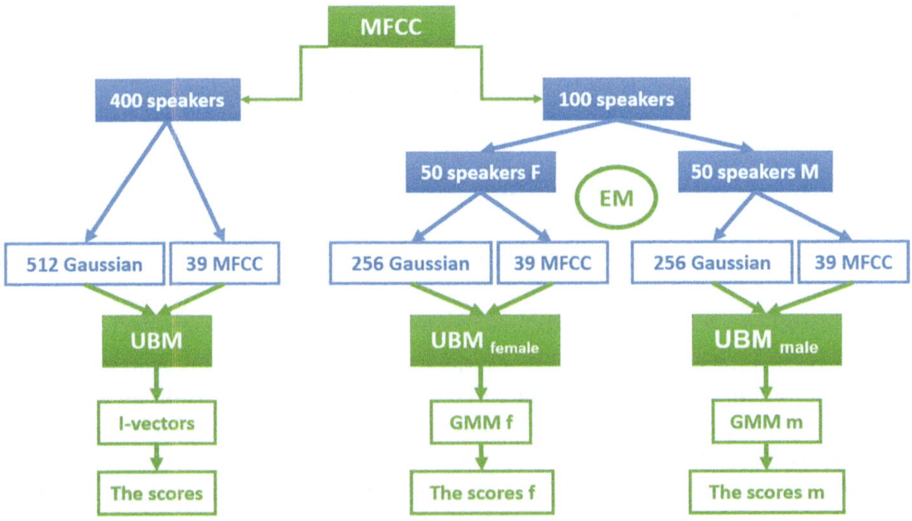

Fig. 2. Diagram of the FVC Task based on ALIZE Toolkit.

On the other hand, the GMM model uses 3 iterations and 256 Gaussians. It is worthy to note, that the adaptation of the GMMs models were made from the UBM_Female.

5 Results

This paper deals with two systems for FVC task, one based on the GMM-UBM model and the other on the i-vector extractor, both implemented under ALIZE platform. After modeling the speakers, we use the file "test" for the comparison of the suspects with the obtained models in order to have scores highlighting the accusations.

Results are illustrated using the Detection Error Tradeoff (DET) curve, plotted by a "GNU detware" evaluation tool. The verification rate is obtained by subtracting the error rate, which is the intersection between the first bisector and the DET curve, by a percentage of 100%. As shown in Fig. 3, we obtained an Equal Error Rate of 9.352% for the GMM-UBM model, and therefore a verification rate of 90.648%. By cons, the i-vectors are more efficient with a verification rate more than 94%. It is important to note that the verification task has been done in mismatched condition, using different types of microphone such us a high fidelity microphone (44.1 kHz, 16 bit), or a mobile phone with Android system (16 kHz, 16 bit), and a mobile phone with iOS system (16 kHz, 16bit). Indeed, regarding the microphone types, the i-vectors seem to be the most discriminant.

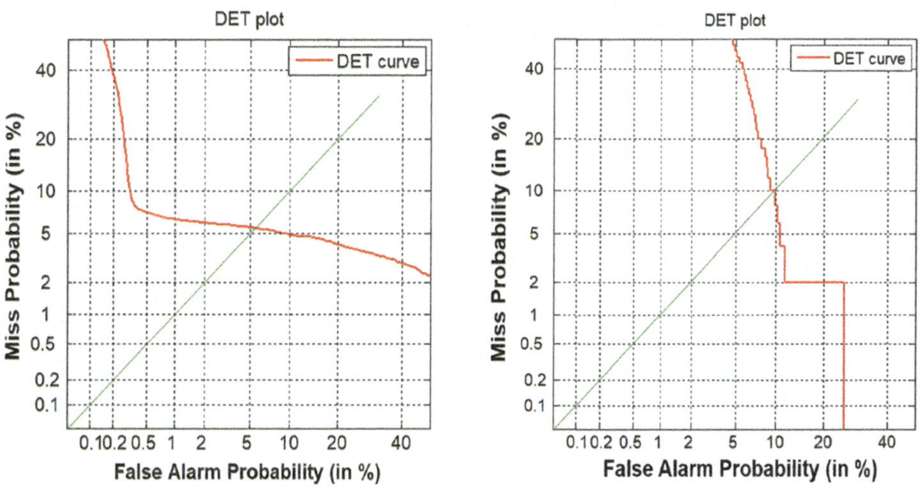

Fig. 3. DET curve for i-vectors (a) and GMM-UBM (b).

For a better understanding, Table 2 presents the GMM scores obtained after comparing speaker S0195 against 4 other speakers.

Table 2. GMMs scores for speaker S0195.

Speaker	S0195	Spkn1	Spkn2	Spkn3	Spkn4
Score	1.1139	−0.0166	−0.2803	−0.1237	−0.1659

The highest score value represents the accused, while the rest of the scores represent the innocents, we can recognize the performance of a system by calculating the difference between the two latter. For this example, the performance of the GMM-UBM system for FVC can be measured as: $1.1139 - (-0.0166) = 1.1305$. As expected, the GMM-UBM gives a small score width, less than 1.2, which is not significant or decisive for a judge decision.

On the other hand, a remarkable difference between the i-vector scores is observed. For example, the obtained scores for speaker S0764 are given in Table 3. The suspect has a score of 17.4130 while it is negative for the innocent. Thus, the score width is more than 23.

Table 3. I-vectors scores for speaker S0764.

Speaker	S0764	Spkn1	Spkn2	Spkn3	Spkn4
Score	17.4130	−6.3010	−13.9659	−23.7629	−33.9229

The difference in the i-vector score is much larger than that of the GMMs. So, a better (more confident) discrimination between the accused and the innocent is made.

6 Conclusion

This paper presents the Forensic Voice Comparison (FVC) of 400 Mandarin speakers using i-vector and GMM-UBM implemented under ALIZE. For recognition, results using GMM-UBM show that the performance is equal to 90%. An increase of accuracy is obtained when using the i-vector extractor approximately by 4% which may affect the decision of the forensic practitioner. While the results are encouraging, there is scope for improvement. For future work, we intend to explore using a robust approach by including phonetic information.

References

1. Drygajlo, A., Haraksim, R.: Biometric evidence in forensic automatic speaker recognition. In: Handbook of Biometrics for Forensic Science, pp. 221–239. Springer (2017). https://doi.org/10.1007/978-3-319-50673-9_10
2. Drygajlo, A., Jessen, M., Gfroerer, S., Wagner, I., Vermeulen, J., Niemi, T.: Methodological guidelines for best practice in forensic semiautomatic and automatic speaker recognition. Verlag für Polizeiwissenschaft (2016)

3. Morrison, G.S.: The impact in forensic voice comparison of lack of calibration and of mismatched conditions between the known-speaker recording and the relevant-population sample recordings. Forensic Sci. Int. **283**, e1–e7 (2018)
4. Reynolds, D.A., Quatieri, T.F., Dunn, R.B.: Speaker verification using adapted gaussian mixture models. Digit. Signal Process. **10**(1–3), 19–41 (2000)
5. Burget, L., Plchot, O., Cumani, S., Glembek, O., Matějka, P., Brümmer, N.: Discriminatively trained probabilistic linear discriminant analysis for speaker verification. In: 2011 IEEE International Conference on Acoustics, Speech and Signal Processing (ICASSP), pp. 4832–4835. IEEE (2011). https://doi.org/10.1109/ICASSP.2011.5947437
6. Lei, L., Kun, S.: Speaker recognition using wavelet cepstral coefficient, i-vector, and cosine distance scoring and its application for forensics. J. Electr. Comput. Eng. **2016** (2016)
7. Tsuge, S., Ishihara, S.: Text-dependent forensic voice comparison: likelihood ratio estimation with the hidden Markov model (HMM) and Gaussian mixture model – universal background model (GMM-UBM) approaches. In: Proceedings of the Australasian Language Technology Association Workshop, pp. 17–25 (2018)
8. Larcher, A., Bonastre, J.F., Fauve, B., Lee, K.A., Levy, C., Li, H., Mason, J., Parfait, J.Y.: ALIZE 3.0-open source toolkit for state-of-the-art speaker recognition. In: Annual Conference of the International Speech Communication Association, Lyon, France (2013)
9. Kenny, P., Stafylakis, T., Ouellet, P., Alam, M.J.: JFA-based front ends for speaker recognition. In: 2014 IEEE International Conference on Acoustics, Speech and Signal Processing (ICASSP), pp. 1705–1709. IEEE (2014). https://doi.org/10.1109/ICASSP.2014.6853889
10. Dehak, N., Kenny, P.J., Dehak, R., Dumouchel, P., Ouellet, P.: Front-end factor analysis for speaker verification. IEEE Trans. Audio Speech Lang. Process. **19**, 788–798 (2010)
11. Sadjadi, S.O., Slaney, M., Heck, L.: MSR identity toolbox v1. 0: a MATLAB toolbox for speaker-recognition research. In: Speech and Language Processing Technical Committee Newsletter, Piscataway, NJ, USA, pp. 1–32. IEEE Signal Processing Society (2013)
12. Prince, S.J.D., Elder, J.H.: Probabilistic linear discriminant analysis for inferences about identity. In: IEEE 11th International Conference on Computer Vision 2007, Rio de Janeiro, pp. 1–8. IEEE (2007). https://doi.org/10.1109/ICCV.2007.4409052
13. Dehak, N., Dehak, R., Glass, J.R., Reynolds, D.A., Kenny, P.: Cosine similarity scoring without score normalization techniques. In: Odyssey, p. 15 (2010)
14. Bu, H., Du, J., Na, X., Wu, B., Zheng, H.: Aishell-1: an open-source mandarin speech corpus and a speech recognition baseline. In: 20th Conference of the Oriental Chapter of the International Coordinating Committee on Speech Databases and Speech I/O Systems and Assessment (O-COCOSDA) 2017, Seoul, pp. 1–5. IEEE (2017). https://doi.org/10.1109/ICSDA.2017.8384449

Using Machine Learning for Heart Disease Prediction

Dhai Eddine Salhi[1(✉)], Abdelkamel Tari[2], and M-Tahar Kechadi[3]

[1] LIMOSE Laboratory, University of Mhamed Bougara, Boumerdes, Algeria
d.salhi@univ-boumerdes.dz

[2] LIMED Laboratory, University of Abderrahmane Mira, Bejaia, Algeria
tarikamel59@gmail.com

[3] Insight Centre for Data Analytics, University College Dublin (UCD), Dublin, Ireland
tahar.kechadi@insight-centre.org

Abstract. In this paper we carried out research on heart disease from data analytics point of view. Prediction of heart disease is a very recent field as the data is becoming available. Other researchers have approached it with different techniques and methods. We used data analytics to detect and predict disease's patients. Starting with a pre-processing phase, where we selected the most relevant features by the correlation matrix, then we applied three data analytics techniques (neural networks, SVM and KNN) on data sets of different sizes, in order to study the accuracy and stability of each of them. Found neural networks are easier to configure and obtain much good results (accuracy of 93%).

Keywords: Machine Learning · Heart disease · Prediction · Detection

1 Introduction

Of an estimated 58 million deaths from all causes worldwide in 2015, cardiovascular disease (CVD) accounted for 30%. This ratio is equal to infectious diseases, nutritional deficiencies and maternal and perinatal conditions combined [12]. It is important to recall that a significant proportion of these deaths (46%) is attributable to people under 70 years of age, in the most productive period of life. Furthermore, 79% of the burden of disease attributed to CVD is in this age group [11]. Between 2015 and 2021, deaths due to non-communicable diseases (half of which will be due to cardiovascular disease) is expected to increase by 17%, while deaths due to infectious diseases, nutritional deficiencies and maternal and perinatal conditions combined is expected to decrease by 3% [12]. Nearly half of the burden of disease in low- and middle-income countries is already due to non-communicable diseases [9].

Heart disease is the leading cause of death in the world. More than 1.6 million people die of heart disease each year. The term "heart disease" refers to several

M. R. Senouci et al. (Eds.): CSA 2020, LNNS 199, pp. 70–81, 2021.
https://doi.org/10.1007/978-3-030-69418-0_7

types of heart problems. The most common type is coronary artery disease, which can cause a heart attack. Other types of heart disease may involve the valves in the heart, or the heart may not pump well and cause heart failure. Some people are born with heart disease. Anyone, including children, can develop heart disease. It happens when a substance called plaque builds up in your arteries. Smoking, unhealthy eating and lack of exercise increase your risk of heart disease. High cholesterol, high blood pressure or diabetes can also increase your risk of heart disease.

To deal with this disease, there are several methods of prevention, such us natural methods, like stopping smoking, maintaining a healthy weight, adopting a healthy diet and practicing sports regularly. We also have the scientific methods such as drugs and surgeries. The prediction of this disease before being infected is part of the prevention methods, where the computer tools are the most used means in it, more precisely Machine Learning algorithms.

Our study about this problem is part of data science applications, where we detect cardiac patients based on well-defined attributes, such as: age, sex, cholesterol and blood pressure. The use of data collected from patients, is very important to train the learning algorithms. In our case, we use a dataset collected from Algerian hospitals in which a certain group of people are considered as sick, and others are not. Before describing our study, we present a state of the art on the most recent research work in this field. After that, we followed by the pre-processing where we select the most relevant attributes that give the best results, using the correlation matrix. Finally, we apply learning algorithms on different sizes of the dataset (600, 800, 1000, 1200 lines), to develop the most appropriate and stable prediction approach.

2 Related Work

The detection and prediction of heart disease are an important issue. For this reason, many research work has been done about this topic. We divide existing work into two categories. The first presents approaches that select the most relevant patient by using features selection; and the second presents approaches that explore the learning algorithms.

2.1 Features Selection

The choice of features is one of the major challenges to train a predictive learning algorithm. Several studies have been carried out. In this section, we present the most recent work in the literature. Hasan et al. [6] made a kind of comparison of the different Machine Learning algorithms on two classes of attributes. The first one contains 10 attributes and the second one contains 14 attributes. The problem with this study is that the authors did not mention the attributes used. Gavhane et al. [5] developed the same approach as [6] with only six attributes, namely, age, sex, blood pressure, heart Rate, diabetes and hyper cholesterol. Shanmugasundaram et al. [18] surveyed the most commonly used attributes in cardiac disease detection in different papers, and returned three classes, mentioned in Table 1:

Table 1. The classes of the attributes used

Papers	Attributes
[14] T John Peter et al. (2012) [7] I.S Jenzi (2013) [16] S. Radhimeenakshi (2016)	Number of attributes: 13 Age, Gender, CPT, FBS, RECG, Ex-Ang, SL, Col-Ves, Thal, SC, Thalach, Old peak, RBP
[3] Chaitrali S et al. (2012) [8] C. Kalaiselvi (2016)	Number of attributes: 10 Age, Gender, CPT, FBS, RECG, SC, Thalach, Smoking, Alchool, Obesity
[13] Shamsher Bahadur et al. (2013) [10] Hlaudi Daniel et al. (2014)	Number of attributes: 6 CPT, Ex-Ang, Col-Ves, RBP, Num, Smoking

2.2 Machine Learning Algorithms

The best Machine Learning algorithm is the one with the high accuracy. The prediction of heart patients is very critical, because a simple mistake can lead to death of a human being. This section consists of evaluating and selecting the most frequently used algorithms with high accuracy. Hasan et al. [6] implemented several learning machine algorithms. Logistic Regression has given an accuracy of 93%, Random Forest 92% and Gaussian Naïve Bayes 90%. Babu et al. [1] tested the diagnosis of heart patients by applying two techniques, namely, genetic algorithms and KNN. The best results were obtained by KNN. Priyanka and RaviKumar [15], made a comparative study between two algorithms, namely, Naïve bayes and KNN. The obtained results show that KNN gives much more accuracy (85%) compared to Naïve bayes). Raju et al. [17], conducted a study on the prediction of heart disease, using several algorithms. The obtained results show that SVM reached the best accuracy with 99.3% compared to neural networks, KNN and decision trees with 91.1%, 87.2% and 82.3% respectively. Shanmugasundaram et al. [18] conducted a survey on the most used Machine Learning techniques for this topic. Table 2 illustrates the algorithms, the papers in which these algorithms are used and the obtained accuracy values.

After analyzing the techniques used in these papers, we decided to select the most important attributes of the three classes [18], and the algorithms to implement, namely, neural networks (NN), KNN and SVM. These algorithms and attributes are used to test our dataset containing information on Algerian patients.

Table 2. The most frequently cited algorithms in the papers

Algorithms	Papers	Accuracy
Neural networks	[14] T John Peter et al. (2012)	78%
	[3] Chaitrali S et al. (2012)	100%
KNN	[14] T John Peter et al. (2012)	75%
	[8] C. Kalaiselvi 2016	87%
SVM	[14] T John Peter et al. (2012)	76%
	[3] Chaitrali S et al. (2012)	99%
	[13] Shamsher Bahadur et al. (2013)	99%
	[19] B.Venkata-lakshmi et al. (2014)	84%

3 Proposed Approach

Our proposal is based on two main phases, namely, the pre-processing phase and Machine Learning algorithm application. In the first phase, we choose the most relevant attributes. In the second phase we apply Machine Learning algorithms in order to select the algorithm that gives a best accuracy value. The proposed phases of our approach are described in Fig. 1.

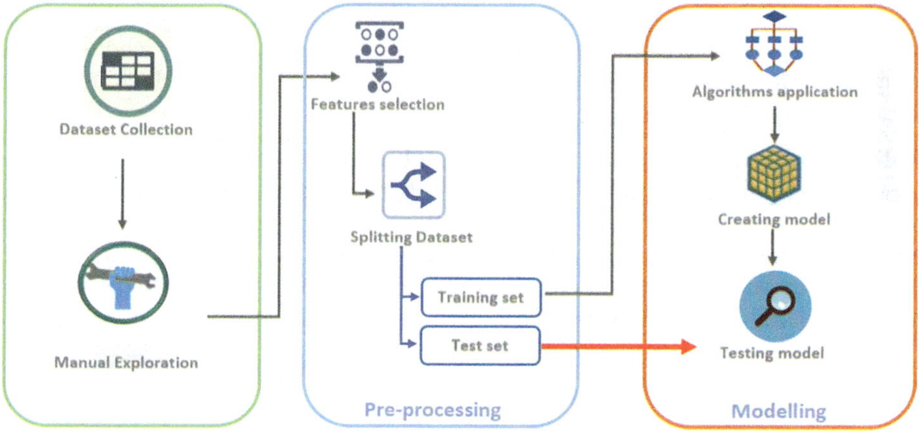

Fig. 1. Approach proposal

3.1 Dataset Collection

Data collection is defined as the procedure of collecting, measuring and analyzing accurate insights for research. Researchers can evaluate their hypothesis on the basis of collected data. In our case, we use a dataset of persons who have performed analyzes and tests to detect heart disease. The dataset is represented as a matrix, where rows represent the patients and columns represent the attributes (features) to be tested.

3.2 Manual Exploration

Data exploration or Manual Exploration is the initial step in data analysis, where users explore large unstructured datasets to uncover initial patterns, characteristics, and points of interest. This process does not involve revealing every bit of information a dataset holds, but rather to help create a broad picture of important trends and major points to study in more detail. This phase is very important in the process of Machine Learning algorithms application. Dataset is manually analyzed to label each person as sick or not sick. This annotated data is used by the different algorithms in the training step.

3.3 Data Pre-processing

Data pre-processing is performed to ensure the quality and the usefulness of data. This step is extremely important, because it affects directly the learning ability of our model.

Features Selection. In a dataset where we have a large set of features [2], we choose the most relevant ones using Pearson Correlation Method (Correlation matrix) [4]. To detect the links between attributes, we only choose those attributes that are highly dependent each other in order to apply Machine Learning algorithms and achieve better accuracy.

Splitting Data Set. To train the Machine Learning algorithm, we mention the target column in the dataset. Then, we divide the dataset into two sub-datasets, namely, training-set and test-set. Figure 2 explains the method we followed to apply Machine Learning algorithms. First, we split our dataset into two parts (training-set and test-set). Then, the training-set is splitted into two other parts, namely, training and validation.

```
library(caTools)
set.seed(123)
split = sample.split(maxmindf$target, SplitRatio = 0.8)
trainset = subset(maxmindf, split == TRUE)
testset = subset(maxmindf, split == FALSE)
```

Fig. 2. Splitting dataset

3.4 Modelling

This phase represents the application and test of the three chosen algorithms, namely, neural networks, SVM and KNN.

Algorithms Application. Our approach is based on the application of the three algorithms on datasets of various sizes in order to compute the accuracy values of each algorithm and defining the best and the more stable one in training.

Testing Algorithms. We used the confusion matrix and the accuracy ratio to test the three algorithms on the test set versus manual exploration. A confusion matrix is a table that is often used to describe the performance of a classification model "classifier" on a set of test data for which the true values are known. After this test phase, we will consider the best algorithm to use in our context.

4 Results and Discussion

In this section, we describe the different phases of our approach, namely: data collection, manual exploration, data pre-processing, modelling and testing.

4.1 Data Collection

In our study, we use a structured dataset of Algerian persons who have done analyzes at the Mohand Amokrane EHS Hospital, previously known as ex CNMS, located Algiers, Algeria, with a size of 1200 rows and 20 columns. These columns are represented by the following features: age, sex, cp, trestbps, chol, Ex-Ang, Col-Ves, fbs, restecg, thalach, exang, oldpeak, slope, RBP, ca, thal, Smoking, Alchool, Obesity and OilyCough (Fig. 3).

	age	sex	cp	trestbps	chol	fbs	restecg	thalach	exang	oldpeak	slope	ca	thal
1	29	1	1	130	204	0	0	202	0	0.0	2	0	2
2	29	1	1	130	204	0	0	202	0	0.0	2	0	2
3	29	1	1	130	204	0	0	202	0	0.0	2	0	2
4	29	1	1	130	204	0	0	202	0	0.0	2	0	2
5	34	1	3	118	182	0	0	174	0	0.0	2	0	2
6	34	0	1	118	210	0	1	192	0	0.7	2	0	2
7	34	1	3	118	182	0	0	174	0	0.0	2	0	2
8	34	0	1	118	210	0	1	192	0	0.7	2	0	2
9	34	1	3	118	182	0	0	174	0	0.0	2	0	2
10	34	0	1	118	210	0	1	192	0	0.7	2	0	2
11	35	0	0	138	183	0	1	182	0	1.4	2	0	2
12	35	1	1	122	192	0	1	174	0	0.0	2	0	2
13	35	1	0	120	198	0	1	130	1	1.6	1	0	3

Fig. 3. Dataset collection representation

4.2 Manual Exploration

In our study, We added a column in the used dataset (Target) which contains zero or one (0 = not sick, 1 = sick), to start the pre-processing (Fig. 4).

fbs	restecg	thalach	exang	oldpeak	slope	ca	thal	target
0	1	192	0	0.7	2	0	2	1
0	0	174	0	0.0	2	0	2	1
0	1	192	0	0.7	2	0	2	1
0	1	182	0	1.4	2	0	2	1
0	1	174	0	0.0	2	0	2	1
0	1	130	1	1.6	1	0	3	0
0	0	156	1	0.0	2	0	3	0
0	1	182	0	1.4	2	0	2	1
0	1	174	0	0.0	2	0	2	1
0	1	130	1	1.6	1	0	3	0
0	0	156	1	0.0	2	0	3	0
0	1	182	0	1.4	2	0	2	1
0	1	174	0	0.0	2	0	2	1
0	1	130	1	1.6	1	0	3	0
0	0	156	1	0.0	2	0	3	0

Fig. 4. Manual exploration

Table 3. Description and values of selected features

Attribute	Description	Values
Age	Age	29 to 62 years
Sex	Sex	1 - male 2 - female
CP	Chest pain type	1 - typical angina pectoris 2 - atypical angina 3 - non-anginal pain 4 - asymptomatic
trestbps	Resting blood pressure in mm/Hg	Numeric value: Example: 140 mm/Hg
Chol	Serum cholesterol in mg/dl	Numeric value: Example: 289 mg/hg
Fbs	Fasting blood pressure >120 mg/dl	Numeric value: Example: 129 mm/Hg
Restecg	Resting electrocardiographic results	0 - normal, 1 - have the ST-T 2 - hypertrophy
thalach	Maximum heart rate achieved	Numeric value: Example: 140,173
Exang	Exercise induced angina	1 - Yes 2 - No
Oldpeak	ST depression induced by exercise relative to rest	Numeric Value
Slope	The slope of the peak exercise ST segment	1 - on the rise 2 - flat 3 - the downhill slope
Ca	Number of major vessels colored by fluoroscopy	0 to 3 vessels
Thal	Thalassemia	3 - normal, 6 - defect repaired, 7 - reversible defect

4.3 Data Pre-processing

The preparation of the data to be implemented is carried out in two phases, namely: Features selection and Splitting Dataset.

Features Selection. Initially we had 20 attributes. After applying Pearson correlation matrix, we detected 13 attributes, for instance: age, sex, cp, trestbps, chol, fbs, restecg, thalach, exang, oldpeak, slope, ca and thal. These features, detailed in Table 3, are related and dependent each other.

This selection of features has been proven by the correlation matrix presented in Fig. 5.

	age	sex	cp	trestbps	chol	fbs	restecg	thalach	exang
age	1.00000000	-0.06776463	-0.07384384	0.29690194	0.23530980	0.12986009	-0.15493507	-0.37821813	0.08894420
sex	-0.06776463	1.00000000	-0.03294819	-0.09076869	-0.21521726	0.03667062	-0.04419083	-0.06906782	0.13054375
cp	-0.07384384	-0.03294819	1.00000000	0.03685857	-0.08670788	0.06890690	0.03984093	0.31649274	-0.40416943
trestbps	0.29690194	-0.09076869	0.03685857	1.00000000	0.12506180	0.19897193	-0.14364846	-0.04781646	0.06037417
chol	0.23530980	-0.21521726	-0.08670788	0.12506180	1.00000000	0.02343684	-0.12984956	-0.03248069	0.06413524
fbs	0.12986009	0.03667062	0.06890690	0.19897193	0.02343684	1.00000000	-0.09413557	-0.00787552	0.05907449
restecg	-0.15493507	-0.04419083	0.03984093	-0.14364846	-0.12984956	-0.09413557	1.00000000	0.05919509	-0.05699088
thalach	-0.37821813	-0.06906782	0.31649274	-0.04781646	-0.03248069	-0.00787552	0.05919509	1.00000000	-0.39283739
exang	0.08894420	0.13054375	-0.40416943	0.06037417	0.06413524	0.05907449	-0.05699088	-0.39283739	1.00000000

Fig. 5. The correlation matrix of selected features

Splitting Dataset. The used dataset has been divided into two parts: the first part represents the training dataset with 80%, and the second part represents the test dataset with a size of 20%.

4.4 Modelling

In this phase of experimental study, we apply the selected algorithms, namely, neural networks, SVM, KNN; and discussed the obtained results. The application of the three algorithms is carried out on the dataset. We tested the algorithms on the same dataset using different sizes (600, 800, 1000 and 1200) in order to detect the algorithm which gives more accuracy and at the same time insures stability. Figure 6, Fig. 7 and Fig. 8 shows the obtained results by the three algorithms, NN, SVM and LR respectively.

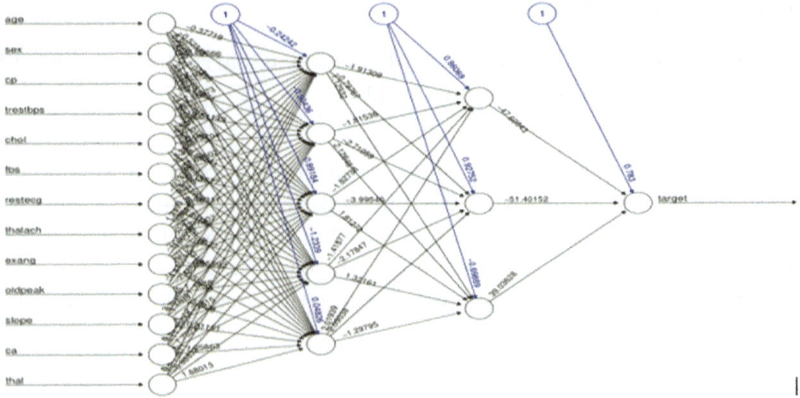

Fig. 6. Neural networks application

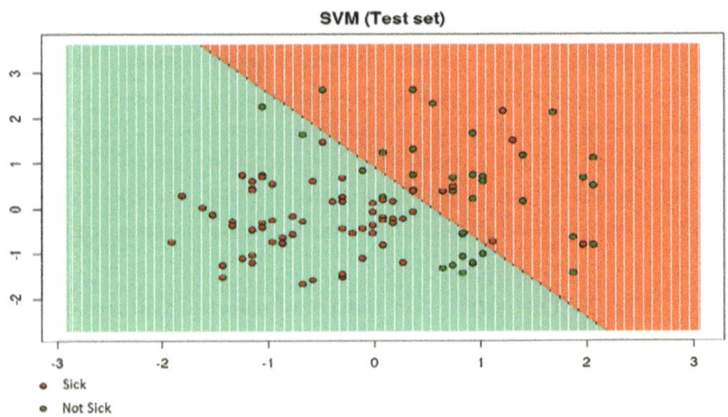

Fig. 7. SVM application

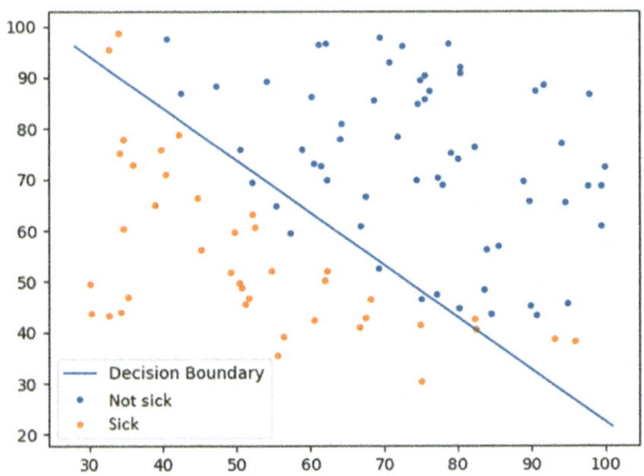

Fig. 8. Logistic regression application

4.5 Testing Algorithms

After training the 3 algorithms using the 4 sub datasets (600, 800, 1000 and 1200 lines), we test each algorithm and compute the accuracy values obtained for these sub datasets. Table 4 represents the confusion matrix, including the accuracy values obtained by each of the three algorithms (using the sub dataset containing 1200 lines).

Table 4. The confusion matrix with accuracy values

	Neural network dataset (1200 lines)		SVM dataset (1200 lines)		KNN dataset (1200 lines)	
	Sick	*Not sick*	*Sick*	*Not sick*	*Sick*	*Not sick*
Sick	94	8	90	11	84	18
Not sick	6	92	9	90	11	87
Accuracy	**93%**		**90%**		**85,5%**	

Table 5 shows the accuracy values obtained by the three algorithms for the four configurations (sub datasets).

Table 5. The accuracy of the different algorithms

Algorithm	Accuracy			
	Data set (600 lines)	Data set (800 lines)	Data set (1000 lines)	Data set (1200 lines)
Neural network	91.8%	92%	92.6%	93%
SVM	89.7%	89%	90.1%	88%
KNN	85.1%	85.3%	86%	85.5%

By considering the obtained results, showed in Table 5, we can clearly observe that neural networks algorithm outperforms the two other algorithm with an accuracy value greater than 91% in all tested configurations.

5 Conclusion

Heart diseases have become more and more frequent among people, including our country (Algeria). Therefore, predicting the disease before becoming affected decreases the risk of death. This topic is of certain importance and attracts a lot of attention from the scientific community. Our paper is part of the research on the detection and prediction of heart disease. It is based on the application of Machine Learning algorithms, from which we selected the 3 most used algorithms, namely, neural network, SVM and KNN. We applied our approach to real dataset of Algerian persons. The initial results showed a very interesting accuracy value of 93% obtained using neural network. Also, a study on the features selection have been carried out using the correlation matrix to detect the features' dependencies. As future work, we propose the use of deep Learning techniques, as well as the building of more large dataset.

References

1. Babu, S., Vivek, E., Famina, K., Fida, K., Aswathi, P., Shanid, M., Hena, M.: Heart disease diagnosis using data mining technique. In: 2017 International conference of Electronics, Communication and Aerospace Technology (ICECA), vol. 1, pp. 750–753. IEEE (2017)
2. Cai, J., Luo, J., Wang, S., Yang, S.: Feature selection in machine learning: a new perspective. Neurocomputing **300**, 70–79 (2018)
3. Dangare, C.S., Apte, S.S.: Improved study of heart disease prediction system using data mining classification techniques. Int. J. Comput. Appl. **47**(10), 44–48 (2012)
4. Fang, X., Hodge, B.M., Du, E., Zhang, N., Li, F.: Modelling wind power spatial-temporal correlation in multi-interval optimal power flow: a sparse correlation matrix approach. Appl. Energy **230**, 531–539 (2018)
5. Gavhane, A., Kokkula, G., Pandya, I., Devadkar, K.: Prediction of heart disease using machine learning. In: 2018 Second International Conference on Electronics, Communication and Aerospace Technology (ICECA), pp. 1275–1278. IEEE (2018)
6. Hasan, S., Mamun, M., Uddin, M., Hossain, M.: Comparative analysis of classification approaches for heart disease prediction. In: 2018 International Conference on Computer, Communication, Chemical, Material and Electronic Engineering (IC4ME2), pp. 1–4. IEEE (2018)
7. Jenzi, I., Priyanka, P., Alli, P.: A reliable classifier model using data mining approach for heart disease prediction. Int. J. Adv. Res. Comput. Sci. Softw. Eng. **3**(3), 20–24 (2013)
8. Kalaiselvi, C.: Diagnosing of heart diseases using average k-nearest neighbor algorithm of data mining. In: 2016 3rd International Conference on Computing for Sustainable Global Development (INDIACom), pp. 3099–3103. IEEE (2016)
9. Lopez, A.D., Mathers, C.D., Ezzati, M., Jamison, D.T., Murray, C.J.: Global and regional burden of disease and risk factors, 2001: systematic analysis of population health data. Lancet **367**(9524), 1747–1757 (2016)
10. Masethe, H.D., Masethe, M.A.: Prediction of heart disease using classification algorithms. In: Proceedings of the world Congress on Engineering and Computer Science, vol. 2, pp. 22–24 (2014)
11. World Health Organization: The world health report 2002: reducing risks, promoting healthy life. World Health Organization (2016)
12. World Health Organization of Canada, Public Health Agency of Canada, Canadian Pharmacists Association: Preventing chronic diseases: a vital investment. World Health Organization (2015)
13. Patel, S.B., Yadav, P.K., Shukla, D.: Predict the diagnosis of heart disease patients using classification mining techniques. IOSR J. Agric. Vet. Sci. (IOSR-JAVS) **4**(2), 61–64 (2013)

14. Peter, T.J., Somasundaram, K.: An empirical study on prediction of heart disease using classification data mining techniques. In: IEEE-International Conference on Advances in Engineering, Science and Management (ICAESM 2012), pp. 514–518. IEEE (2012)
15. Priyanka, N., RaviKumar, P.: Usage of data mining techniques in predicting the heart diseases–Naïve Bayes & decision tree. In: 2017 International Conference on Circuit, Power and Computing Technologies (ICCPCT), pp. 1–7. IEEE (2017)
16. Radhimeenakshi, S.: Classification and prediction of heart disease risk using data mining techniques of support vector machine and artificial neural network. In: 2016 3rd International Conference on Computing for Sustainable Global Development (INDIACom), pp. 3107–3111. IEEE (2016)
17. Raju, C., Philipsy, E., Chacko, S., Suresh, L.P., Rajan, S.D.: A survey on predicting heart disease using data mining techniques. In: 2018 Conference on Emerging Devices and Smart Systems (ICEDSS), pp. 253–255. IEEE (2018)
18. Shanmugasundaram, G., Selvam, V.M., Saravanan, R., Balaji, S.: An investigation of heart disease prediction techniques. In: 2018 IEEE International Conference on System, Computation, Automation and Networking (ICSCA), pp. 1–6. IEEE (2018)
19. Venkatalakshmi, B., Shivsankar, M.: Heart disease diagnosis using predictive data mining. Int. J. Innov. Res. Sci. Eng. Technol. **3**(3), 1873–1877 (2014)

Operational Research and Optimization

Ant Lion Optimizer for the Estimation of Photovoltaic (PV) Cells Parameters

Dallel Nasri(✉) and Diab Mokeddem

Automatic Laboratory (LAS), Department of Electrical Engineering,
Faculty of Technology, University of Ferhat Abbas Setif-1, Setif, Algeria
dallelnasri@gmail.com, mokeddem_d@yahoo.fr

Abstract. Recently, solar energy as a clean source is widely used. In this way, the design of photovoltaic cells has attracted the attention of researchers over the world. Two main problems are faced in this field: having a useful model to characterize the solar cells and the absence of data about photovoltaic cells. This situation even affects the performance of the photovoltaic modules (panels). The characteristics of the current vs. voltage are used to describe the behaviour of solar cells. Two metaheuristic algorithms are proposed to identify the parameters of the photovoltaic cells and panels namely Ant Lion Optimizer (ALO) and particle swarm optimization (PSO). Experimental results demonstrate the high performance of the proposed ALO method compared to PSO.

Keywords: Optimization · Ant Lion Optimizer - ALO · Particle Swarm Optimization · Photovoltaic cell

1 Introduction

In recent decades, renewable energy sources have attracted the attention of researchers and scientists. Among these sources, solar energy is the most promising as for Solar PV (Photovoltaic) cell has been reported to be the most important. Therefore, a distributed generation unit that would provide accurate modelling of solar cells at the charging site is very important. The modelling of PV cells consists of two steps: the formulation of the mathematical model and the precise estimation of their parameter values. For the mathematical model, the current vs. the voltage characteristics govern the behaviour of a solar cell. There are two equivalent electronic circuits modelling the behaviour of a solar cell. Such circuits are called single diode (SD) and double diode (DD) models. Whatever the model selected, it is necessary to estimate all its parameters, the main problem is to identify the optimal values of the parameters applied to the selected models that make it possible to obtain the best possible approximation of the experimental data obtained by the true solar cell. Many optimization technics dealt with the problem of identification of PV cell parameters such as ABC (Artificial Bee Colony) [1], GA (Genetic Algorithm) [2–4], PSO (Particle Swarm Optimization) [5,6], SA (Simulated Annealing) [7], HS (Harmony Search) [8],

M. R. Senouci et al. (Eds.): CSA 2020, LNNS 199, pp. 85–95, 2021.
https://doi.org/10.1007/978-3-030-69418-0_8

BFA (Bacterial Foraging Algorithm) [9] and TLBO (Teaching Learning Based Optimization) [10]. Recently, Kanimozhi and Kumar [11] propose the modeling of solar cell with single diode by Ant Lion Optimizer. In the proposed work, the intelligent technique ant lion optimizer (ALO) [12] is successfully applied to identify the parameters of PV cells with both single diode and double diode models. In order to illustrate efficiency of the proposed approach, it is compared with PSO algorithm. Experimental results proved the superiority of the proposed model obtained by ALO algorithm.

2 Ant Lion Algorithm

Ant lion optimizer (ALO) emerged as a successful nature-inspired algorithm, which mimics the hunting activities of ant lions in nature, 2015 Mirjalili [12]. As their name indicates, ant lions have exceptional hunting tactics to catch their prey (ants). Firstly, an ant lion larva digs a cone-shaped pit by throwing out sands along a circular path with their powerful massive jaw. Two main factors affect the dimension of the trap: ant lion hunger level and moon size. After building the trap, hunters will hide themselves underneath the bottom of the cone and wait for ants to trip over their trap. Once the prey is on the pit surface, ant lions start to toss sands just before the edges to slide the frightened victim. At this stage, ants attempt to survive by performing some abrupt movements to increase their safe distance to the jaw or it will be dragged under the surface and killed. Referring to the above behaviours, in ALO algorithm there are mainly five operations namely random movement of ants, construction of trap, trapping of ants in traps, catching preys and re-construction of traps. Thus, two classes of search agents are introduced namely ants and ant lion. Ant lion is selected as the best search agent that never change its position except when substituting a particular ant. Ant agents can carry out random walks in the solution space. The random positions of the ants are saved in the matrix Aant.

$$A_{\text{ant}} = \begin{bmatrix} A_{1,1} & A_{1,2} & \cdots\cdots & A_{1,d} \\ A_{2,1} & A_{2,2} & \cdots\cdots & A_{2,d} \\ \vdots & \vdots & \vdots \quad \vdots & \vdots \\ A_{n,1} & A_{n,2} & \cdots\cdots & A_{n,d} \end{bmatrix} \tag{1}$$

$A_{m_n}$is the value of m-th variable (dimension) of n-th ant, n = number of ants (population size). Fitness of all ants will be stored in the matrix A_{OA} in terms of objective function f:

$$\mathrm{A_{OA}} = \begin{bmatrix} \mathrm{f}\,(\mathrm{A_{1,1}} & \mathrm{A_{1,2}} & \cdots\cdots & \mathrm{A_{1,d}}) \\ \mathrm{f}\,(\mathrm{A_{2,1}} & \mathrm{A_{2,2}} & \cdots\cdots & \mathrm{A_{2,d}}) \\ \vdots & \vdots & \vdots \quad \vdots & \vdots \\ \mathrm{f}\,(\mathrm{A_{n,1}} & \mathrm{A_{n,2}} & \cdots\cdots & \mathrm{A_{n.d}})) \end{bmatrix} \tag{2}$$

The position and fitness of ant lion are represented by the matrices $A_{antlion}, A_{OAL}$ respectively.

$$A_{antlion} = \begin{bmatrix} AL_{1,1} & AL_{1,2} & \cdots\cdots & AL_{1,d} \\ AL_{2,1} & AL_{2,2} & \cdots\cdots & AL_{2,d} \\ \vdots & \vdots & \vdots \; \vdots & \vdots \\ AL_{n,1} & AL_{n,2} & \cdots\cdots & AL_{n,d} \end{bmatrix} \quad (3)$$

$$\mathrm{A_{OAL}} = \begin{bmatrix} \mathrm{f\,(AL_{1,1}} & \mathrm{AL_{1,2}} & \cdots\cdots & \mathrm{AL_{1,d}])} \\ \mathrm{f\,([AL_{2,1}} & \mathrm{AL_{2,2}} & \cdots\cdots & \mathrm{AL_{2,d}])} \\ \vdots & \vdots & \vdots \; \vdots & \vdots \\ \mathrm{f\,(AL_{n,1}} & \mathrm{AL_{n,2}} & \cdots\cdots & \mathrm{AL_{nd}])} \end{bmatrix} \quad (4)$$

2.1 Random Movement of Ants

Ants randomly move to search for food. This random movement of ants is given by Eq. (5)

$$X(t) = [0, cusum(2r(t_1) - 1, cusum(2r(t_2) - 1), ..., cusum(2r(t_n) - 1)] \quad (5)$$

where $cusum$ represents cumulative sum, $n =$ maximum number of ants, $t =$ step of random walk (iteration) and

$$r(t) = \begin{cases} 1, & rand > 0.5 \\ 0, & rand <= 0.5 \end{cases} \quad (6)$$

To restrict the random movement within the search space, normalized form is used, which is based on min–max normalization. Position of ants can be updated by Eq. (7).

$$X_i^t = \frac{(X_i^t - a_i) \times (d_i - c_i^t)}{(b_i - a_i)} + c_i \quad (7)$$

where a_i, b_i are minimum and maximum of random walk of ants c_i^t, d_i represents minimum and maximum $i - th$ variable at $t - th$ iteration.

2.2 Trapping of Ants

Equations (8) and (9) give the mathematical expression of trapping ants to the ant lion's pits.

$$c_i^t = Antlion_j^t + c^t \quad (8)$$

$$d_i^t = Antlion_j^t + d^t \quad (9)$$

where:
c^t, d^t are the minimum and maximum of all variables at t-th iteration, and $Antlion_j^t$ shows the position of the selected $j - th$ ant lion at $t - th$ iteration.

2.3 Construction of Trap

Equations (10) and (11) shrink the radius of updating ant's positions and mimics sliding process of ant inside the pits. This guarantees exploitation of search space

$$c^t = \frac{c^t}{l} \tag{10}$$

$$d^t = \frac{d^t}{l} \tag{11}$$

where I $= 10^w \frac{t}{maxiter}$, t is current iteration, $maxiter$ is the maximum number of iterations and w is a constant whose value is given by:

$$w = \begin{cases} 2 \text{ if } t > 0.1 \text{ maxiter} \\ 3 \text{ if } t > 0.5 \text{ maxiter} \\ 4 \text{ if } t > 0.75 \text{ maxiter} \\ 5 \text{ if } t > 0.9 \text{ maxiter} \\ 6 \text{ if } t > 0.95 \text{ maxiter} \end{cases} \tag{12}$$

2.4 Catching the Prey and Reconstruction of the Pit

Ant lion catches the ant when it reaches the pit bottom and consumes it. After this, ant lion has to update the position in order to catch new prey. This process is represented by Eq. (13).

$$\text{Antlion}_j^t = Ant_i^t \text{ if } f\left(Ant_i^t\right) > f\left(\text{ Antlion }_j^t\right) \tag{13}$$

Elitism is used to keep the best solutions in every stage. Best ant lion obtained is treated as elite, which is the fittest ant lion. Elite should affect the ant lion in every stage (random movement). For this, every ant is assumed to associate with an ant lion and elite by Roulette wheel is given by Eq. (14).

$$Ant_i^t = \frac{R_A^t + R_E^t}{2} \tag{14}$$

R_A, R_E represents the random walk around the selected ant lion and elite at t_th iteration simultaneously and Ant_i^t indicate the position of i_th ant at t_th iteration.

3 Photovoltaic Models

To generate a precise design of solar cells, it is necessary to have a good mathematical model formulation and the accurate estimation of their parameter values. In general, there exist two models: single diode (SD) and double diode (DD) [14]. In this section, these models are described and their objective functions are formulated.

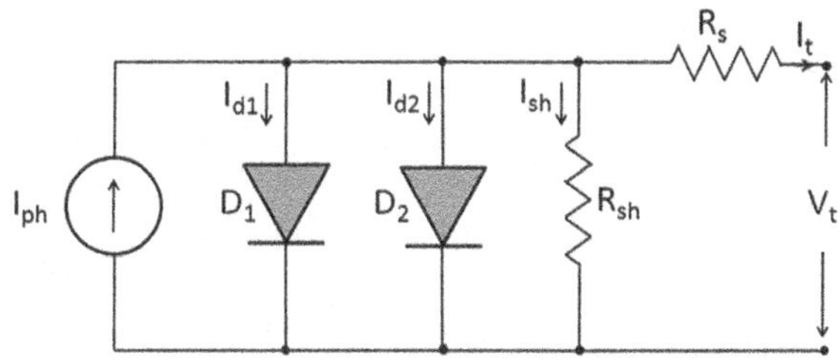

Fig. 1. Double diode model of solar cells.

3.1 Double Diode Model (DD)

Figure 1 defines the equivalent circuit for a double diode model. The two diodes are used to shunt the photo-generated (Iph) current source. One of these diodes is set as a rectifier and the second diode is used to model the charge recombination current and some non-idealities. The double diode model also requires a resistor connected in series with the diodes [5,14]. The cell terminal current is computed as follows:

$$I_t = I_{ph} - I_{d1} - I_{d2} - I_{sh} \tag{15}$$

where the terminal current is I_t, I_{ph} is the photo-generated current, I_{d1}, I_{d2} are the first and second diode currents, respectively, whereas Ish is the shunt resistor current. For the proper modelling of PV cells, it used the Shockley diode equation. Hence, Eq. (15) is rewritten as shown in Eq. (16):

$$\begin{aligned} I_t = I_{ph} - I_{sd1} \left[\exp\left(\left(\frac{q(V_t+R_s.I_t)}{n_1.k.T}\right) - 1\right]\right. \\ - I_{sd2}\left[\exp\left(\left(\frac{q(V_t+R_s.I_t)}{n_2.k.T}\right) - 1\right] - \frac{V_t+I_tR_s}{R_{sh}}\right. \end{aligned} \tag{16}$$

where V_t is the terminal voltage whereas the series and shunt resistances are represented by R_s and R_{sh} respectively. I_{sd1} and I_{sd2} are the diffusion and saturation current, respectively. n_1 and n_2 are the diffusion and recombination diode ideality factors, respectively. In the Shockley diode equation, $q = 1.602.10^{-19}$ (coulombs) is the magnitude of charge on an electron, $k = 1.381.10^{-23}(J/K)$ is the Boltzmann constant and T is the cell temperature (K).

Therefore, ($R_s, R_{sh}, I_{ph}, I_{sd1}, I_{sd2}, n_1$ and n_2) are the seven unknown parameters in Eq. (16). The identification of these parameters improves the optimal performance of a solar cell.

3.2 Single Diode Model (SD)

In single diode model the diffusion and saturation currents, I_{sd1} and I_{sd2} respectively, are represented by a non-physical ideality factor n [1,14]. Figure 2 represents the equivalent circuit of single diode model. In this approach, only five parameters are to be estimated. The SD model is adapted in Eq. (17) as follows:

$$I_t = I_{ph} - I_{sd}\left[\exp\left(\frac{q(V_t + I_tR_S)}{n.k.T}\right) - 1\right] - \frac{V_t + I_tR_S}{R_{sh}}. \tag{17}$$

Therefore, the parameters to be identified are $R_s, R_{sh}, I_{ph}, I_{sd}$ and n.

4 Formulation of the Optimization Problem

Firstly, it is necessary to define the optimization problem. The parameters of each equivalent circuit (DD and SD) must be modified as variable vectors. The dimensionality of optimization space has been determined. The values of some variables must be limited to define the search space. These limits are presented in Eq. (18)

$$\min f(x), x = [x_1, x_2 \ldots \ldots x_d], d\epsilon[5, 7] \tag{18}$$

with

$$\begin{array}{cc}
d = 5(SD) & d = 7(DD) \\
0 \leq x_1 (R_S) \leq 0.5 & 0 \leq x_1 (R_S) \leq 0.5 \\
0 \leq x_2 (R_{Sh}) \leq 100 & 0 \leq x_2 (R_{Sh}) \leq 100 \\
0 \leq x_3 (I_{ph}) \leq 1 & 0 \leq x_3 (I_{ph}) \leq 1 \\
0 \leq x_4 (I_{sd}) \leq 1 & 0 \leq x_4 (I_{sd1}) \leq 1 \\
1 \leq x_5(n) \leq 2 & 0 \leq x_5 (I_{sd2}) \leq 1 \\
 & 1 \leq x_6 (n_1) \leq 2 \\
 & 1 \leq x_7 (n_2) \leq 2
\end{array}$$

The evaluation consists of measuring the performance of each individual. For this, we use a function directly related to the objective function (function to optimize) called fitness function. Performance indices are a quantitative measures usually used in the search environment to reduce the minimum value of the error between the input and the desired output [14].

It is possible to estimate the parameters of the model of photovoltaic cells using Eqs. (19) and (20). However, in the presented approach, it is necessary to define an objective function to determine the quality of the estimated parameters. In other words, a good set of parameters produces a precise approximation between the measurements of the current with respect to the voltage (I-V) of the physical cell and the values of the mathematical model. In case of a photovoltaic cell with a single diode the error is defined by [14]:

$$f_{SD} (V_t, I_t, X) = I_t - x_3 + x_4 \left[\exp \left(\frac{q (V_t + x_1 I_t)}{x_5 \cdot k.T} \right) - 1 \right] + \frac{V_{t+} x_1 \cdot I_t}{x_2} \tag{19}$$

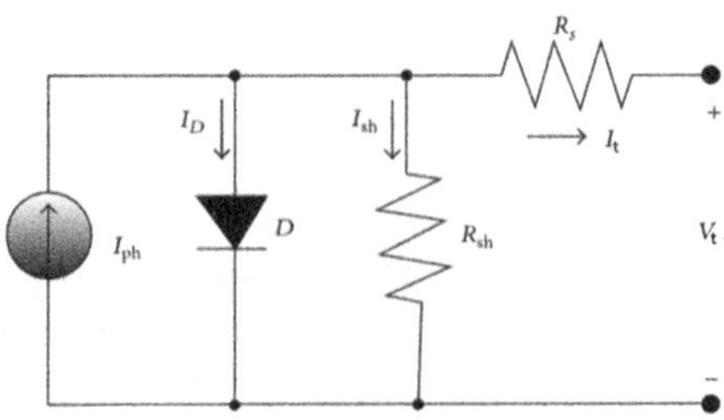

Fig. 2. Single diode model of solar cells.

Meanwhile, the error in case of double diode is given by:

$$f_{DD}(V_t, I_t, X) = I_t - x_3 + x_4\left[\exp\left(\frac{q(V_t + x_1 \cdot I_t)}{x_6 \cdot k.T}\right) - 1\right] + x_5\left[\exp\left(\frac{q(V_t + x_1 \cdot I_t}{x_{7.k.T}}\right) - 1\right] + \frac{V_t + x_1 \cdot I_t}{x_2} \tag{20}$$

In both functions f_{DD} and f_{SD}, the values of V_t and I_t are experimentally collected from the solar cell. x is a vector that contains the model parameters, where $x = R_s, R_{sh}, I_{ph}, I_{d1}, I_{d2}, n_1, n_2$ is the model parameters for DD and $x = R_s, R_{sh}, I_{ph}, I_d, n$ for SD.

The objective functions used in this work are the Root Mean Square Error (RMSE) and Mean Absolute Error (MAE):

Root Mean Square Error (RMSE)

$$\text{RMSE}(X) = \sqrt{\frac{1}{N_E}\sum_{C=1}^{N_E}\left(f_M(V_t, I_t, X)\right)^2} \tag{21}$$

From this equation f_M is used to select the model DD or SD.

Mean Absolute Error (MAE)

$$\text{MAE} = \sum_{i=1}^{N_E}\frac{|I_t - I_{te}|}{N_E} \tag{22}$$

Where:

I_t: measures the PV obtained currant.

I_{te}: estimated currant.

N_E: the data set.

5 Results and Discussion

The objective of this part is to apply two strategies Ant lion algorithm and Particle Swarm optimization, to calculate the parameters for solar cell models $R_s, R_{sh}, I_{ph}, I_{sd1}, I_{sd2}, n_1, n_2$ of DD and $R_s, R_{sh}, I_{ph}, I_{sd}, n$ of SD. For this, we have used experimental measurements of a silicon solar cell with a diameter of 57 mm studied by the authors [1]. These authors collected 20 samples (Table 1) measuring current and voltage under the conditions: 1 sun (1000 W/m^2) at $T = 33\,^{\circ}$C and respecting the constraints defined by Eq. (18).

Table 1. Terminal ($V_t - I_t$) measurements.

Data	01	02	03	04	05	06	07	08	09	10
V_t	0.0057	0.0646	0.1185	0.1678	0.2132	0.2545	0.2924	0.3269	0.3585	0.3873
I_t	0.7605	0.7600	0.7590	0.7570	0.7570	0.7555	0.7540	0.7505	0.7465	0.7385
Data	11	12	13	14	15	16	17	18	19	20
V_t	0.4137	0.4373	0.4590	0.4784	0.4960	0.5119	0.5265	0.5398	0.5521	0.5633
I_t	0.7280	0.7065	0.6755	0.6320	0.5730	0.4990	0.4130	0.3165	0.2120	0.1035

The parameters of the algorithm ALO and PSO used for this example are given in Table 2.

Table 2. Parameters values of ALO and PSO.

	Population size	N° of iterations	w_{min}, w_{max}	C_1, C_2
PSO parameters	30	200	0.1, 0.6	1.5
ALO parameters	30	200	/	/

The parameters of the identified photovoltaic cells, obtained by the methods of ALO and PSO are given by the Tables 3 and 4 for the SD and DD models using different objective functions.

Table 3. Comparison results for the *SD* model.

Parameters	$I_{ph}(A)$	$R_s(\Omega)$	$R_{sh}(\Omega)$	$MAE/RMSE$	$I_{sd}(A)$	n
ALO	0.761167	0.0399823	54.1418	0.0199/0.0028	$3.46563e^{-007}$	1.49291
PSO	0.747673	0.03	55	0.17112/0.043707	$2.1e^{-07}$	1.5

From Tables 3 and 4, we observe that the algorithm ALO with the RMSE performance criterion has a better performance (smaller value) to identify the parameters of the SD and DD cells.

Figure 3 shows the evolution of the objective function (RMSE) by the two algorithms ALO and PSO vs the number of iterations in the case of the single and double diode model. This figure shows that the ALO algorithm gives a faster convergence compared to PSO algorithm.

The current vs the voltage and the power vs the voltage are given by Fig. 4 and Fig. 5 for the case of the model *SD* and *DD*, respectively. These curves are calculated using the parameters estimated by each optimization algorithm.

Table 4. Comparison results for the *DD* model.

Parameters	$I_{ph}(A)$	$I_{sd1}(A)/I_{sd2}(A)$	n_1/n_2	$R_s(\Omega)$	$R_{sh}(\Omega)$	$MAE/RMSE$
ALO	0.747017	$2.1e^{-07}/2.1e^{-07}$	1.5/1.5	0.0315244	54.999	0.17557/0.23848
PSO	0.747073	$2.1e^{-07}/2.1e^{-07}$	1.5/1.5	0.03	55	0.17566/0.25899

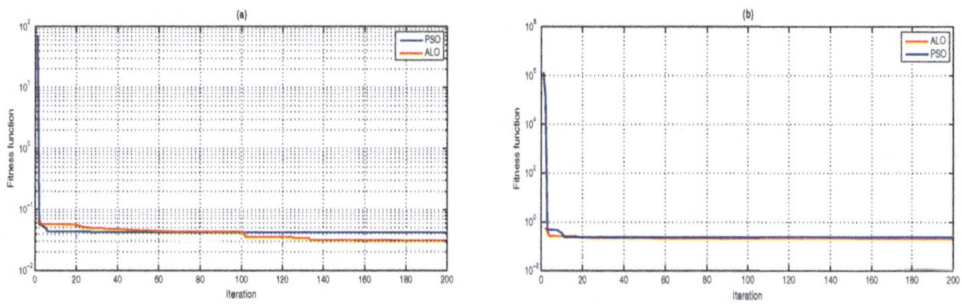

Fig. 3. RMSE evolution of ALO and PSO for (a) Single diode model and (b) double diode model.

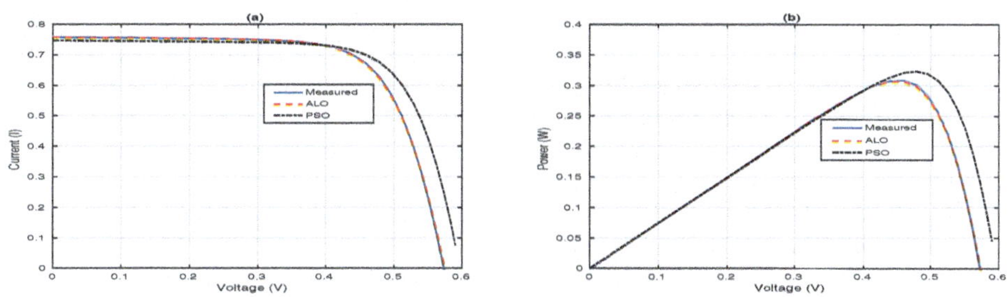

Fig. 4. (a) Current V_s measured voltage, (b) power V_s the measured voltage for the *SD* model computed using ALO and PSO algorithms.

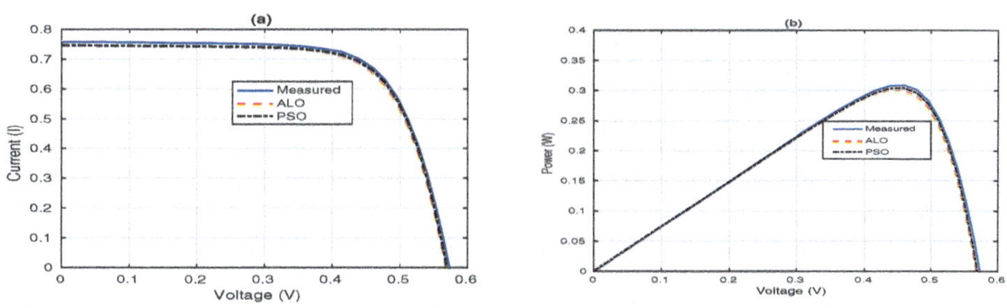

Fig. 5. (a) Current V_s measured voltage, (b) power V_s the measured voltage for the DD model computed using ALO and PSO algorithms for estimating the parameters of SD and DD models.

6 Conclusion

This work is concerned with the application of new global optimization methods based on metaheuristic algorithms for estimating the parameters of a single diode and double diode models of a solar cell. The algorithms ALO, inspired by the hunting behaviour of ant lion larva and PSO, Particle swarm optimization, were proposed. The proposed algorithms are introduced to estimate the parameters of the solar cells, while minimizing certain objective functions such as RMSE and MAE. The ALO and PSO approaches are compared and the experimental results proved evidently performance superiority of ALO. In future work, the proposed method will be implemented to other interesting models of solar cells. Also, some other modification will be introduced, extending the use of optimization algorithms for renewable energy problems.

References

1. Oliva, D., Cuevas, E., Gonzalo, P.: Parameter identification of solar cells using artificial bee colony optimization. Appl. Energy 141–154 (2014). https://doi.org/10.1016/j.energy.2014.05.011
2. AlRashidi, M.R., AlHajri, M.F., El-Naggar, K.M., Al-Othman, A.K.: A new estimation approach for determining the I-V characteristics of solar cells. Sol. Energy **85**(7), 1543–1550 (2011). https://doi.org/10.1016/j.solener.2011.04.013
3. Jervase, J.A., Bourdoucen, H., Al-Lawati, A.: Solar cell parameter extraction using genetic algorithms. Meas. Sci. Technol. **12**(11), 1922–1925 (2001)
4. Appelbaum, J., Peled, A.: Parameters extraction of solar cells: a comparative examination of three methods. Sol. Energy Mater. Sol. Cells **122**, 164–173 (2014)
5. Ye, M., Wang, X., Xu, Y.: Parameter extraction of solar cells using particle swarm optimization. J. Appl. Phys. **105**(9), 094502–8 (2009). https://doi.org/10.1063/1.2895396
6. Wei, H., Cong, J., Lingyun, X., Deyun, S.: Extracting solar cell model parameters based on chaos particle swarm algorithm. In: International Conference on Electric Information and Control Engineering (ICEICE), pp. 398–402 (2011)
7. El-Naggar, K.M., AlRashidi, M.R., AlHajri, M.F., Al-Othman, A.K.: Simulated annealing algorithm for photovoltaic parameters identification. Sol. Energy **86**(1), 266–274 (2012). https://doi.org/10.1016/j.solener.2011.09.032
8. Askarzadeh, A., Rezazadeh, A.: Parameter identification for solar cell models using harmony search-based algorithms. Sol. Energy **86**(11), 3241–3249 (2012)
9. Rajasekar, N., Krishna, K.N., Venugopalan, R.: Bacterial Foraging Algorithm based solar PV parameter estimation. Sol. Energy **97**, 255–265 (2013)
10. Niu, Q., Zhang, H., Li, K.: An improved TLBO with elite strategy for parameters identification of PEM fuel cell and solar cell models. Int. J. Hydrogen Energy **39**(8), 3837–3854 (2014). https://doi.org/10.1016/j.ijhydene.2013.12.110
11. Kanimozhi, G., Kumar, H.: Modeling of solar cell under different conditions by Ant Lion Optimizer with LambertW function. Appl. Soft Comput. **71**, 141–151 (2018)

12. Mirjalili, S.: The ant lion optimizer. Adv. Eng. Softw. **83**, 80–98 (2015)
13. Oliva, D., Aziz, M.A., Hassanien, A.: Parameter estimation of photovoltaic cells using an improved chaotic whale optimization algorithm. Appl. Energy **200**, 141–154 (2017). https://doi.org/10.1016/j.apenergy.2017.05.029
14. Mokeddem, D., Mirjalili, S.: Improved whale optimization algorithm applied to design PID plus second-order derivative controller for automatic voltage regulator system. J. Chin. Inst. Eng. **43**(6), 541–552 (2020)

Discrete Elephant Algorithms for Target Detection in Complex and Unknown Environments

Naila Aziza Houacine(✉) and Habiba Drias

LRIA, USTHB, Algiers, Algeria
{nhouacine,hdrias}@usthb.dz

Abstract. Multi-agent systems as an active research axis in artificial intelligence (AI) have known a tremendous progress these last decades. Their main strength resides in the cooperation that makes agents intelligent as a group. The paradigm of multi-robot systems derives from the agent technology field. In this paper, we are interested in developing a multi-robot system for target search in complex and unknown environments because of the numerous applications that exist for this issue and that are present in the modern life. As AI tools, we exploit recent swarm intelligent algorithms such as EHO and ESWSA, as they have shown effectiveness and efficiency for several applications. As a first contribution, we propose discrete versions of these algorithms as the addressed issue is a discrete problem. Then based on these algorithms, we will build a robust multi-robot system capable to find a maximum number of targets by avoiding potential obstacles of complex and unknown environments. Experiments have been performed on public datasets and the results are very promising.

Keywords: Multi-robots system · Swarm intelligence · Complex and unknown environment · Target searching · DEHO · DESWSA

1 Introduction

Because of the numerous real applications in the industrial, civil and military fields, the target detection problem in a complex and unknown environment has recently been the center of interest of many researchers. Different approaches of resolution have been proposed such as anti-flocking algorithms [8], the potential fields approach [6] and some swarm intelligence approaches. On the other hand, multi-robot systems (MRSs) are the subject of many applications, including the problem of finding targets [10]. For example, for planetary exploration, the search for victims in a disaster scenario, the detection of radioactive sources or the location of antipersonnel mines or bombs [1]. Swarms of Robots (SRs) are a special category of MRSs. Many recent works exploit the cooperation of SRs for the particular target search problem because of their robustness, flexibility and scalability [12]. SRs can use swarm intelligence algorithms, knowing that

M. R. Senouci et al. (Eds.): CSA 2020, LNNS 199, pp. 96–106, 2021.
https://doi.org/10.1007/978-3-030-69418-0_9

in recent decades several swarms and meta-heuristics based algorithms for this particular problem have emerged. For example, GSO (Glowworm Swarm Optimization) [5] has been applied to multi-robot systems for locating the source of a signal-emitting target. A hybrid of modified PSO and Local Search on a multi-robots search system [9] that guarantees the global convergence. The A-RPSO (Adaptive Robotic Particle Swarm Optimization) [1] that provides an obstacle avoidance strategy included in the update equation of velocity, as well as the adaptive inertia weight to avoid local optima. The MFPSO (Multi-swarm hybrid FOA-PSO) [13] is a hybrid multi-swarm algorithm based on FOA and PSO for finding a target in unknown environments, this meta-heuristic avoids premature convergence by the independence of the swarms of robots and allow obstacles avoidance through the escape mechanism (MSCM). We noticed that these solving methods assumed that the target has a very large signal emission range and did not take into account its potential variation. Also, these algorithms have only been developed for the search of a unique-target.

In this article, we are interested in a more complex issue, which is the search for targets in a complex and unknown environment. Our SRs ignore the positions of targets and obstacles and only know the limits of the search area. We suppose that the targets emit signals on a certain radius. These signals are perceptible by our robots through adequate sensors. To tackle such a question, two elephant-inspired approaches, namely, Elephant Herding Optimization (EHO) [14] and Elephant Swarm Water Search Algorithm (ESWSA) [7] were chosen because of the similarity found between elephants' behavior searching for food and water, and robots searching for targets, in addition to the encouraging results these recent methods obtained in optimization tasks. Although, these techniques are based on elephant herding, their design are completely different. As these swarms' algorithms are developed for continuous applications, discrete versions of EHO have recently been published to solve different problems such as [3] for the TSP problem and [11] for the QoS aware web service composition. In this article, we propose Discrete version of both EHO and ESWSA a prior to their adaptation to our problem.

The article is organized around 6 sections. In the next one, we present the problem formulation. In the third section, we propose a discrete version of EHO and adapt it to the issue of target detection in a complex and unknown environment. Then in the fourth section, a discrete version of ESWSA is developed for the same issue. The performed experiments and the achieved results are exposed in the fifth section. At last, a conclusion summarizes the contributions of this study and suggests some future works.

2 Problem Formulation

The search environment is a 2D grid-based representation, each box has a value and is defined by a coordinate (x, y). To distinguish the different objects on the discrete search environment. The boxes take several values, the target takes the value 1, the values of the target emission range are bounded by 0 and

1, obstacles are assigned a value of −1 and the rest of the area is neutral and takes the value 0.

A robot ignores the positions of targets and obstacles. It occupies a position (x, y) of the environment and only has visibility on a radius of 10 squares around him. Despite the discrete nature of the environment, our robots move continuously like real robots.

A target takes a single random position and has a fixed signal emission range. The range decreases from the value "1" of the target position to "0" when the distance to the target exceeds the range size.

An obstacle is a fixed object occupying a rectangular surface of the environment and blocks robots from positioning themselves there or cross through. Their position is randomly generated. There are three environments types according to their obstacles density, simple environments (without obstacles), environments with obstacles (low density) and complex environments (high density).

The research space is our 2D environment composed of positions.

A solution is a geographic position of coordinate (x, y) and it's optimal when it maximizes the fitness function. A solution must be included in the boundaries of the environment and mustn't match the position of an obstacle. For multiple targets, we will have to find as many optimal solutions as there are targets.

The fitness function takes its values in the interval [0, 1]. It is equal to the intensity of the target's signal emissions captured by the robots' sensors. The evaluation includes all the positions in the visibility area of the robot.

3 Discrete EHO for the Target Detection Problem

In this section, a Discrete EHO algorithm called DEHO is proposed for the target search problem. The framework for both EHO and DEHO is similar. However, the discrete version requires us to convert the arithmetic operators of the continuous functions into discrete 2D ones. Furthermore, an obstacle avoidance strategy is implemented by sampling the robot's input space [4]. In DEHO, a solution is a position where an elephant robot is located and is responsible of evaluating its solution through the fitness function. Elephants are organized into clans grouped in one area, with random initial positions. From these positions are calculated the elephants' positions on a round surface radius of 5 squares. The positions' update of each elephant relative to its clan follows Eq. (1).

$$\begin{cases} x_{c,i}^{t+1} = x_{c,i}^{t} + \alpha * (x_{Best,c}^{t} - x_{c,i}^{t}) * r \\ y_{c,i}^{t+1} = y_{c,i}^{t} + \alpha * (y_{Best,c}^{t} - y_{c,i}^{t}) * r \end{cases} \tag{1}$$

where r ∈ [0, 1] is a uniform distribution, $(x_{c,i}^{t}, y_{c,i}^{t})$ and $(x_{c,i}^{t+1}, y_{c,i}^{t+1})$ are positions before and after the update of the ith elephant in the clan c and $\alpha \in$ [0,1] represents the influence rate of the matriarch of the clan on the ith elephant.

The elephant who has the best solution of his clan is called "Matriarch". Its next position is calculated using Eq. (2).

$$\begin{cases} x_{Best,c} = \beta * x_{center,c} \\ y_{Best,c} = \beta * y_{center,c} \end{cases} \quad (2)$$

where $\beta \in [0, 1]$ is the influence rate of the center of gravity $(x_{center,c}, y_{center,c})$ of the clan c on the matriarch.

In each clan, the elephant evaluated as owning the worst solution called "male elephant" updates its position following Eq. (3).

$$\begin{cases} x_{worst,c} = x_{min} + (x_{max} - x_{min}) * rand \\ y_{worst,c} = y_{min} + (y_{max} - y_{min}) * rand \end{cases} \quad (3)$$

where rand $\in [0, 1]$ is a uniform stochastic distribution and (x_{min}, y_{min}) and (x_{max}, y_{max}) are lower and upper limits of the environment.

The algorithm DEHO is outlined in Algorithm 1. Its empirical parameters are: nClan: the number of clans, N: the number of elephants in a clan, α: the influence rate of the matriarch on an elephant and β: the influence rate of the center of gravity of the clan on the matriarch.

```
Input: Empirical parameters (nClan,N,α,β), Iter_max : maximum number
       of iterations, nTarget : targets number
Output: TargetsPosition
begin
    t ←1;/* Initialization of the generation counter          */
    c ←0;/* Initialization of the detected targets number     */
    Initialization of the elephants positions.
    while t < Iter_max And c < nTarget do
        Sort Elephants of each clan;
        Update Elephants' Positions; using Eq 1
        Update Best Elephants' Positions; using Eq 2
        Update Worst Elephants' Positions; using Eq 3
        EvaluatePopulation
        if TargetFound then
            TargetsPosition[c] ← Position;
            c ← c +1;
        t ← t + 1;
    Return TargetsPosition;
```

Algorithm 1: DEHO for target detection

4 Discrete ESWSA for the Target Detection Problem

In this section, a Discrete ESWSA (DESWSA) algorithm is proposed and adapted to our problem as a second contribution of this paper. Same observation as for the previous section about the original ESWSA and the obstacle avoidance strategy.

The design of DESWSA is completely different from that of DEHO. DESWSA simulates the movements of the elephants in the swarm as does PSO, which was the first swarm approach based on movement evolution. However, the strategy of movements is different from these two swarm algorithms. In DESWSA, a solution and its evaluation are defined in the same way as for DEHO. Each clan of elephants is represented by a single elephant i. Equations (4) and (5) show how to update the velocity $(v_{x,i}^t, v_{y,i}^t)$ of the ith elephant.

$$v_{x,i}^{t+1} = \begin{cases} v_{x,i}^t * w^t + rand1 * (Gbest_x^t - x_i^t) \text{ if } random > p \\ v_{x,i}^t * w^t + rand1 * (Pbest_{x,i}^t - x_i^t) \text{ if } random \leq p \end{cases} \tag{4}$$

$$v_{y,i}^{t+1} = \begin{cases} v_{y,i}^t * w^t + rand2 * (Gbest_y^t - y_i^t) \text{ if } random > p \\ v_{y,i}^t * w^t + rand2 * (Pbest_{y,i}^t - y_i^t) \text{ if } random \leq p \end{cases} \tag{5}$$

where $(Pbest_{x,i}^t, Pbest_{y,i}^t)$ is the best solution of the ith elephant group at the current iteration and $(Gbest_x^t, Gbest_y^t)$ is the best global solution at the current iteration. (rand1, rand2, random) $\in [0, 1]^3$ are random values. From this velocity, the elephants' new positions (x_i^t, y_i^t) are calculated using Eq. (6).

$$\begin{cases} x_i^{t+1} = v_{x,i}^{t+1} + x_i^t \\ y_i^{t+1} = v_{y,i}^{t+1} + y_i^t \end{cases} \tag{6}$$

The inertia weight is updated according to Eq. (7), using $Iter_{max}$ the maximum iterations number, W_{min} and W_{max} its min and max possible values.

$$W^t = W_{max} - \frac{W_{max} - W_{min}}{Iter_{max}} * t \tag{7}$$

The ESWSA's empirical parameters contain, N: the groups number, p: the switching constant (between global and local search) and W: the initial inertia weight.

```
Input: Empirical parameters (N, p, W), Iter_max : maximum number of
       iterations, nTarget : targets number.
Output: TargetsPosition
begin
    t ←1;/* Initialization of the iteration counter              */
    c ←0;/* Initialization of the detected targets number        */
    Initialize Position, Velocity, Pbest and Gbest
    while t < Iter_max And c < nTarget do
        for i ← 1 to N do
            Update velocity (v^t_{x,i}, v^t_{y,i}) according to Eq. (4) and (5)
            Update position (x^t_i, y^t_i) using Eq. (6)
            Evaluation f(x^t_i, y^t_i);
            Update Pbest and Gbest;
            Update weight inertia W^t using Eq.(7)
        if TargetFound(Gbest^t_x,Gbest^t_y) then
            TargetsPosition[c] ← (Gbest^t_x,Gbest^t_y);
            c ← c + 1;
        t ← t + 1;
    Return (TargetsPosition)
```

Algorithm 2: DESWSA for target detection

5 Experiments and Results

We conducted a series of experiments to validate the effectiveness of our adaptation of DEHO and DESWSA approaches to this problem. We created our own simulator by following the description of the BRobotic Target Search Simulator in [1]. So our simulator and our targets search approaches have all been implemented in Java. In addition to the parameters tuning of each algorithm, we experimented the behavior and effectiveness of our approaches to different variations of external parameters relative to the environment and the targets, organized as shown in Fig. 1.

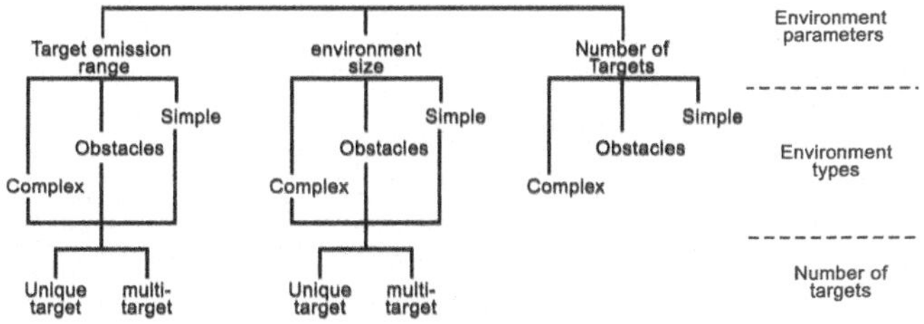

Fig. 1. Organization of the experiments related to external parameters.

In order to compare the results with state-of-the-art algorithms, two Bee Swarm Optimization (BSO) [2] inspired approaches (Mono-BSO, Multi-BSO) have been adapted to the same target search problem, exploiting the same obstacles avoidance strategy as for DEHO and DESWSA.

5.1 Influence of Target Emission Range

Initially, we set the size of the environment to 500×500 squares and vary the range of a 10-squares radius to 100 squares with a step of 10. We tested 4 different environments and executed 10 times each one (which makes 400 executions for the unique-target mode and 400 others for the multi-target one, for each approach and for each environment type). The results we obtained show that even for small targets emission range, all our algorithms find all the targets (100% of success). The gradual widening of the target emission range has caused the number of iterations to decline. And despite the acceptable execution times, they are moderately more important for the multi-target mode.

5.2 Influence of the Environment Size

For these experiments, we adapted the signal emission range of the targets to each environment size (range = 10% of the environment size). For this we have selected some environment sizes between 50×50 and 5000×5000. We undertook 40 executions per environment size, which amounts to a total of 400 executions for the unique-target and 400 others for the multi-target for each approach and for each environment type. In Table 1 we present some experiments results for environments with obstacles in unique-target mode.

Table 1. Comparing average number of iterations, average execution time and success rate in environments with obstacles with different environment sizes.

Size	DEHO			DESWSA			Mono-BSO			Multi-BSO		
	Iter	Time	Succ%	Iter	Time	Succ%	Iter	Time	Succ%	Iter	Time	Succ%
50	1.08	0.00	100	1.08	0.00	100	2.55	0.00	100	2.03	0.00	100
100	1.35	0.00	100	1.10	0.00	100	4.40	0.00	100	2.43	0.00	100
200	1.75	0.00	100	2.05	0.00	100	4.65	0.00	100	2.75	0.00	100
400	3.33	0.01	100	2.03	0.01	100	9.18	0.00	100	4.58	0.00	100
600	1.68	0.01	100	2.38	0.03	100	5.73	0.00	100	4.23	0.01	100
800	2.05	0.02	100	1.98	0.08	100	15.48	0.01	100	4.50	0.01	100
1000	6.43	0.03	100	28.83	0.29	100	77.00	0.03	100	21.40	0.03	100
1500	1.58	0.13	100	2.38	0.39	100	46.13	0.07	100	10.98	0.07	100
3000	15.63	0.58	100	29.78	3.03	100	441.15	0.37	75	48.45	0.49	100
5000	9.35	3.83	100	2.60	15.56	100	566.18	1.35	55	295.58	2.31	87

The success rates presented in Fig. 2 indicate that the two elephant-inspired algorithms find the target in 100% of cases, unlike Mono-BSO and Multi-BSO. Both DEHO and DESWSA have the lowest number of iterations, not exceeding an average of 29.78 iterations. On the other hand, the number of iterations performed by Mono-BSO and Multi-BSO increases considerably up to 566.18 and 295.58 iterations respectively. Figure 2 also shows that the execution times increase due to the widening of the environment surfaces, such as DEHO, Mono-BSO and Multi-BSO are the fastest with a max of 3.83 s compared to DESWSA reaching 15.56 s.

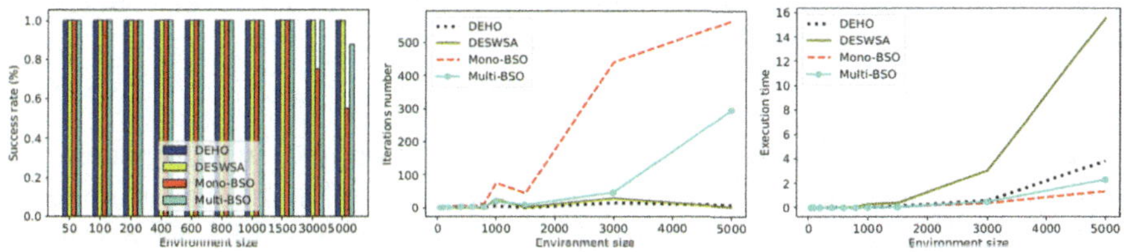

Fig. 2. Comparing average iterations number, average execution time and success rate in environments with obstacles with different environment sizes.

In Table 2, some experiments results for complex environments in unique-target mode are presented. For complex environments also, the size of the search space still does not affect elephant-inspired approaches, whereas BSO-based approaches have difficulty to find the target in large environments. As shown in Fig. 3, we note that the iterations number of Mono-BSO and Multi-BSO increases significantly (up to 747.78 and 444.45 iterations respectively), unlike DEHO and DESWSA where it is much more reasonable (up to 13.83 and 70.45 iterations respectively). We also noticed that all our approaches experience an increasing execution time with the enlargement of the environment sizes. DESWSA reaches 12.32 s while DEHO and the two BSO do not exceed 3.03 s.

Table 2. Comparing average number of iterations, average execution time and success rate in complex environments with different environment sizes.

Size	DEHO			DESWSA			Mono-BSO			Multi-BSO		
	Iter	Time	Succ%	Iter	Time	Succ%	Iter	Time	Succ%	Iter	Time	Succ%
50	1.35	0.00	100	1.05	0.00	100	2.70	0.00	100	2.05	0.00	100
100	3.70	0.00	100	1.43	0.00	100	4.38	0.00	100	2.73	0.00	100
200	3.23	0.00	100	1.73	0.00	100	7.35	0.00	100	3.60	0.00	100
400	4.33	0.00	100	2.03	0.01	100	20.68	0.00	100	3.98	0.00	100
600	14.58	0.02	100	5.75	0.03	100	46.58	0.01	100	15.10	0.01	100
800	9.48	0.02	100	4.48	0.06	100	28.85	0.01	100	7.90	0.01	100
1000	10.30	0.03	100	18.75	0.16	100	171.45	0.03	97	33.73	0.02	100
1500	7.93	0.09	100	33.43	0.34	100	105.60	0.06	100	35.78	0.06	100
3000	8.98	0.52	100	7.33	2.58	100	71.45	0.49	100	21.23	0.48	100
5000	13.83	3.03	100	70.45	12.32	100	747.78	1.00	40	444.45	1.92	80

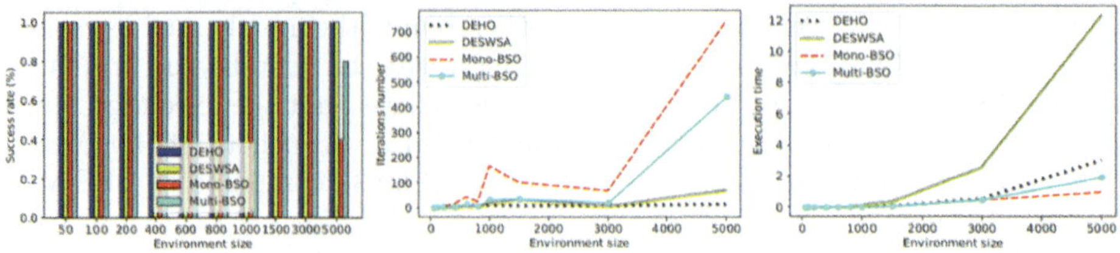

Fig. 3. Comparing average iterations number, Average execution time and success rate in complex environments with different environment sizes.

Following all the results relative to the size of simple, with obstacles and complexes environments and according to the unique and multi-target modes, we note that unlike the two bee-inspired approaches, DEHO and DESWSA achieve the maximum success rate in their search for the target(s) regardless of the environment size. However, for DESWSA the computational time increase significantly comparing to other approaches in both Unique-target and Multi-target modes. For all algorithms, the iterations number increases with increasing environment sizes, but at different speeds. Such as DEHO and DESWSA perform the least iterations, this is due to the large steps between two successive positions that they generate, but it is expensive for them in time, which is the opposite of the two BSO. Finally, for both modes and for all approaches, execution times and number of iterations increase with the increasing complexity of environments.

5.3 Influence of Number of Targets

In order to study the influence of the searched target's number on the behavior of our algorithms, we have set the size of the environments to 500×500 squares and the target's signal emission range to 50. We selected the following set of numbers of targets $\{1, 3, 5, 7, 9, 11, 13, 15\}$. As a consequence, the results of these tests come from executions on 32 environments (4 per target's number), which corresponds to 320 executions for each approach and for each environment type. We realize that all our approaches find the totality (100%) of the target's number, regardless of their number and in all environment types. We also noticed that an increase in the target's number searched leads to a slight growth of execution times and iterations number of all the approaches, with softly better results for DEHO and DESWSA.

5.4 Influence of the Environment Types

The number of obstacles also influences the robots' search performances. Such as the average runtime and the number of iterations increase when moving from simple (without obstacles) environments to environments with obstacles, but they increase dramatically when dealing with complex environments. This is due to the complexity of the environments hindering and making more difficult the robot's movement, so the choice of the right trajectory becomes more and more complex which impacts the execution time. In terms of success rates, they are much more dependent on search methods.

6 Conclusion

In this article, we proposed discrete versions of two swarm-based approaches, namely DEHO and DESWSA to the problem of finding targets in complex and unknown environments. We noticed, after several experiments by varying the parameters of the search algorithms as well as the parameters of the environment, that DEHO and DESWSA outperform state-of-the-art approaches. More precisely, DEHO presents the best performances in terms of search success rate, execution time and the iterations number. These results are due to the fact DEHO is composed of several independent sub-swarms seeking targets, which is obviously more advantageous than a single swarm, preventing this way premature convergence. While, DESWSA solves the problem with adaptive inertia weight and switching between local and global search. Finally, obstacle avoidance strategy by sampling guarantees to find the shortest safe path very quickly. In the future, we plan to develop a GPU version of these approaches in order to improve their efficiency. Also, we think about enhancing them with the cultural dimension in order to improve their effectiveness.

References

1. Dadgar, M., Jafari, S., Hamzeh, A.: A PSO-based multi-robot cooperation method for target searching in unknown environments. Neurocomputing **177**, 62–74 (2016). https://doi.org/10.1016/j.neucom.2015.11.007
2. Drias, H., Sadeg, S., Yahi, S.: Cooperative bees swarm for solving the maximum weighted satisfiability problem. In: Computational Intelligence and Bioinspired Systems, pp. 318–325. Springer, Heidelberg (2005). https://doi.org/10.1007/11494669_39
3. Hossam, A., Bouzidi, A., Riffi, M.E.: Elephants herding optimization for solving the travelling salesman problem. Advances in Intelligent Systems and Computing, pp. 122–130. Springer (2019). https://doi.org/10.1007/978-3-030-12065-8_12
4. Kelly, A., Stentz, A.: Rough terrain autonomous mobility–part 1: a theoretical analysis of requirements. Auton. Robots **5**(2), 129–161 (1998). https://doi.org/10.1023/a:1008801421636
5. Krishnanand, K.N., Ghose, D.: A glowworm swarm optimization based multi-robot system for signal source localization. Studies in Computational Intelligence, pp. 49–68. Springer, Heidelberg (2009). https://doi.org/10.1007/978-3-540-89933-4_3
6. Li, G., Tong, S., Cong, F., Yamashita, A., Asama, H.: Improved artificial potential field-based simultaneous forward search method for robot path planning in complex environment. In: 2015 IEEE/SICE International Symposium on System Integration (SII). IEEE, December 2015. https://doi.org/10.1109/sii.2015.7405075
7. Mandal, S.: Elephant swarm water search algorithm for global optimization. Sādhanā **43**(1) (2018). https://doi.org/10.1007/s12046-017-0780-z

8. Miao, Y.Q., Khamis, A., Kamel, M.S.: Applying anti-flocking model in mobile surveillance systems. In: 2010 International Conference on Autonomous and Intelligent Systems, AIS 2010. IEEE, June 2010. https://doi.org/10.1109/ais.2010.5547036
9. Rastgoo, M.N., Nakisa, B., Nazri, M.Z.A.: A hybrid of modified PSO and local search on a multi-robot search system. Int. J. Adv. Robot. Syst. **12**(7), 86 (2015). https://doi.org/10.5772/60624
10. Robin, C., Lacroix, S.: Multi-robot target detection and tracking: taxonomy and survey. Auton. Robots **40**(4), 729–760 (2016). https://hal.archives-ouvertes.fr/hal-01183372
11. Sadouki, S.C., Tari, A.: Multi-objective and discrete elephants herding optimization algorithm for QoS aware web service composition. RAIRO - Oper. Res. **53**(2), 445–459 (2019). https://doi.org/10.1051/ro/2017049
12. Senanayake, M., Senthooran, I., Barca, J.C., Chung, H., Kamruzzaman, J., Murshed, M.: Search and tracking algorithms for swarms of robots: a survey. Robot. Auton. Syst. **75**, 422–434 (2016). https://doi.org/10.1016/j.robot.2015.08.010
13. Tang, H., Sun, W., Yu, H., Lin, A., Xue, M., Song, Y.: A novel hybrid algorithm based on PSO and FOA for target searching in unknown environments. Appl. Intell. **49**(7), 2603–2622 (2019). https://doi.org/10.1007/s10489-018-1390-0
14. Wang, G.G., Deb, S., Coelho, L.d.S.: Elephant herding optimization. In: 2015 3rd International Symposium on Computational and Business Intelligence (ISCBI). IEEE, December 2015. https://doi.org/10.1109/iscbi.2015.8

An IoT-Based Adaptive Traffic Light Control Algorithm for Isolated Intersection

Tarek Amine Haddad, Djalal Hedjazi(✉), and Sofiane Aouag

LASTIC Laboratory, Department of Computer Science, University of Batna 2, Pole Fesdis 05110 RN, 05110 Fesdis Batna, Algeria
{t.haddad,d.hedjazi}@univ-batna2.dz, djalal.hedjazi@gmail.com, sofiane.aouag@gmail.com

Abstract. In this research work, we focus on optimizing urban traffic at a single intersection. Based on the principle of the Internet of Things (*IoT*), some sensors can be deployed in different locations in the road network and the traffic controller system may collect in real-time the pertinent information about the traffic stat. In this regard, a new approach that dynamically decides the phases sequence and green light times is proposed to improve the traffic quality. Any decision is made according to both Average Waiting Time (*AWT*) and number of vehicles in the different phases' queues. Thus, the priority is given to the phase with highest values of *AWT* and queue size. In order to measure the performances of our approach, we developed a traffic simulation by using *NetLogo* platform. By varying the traffic density, the proposed approach can efficiently improve the quality of the traffic while minimizing the waiting time of vehicles at intersection and enhancing the traffic flow.

Keywords: Intelligent transport systems · Wireless sensor network · Internet of Things · Traffic simulation · Emergency vehicle management

1 Introduction

For many years, the traffic congestion becomes a serious problem for people's daily lives, environment quality and enterprises' economic. It consequences in excess delays, diminished safety conditions and increased pollution [1]. That's why enormous efforts have been investigated to face such problem, notably in urban area. As a result, numerous solutions for Intelligent Transportation Systems (ITS) have been proposed to detect congestions and manage them by re-routing vehicles, while improving traffic flow and optimizing vehicles' waiting time. In general, these solutions are classified into two grand categories. The first category interests to propose algorithms for vehicle routing to guide vehicles away from traffic congestion region [2–4] while the second one focuses on the traffic light control to balance traffic flow [5–8].We inscribe this study in the context of this last category because we judge the implementation of an effective light control algorithm at intersections is crucial, notably when too many vehicles share a common road network with limited capacity. By combining the Artificial Intelligence

M. R. Senouci et al. (Eds.): CSA 2020, LNNS 199, pp. 107–117, 2021.
https://doi.org/10.1007/978-3-030-69418-0_10

(*AI*) and the Internet of Things (*IoT*), many intelligent solutions have been proposed to improve traffic flow at intersections while relying on the measurement of traffic at these lasts by wireless sensor networks.

In this paper, we considered to optimize traffic flow in an isolated intersection by dynamic traffic light control which is based on number of vehicles in phases' queues and phases' Average Waiting Time (*AWT*). Thus, we propose a new adaptive approach which is leant on a model based on *IoT* technologies for collecting real-time data about the traffic stat. These data constitute an essential base for measuring both *AWT* and the number of vehicles in all phases in order to dynamically manage traffic light control. The objective is to improve the effectiveness of traffic light controller at intersections, by reducing waiting time for the green light and avoiding traffic jams.

In order to measure the performances of our proposition, a multi-agent model representing vehicles, roads and intersections is designed and prototyped using *Netlogo* platform. By comparing our approach with some proposed adaptive algorithms [16, 26] as well as with the classical algorithm that fixes the lights' times throughout the use of the road network, the simulation shows our approach with promising results.

The remainder of this paper is organized as follows. In Sect. 2, we briefly discuss some important previous works on traffic light control area. Section 3 presents the problem description and objectives, while Sect. 4 explains the functioning's principle of the proposed adaptive algorithm for dynamically controlling the traffic lights at an isolated intersection. Section 5 discusses the simulation results. Finally, Sect. 6 concludes with a summary and outline for future works.

2 Related Works

The optimization of traffic light control at isolated intersections has been the subject of much work. The contribution of such optimization focuses on two key criteria: maximizing the traffic flow through the intersection during the green light and minimizing vehicles' waiting time by using road network's data which are collected by wireless sensors [9]. The literature on this research area is abundant, exploring several theoretical models proposing various intersection signal control systems. These lasts can be categorized into two classes: fixed-time (pre-timed) systems and adaptive systems [10, 11]. Based on previous traffic data, the pre-timed systems fixe the signal timing settings which persist unchangeable during pre-determined periods of the traffic lights use. Because the implementation of this kind of systems is simple, they are adopted by a large number of cities. Nevertheless, they remain less successful, notably in terms of capturing the uncertainties of traffic demand as well as they require regular updates to keep their capacity. As for adaptive control systems, the signal timing settings is dynamically generated based on current traffic data, and continuously adjust their values according to real-time changes of traffic flow. They are more effective than the pre-timed systems, notably when the traffic's demand increase [12].

Among the most used systems in the world, we cite Split Cycle Offset Optimization Technique (*SCOOT*) system and Sydney Co-ordinated Adaptive Traffic (*SCAT*) System [11, 13]. SCOOT [13], is a traffic-adapted centralized system that coordinates traffic signals according to a fixed sequence of green lights in urban areas. It automatically responses the request for real-time traffic sensors implanted in the urban road network. Through SCOOT, it possible to reduce the time delay up to about 20%. SCATS [14], is another widely used system that can provide predefined intelligent traffic plans for programming traffic signals, covering various traffic situations in order to significantly reduce vehicle delays, especially during peak periods [15]. The system may bring about health and ecology benefits such as a reduction of air pollution and accidents.

Moreover, *Genders et al.* proposed a new intelligent approach based on modern deep reinforcement learning to create a truly adaptive traffic signal control agent in the Simulation of Urban MObility *(SUMO)* simulator [11]. *Yousef et al.* presented a design of an adaptive traffic control system that uses and efficiently manages traffic light controllers in both single and multiple intersections [16]. *Zhou et al.* proposed an adaptive traffic light control algorithm that adjusts both the sequence and length of traffic lights in accordance with the real-time traffic detected [17]. The proposed algorithm considers several traffic parameters such as: traffic volume, waiting time, vehicle density, etc., to define green light sequence and the optimal green light length.

Besides, many others AI-based techniques such as Ant Colonies [18], genetic algorithms [10], fuzzy logic control [19, 20], neural network [21], etc., have been proposed to optimize the signal settings in traffic lights.

The research work presented, in this paper, concerns the optimization of the traffic flow in a simple intersection controlled by lights. The goal is to avoid the congestion problem by analyzing in real-time the phases' queues and attributing the priority to the phase corresponding to minimal *AWT*. The idea is based on adaptive approach that is leant on IoT-based model.

3 Problem Description and Objectives

The traffic light control problem lies in how to adaptively react to dynamic traffic changes in order to improve efficiency control while ensuring equity for all lanes. Efficiency here includes the maximization of the flow rate at an isolated intersection (the number of vehicles crossing the intersection) and the optimization of vehicles' *AWT* [22]. This objective is accomplished by using a dynamic-based strategy that respects some rules such as: (i) Dynamic choice and arranging of the traffic phases; (ii) dynamic reacting to the variations in vehicles' arrival and departure rates; (iii) dynamic control of the green and red times. Indeed, we modeled a simple intersection equipped with traffic lights.

A shown in Fig. 1, our model includes an isolated intersection with four directions *(N (north), S (south), E (east), W (west))*. Each direction is containing two lanes with two possible senses *(FR* (go-*Forward/turn Right)* and *L (turn left))* which are controlled by traffic lights. Thus, each passing vehicle can have a direction *DR* of {*N, S, E* and *W*} and a lane *LN* of {*FR* and *L*}. In our model, we can identify the lane where a vehicle is running by the pair *DR* and *LN*. Indeed, by combining the values of such pair (*DR* and *LN*), we can enumerate at most eight lanes: {*WFR, WL, EFR, EL, NFR, NL, SFR, SL*}. Nevertheless, it is extremely important to consider the conflicts' cases which they seem necessary to avoid the deadlock problems. So, the conflict directions matrix can be summarized in the Table 1. Note that if there is no-conflict between two lanes the value 1 is assigned else the value 0 is assigned. Accordingly, to avoid the conflict situations, when the vehicles are passing across the intersection, the controller must lock all the conflicting lanes by using the traffic lights. Based on the Table 1 values, one of twelve different cases of green lights can be selected for time duration (Fig. 2).

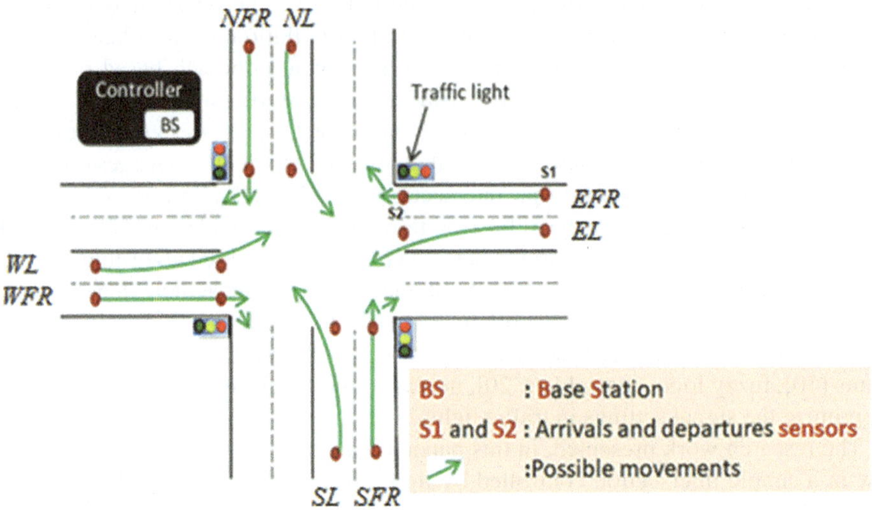

Fig. 1. An isolated intersection model [23].

Moreover, to identify in real-time the lanes' number of vehicles, two sensors (noted *S1* and *S2*) are placed at the both extremities of each lane (Fig. 1). To acquire a correct selection of samples, it is assumed that the distance *D* between *S1* and *S2* should be sufficient.

Table 1. Conflict matrix.

Lane	*WFR*	*WL*	*EFR*	*EL*	*NFR*	*NL*	*SFR*	*SL*
WFR	–	1	1	0	0	1	0	0
WL	1	–	0	1	0	0	1	0
EFR	1	0	–	1	0	0	0	1
EL	0	1	1	–	1	0	0	0
NFR	0	0	0	1	–	1	1	0
NL	1	0	0	0	1	–	0	1
SFR	0	1	0	0	1	0	–	1
SL	0	0	1	0	0	1	1	–

In our model, *T* represents the Traffic cycle duration, *G* represents the green light period of one phase in seconds, and *R* represents the red light period in seconds. As a

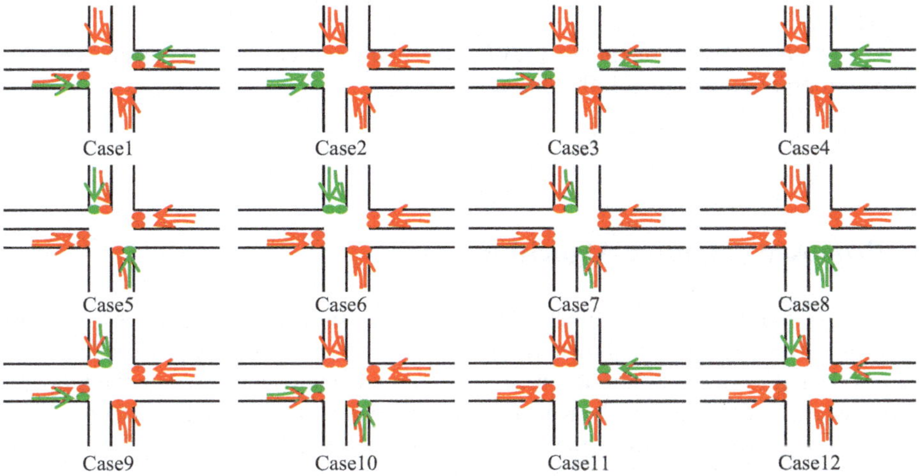

Fig. 2. All possible configurations red green phases.

result, the relation between *T*, *R* and *G* is given by Eq. (4).

$$R = T - G \tag{1}$$

It is assumed that *G* must be inferior to the predetermined maximum time required for the green lights noted T_{max}. The T_{max} has been the subject of several studies as [24, 25]. In addition, *G* is determined according to the both values: *Q* and the required time it takes for a vehicle to cross the intersection (noted t_{pass}).

$$G = Q * t_{pass} \tag{2}$$

All the traffic's information is collected by the controller embedded on the side of the intersection. Thus, it is possible to dynamically measure the number of vehicles (length) in all queues as well as the *AWT* at a junction. The length of queue *j* for a direction at t^{th} cycle is denoted by $Q_j(t)$. *Let* $Rate_{arrival,\,j}(t)$ *and* $Rate_{departure,\,j}(t)$ representing respectively, the arrival rate and the departure rate for the queue *j* at the tth cycle. Therefore, the queue length growth rate for queue*j* at t^{th} cycle which is denoted $\alpha_j(t)$ is given by:

$$a_j(t) = Rate_{arrival,j}(t) - Rate_{departure,j}(t) \tag{3}$$

$$Q_j(t) = Q_j(t-1) + a_j(t) * G + Rate_{arrival,} * R \tag{4}$$

Where: $Q_j(t-1)$ *represents the remaining vehicles from the previous cycle.*

To choice the next phase concerned by the future green light, the queue lengths: *Qi* and Q_j as well as the Waiting Time (*WT*): WT_i and WT_j, for the all two movements *I* and *j* selected from the twelve previous cases, are calculated. Thus, the priority is given to the phase with the highest value the sum that is calculated as:

$$S = Q + WT \tag{5}$$

Where:

The waiting time for the phase k (WT_k) is calculated between the current time and the last stopping time by the following formula:

$$WT_k = T_{current} - T_{stopping} \quad (6)$$

4 Proposed Adaptive Algorithm

We describe in this section the functioning principle of the proposed adaptive algorithm. This last is running, as well shown in Fig. 1, on the *BS* embedded on the roadbed. The data, collected from the different sensors, is used to dynamically calculate the probable queue length, for the next cycle as well as plan the various traffic lights. As previously specified, the target objective of the algorithm is to optimize the the vehicles' *AWT* while enhancing the traffic flow. To achieve this objective, the algorithm executes three steps that are:

(i) Detection in real-time the traffic information;
(ii) Selection of the phase concerned by the following green light cycle;
(iii) Determination of the green light's duration (see Eq. (2)).

As described in Eq. (4), the queue length is calculated, for each phase, after each timing cycle. Furthermore, because the algorithm gives the priority to the emergency vehicles, an operation of checking the presence of these last is extremely important. Therefore, the green light for a TG_{emerg} duration is assigned to the lane on which the emergency vehicle is present. Note that TG_{emerg} represents the green time required to allow the emergency vehicle to across the intersection. It depends on the number of vehicles that precede the emergency vehicle. If many emergency vehicles are present at different lanes, the priority is given to the one with maximum S. The following pseudo code describes the principle of the proposed algorithm:

Algorithm. *Proposed Adaptive Algorithm*

Input:

- $Q_j(t)$: *Total number of vehicles waiting at the traffic signal at* j^{th} *active lane in a* t^{th} *cycle;*
- *Lanes: Set of lanes which are active on the intersection {WFR, WL, EFR, EL, NFR, NL, SFR, SL};*
- *T: Traffic cycle duration;*
- TG_{emerg}: *The green time required to allow the emergency vehicle to across the intersection.*
- *EmergPresent: At minimum one emergency vehicle is present {true, false};*

Output:

- *Green light sequence, phase duration.*

Begin

/ the function of checking the presence of emergency vehicles returns* ***true*** *if at minimum one emergency vehicle is present and the priority is given to the* k^{th} *lane for* TG_{emerg} *duration*/*

EmergPresent = ***Checking the presence of the emergency vehicle*** *(*TG_{emerg}, *k);*
If (EmergPresent) Assign green light to k^{th} *lane for* TG_{emerg} *duration;*
Else {
    Calculate Q for each phase (equation (4))
    Calculate S for each phase (equation (5))
    Calculate the green time G (equation (2))
    If $(G > T_{max})$ $G = T_{max}$;
    Assign the next green light to the phase with maximum S for a G duration;
}

End.

5 Simulation Results and Discussion

To measure the performance of our approach, an implementation using *Netlogo* simulator is accomplished. The experiment is conducted with the objective of comparing the performance of the proposed adaptive algorithm with the fixed algorithm, the algorithm proposed by *Rida et al.* [26] which gives priority to smallest queues, and with algorithm proposed by *Yousef et al.* [16] which prioritizes the largest queues in terms of *AWT*, the traffic *throughput* and the *number of stopped vehicles.* These metrics are calculated during simulation because they indicate the traffic flow models and their consequence on the traffic congestion.

All instances' data are randomly generated. Table 2 describes some simulation data. Thus, to imitate the real life traffic variants, during the use of urban road networks, the dynamic generation of the traffic flow rates is variable from a cycle to other. Besides, a summary of the relative *AWT*, *Throughput* and *Number of stopped vehicles* variation of the both algorithms is graphically depicted in Fig. 3, 4 and 5.

As displayed in Fig. 3, it may be perceived that, on the *AWT*, the proposed adaptive algorithm performs better than the fixed algorithm and the *Rida et al.*'s algorithm significantly. However, it gives close similar performances than *Youcef et al.'s* algorithm. In addition, it is useful to perceive that the gaps' values augment significantly, notably

Table 2. Simulation data.

Parameters	Values
Intersection area	30 × 30 patches
Unit of time	Ticks
Execute time	2000 ticks
Max-speed of vehicle	0.5 patch/ticks
Acceleration of vehicle	0.006 patch/ticks
Density of vehicle	10%–60% per each direction
Lane length	24 patches

when time values vary between 666 and 1661 (Fig. 3). In addition, the Fig. 4 displays that the qualities of both algorithms (our algorithm and fixed one) are almost similar. However, the adaptive algorithm shows a relative improvement over throughput provided by the fixed algorithm. Similarly to the previous analysis, the evolution of the number of stopped vehicles illustrated in Fig. 5 which shows that the four algorithms present more or less the same performances.

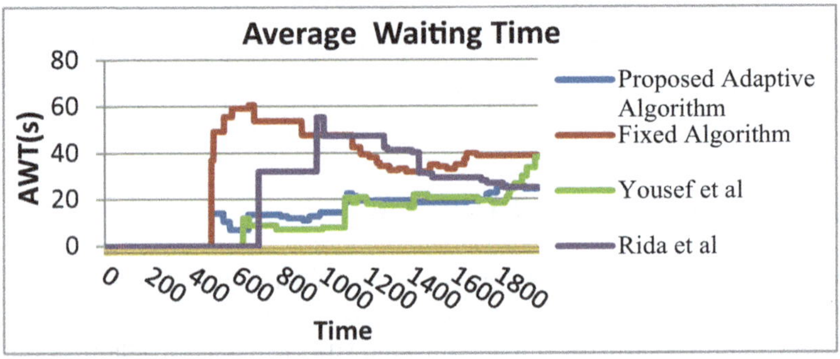

Fig. 3. Average waiting time of lights control algorithm (Fixed vs. Proposed adaptive algorithm, Rida et al. [26] algorithm and Yousef et al. [16] algorithm).

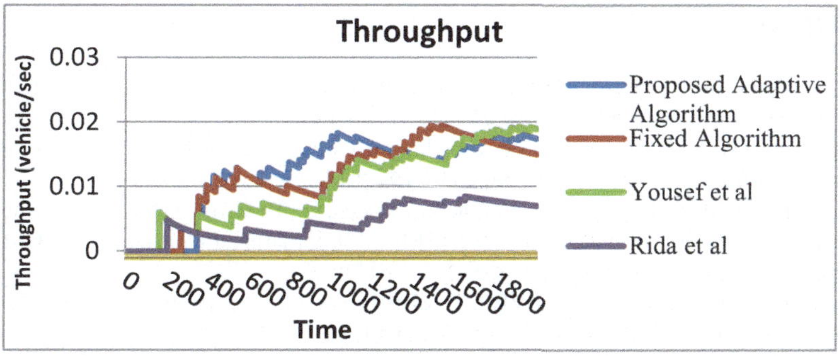

Fig. 4. Throughput of lights control algorithm (Fixed vs. Proposed adaptive algorithm, Rida et al. [26] algorithm and Yousef et al. [16] algorithm).

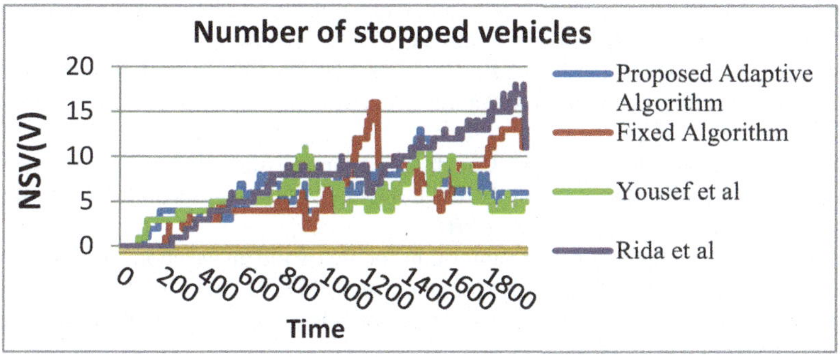

Fig. 5. Number of stopped vehicles at an intersection (Fixed vs. Proposed adaptive algorithm, Rida et al. [26] algorithm and Yousef et al. [16] algorithm).

6 Conclusion

In this paper, we have presented a new algorithm for dynamic control of traffic lights. Our objective is to achieve higher *Throughput* and *AWT* at an isolated intersection. The main contribution of the proposed algorithm is that it can effectively select the best possible setting for the green light sequence. In order to measure the performances of our proposition, the both fixed and Adaptive algorithms are implemented and simulated through *NetLogo* platform.

The analysis of the performances based on a simulation demonstrates that the proposed adaptive algorithm outperforms the fixed algorithm in terms of *AWT*. The results show that the new adaptive algorithm has a significant improvement that can be up to 12 s in terms of *AWT*. As for the *Throughput* and the *Number of stopped vehicles* metrics, the results demonstrate that the both algorithms produce almost the same performances.

Finally, in terms of perspective, we plan to extend or adapt the proposed adaptive algorithm to be able to dynamically control several intersections.

References

1. Papageorgiou, M., Kiakaki, C., Dinopoulou, V., Kotsialos, A., Wang, Y.: Review of road traffic control strategies. Proc. IEEE **91**, 2043–2067 (2003)
2. Jabbarpour, M.R., et al.: Ant-based vehicle congestion avoidance system using vehicular networks. Eng. Appl. Artif. Intell. **36**, 303–319 (2014)
3. Pan, J., Khan, M.A., Popa, I.S., Zeitouni, K., Borcea, C.: Proactive vehicle re-routing strategies for congestion avoidance. In: 2012 IEEE 8th International Conference on Distributed Computing in Sensor Systems, pp. 265–272 (2012)
4. Pan, J., Popa, I.S., Borcea, C.: DIVERT: a distributed vehicular traffic re-routing system for congestion avoidance. IEEE Trans. Mob. Comput. **16**, 58–72 (2017)
5. Jovanović, A., Nikolić, M., Teodorović, D.: Area-wide urban traffic control: a bee colony optimization approach. Transp. Res. Part C Emerg. Technol. **77**, 329–350 (2017)
6. Li, X., Sun, J.-Q.: Signal multiobjective optimization for urban traffic network. IEEE Trans. Intell. Trans. Syst. **19**, 3529–3537 (2018)
7. McKenney, D., White, T.: Distributed and adaptive traffic signal control within a realistic traffic simulation. Eng. Appl. Artif. Intell. **26**, 574–583 (2013)
8. Soon, K.L., Lim, J.M.-Y., Parthiban, R.: Coordinated traffic light control in cooperative green vehicle routing for pheromone-based multi-agent systems. Appl. Soft Comput. **81**, 105486 (2019)
9. Rida, N., Ouadoud, M., Hasbi, A., Chebli, S.: Adaptive traffic light control system using wireless sensors networks. In: 2018 IEEE 5th International Congress on Information Science and Technology (CiSt), pp. 552–556 (2018)
10. Guo, J., et al.: A model and genetic algorithm for area-wide intersection signal optimization under user equilibrium traffic. Math. Comput. Simul. **155**, 92–104 (2019)
11. Genders, W., Razavi, S.: Using a deep reinforcement learning agent for traffic signal control. arXiv:1611.01142*[cs]* (2016)
12. Zhao, D., Dai, Y., Zhang, Z.: Computational intelligence in urban traffic signal control: a survey. IEEE Trans. Syst. Man Cybern. C **42**, 485–494 (2012)
13. Robertson, D.I., Bretherton, R.D.: Optimizing networks of traffic signals in real time-the SCOOT method. IEEE Trans. Veh. Technol. **40**, 11–15 (1991)
14. Sims, A.G., Dobinson, K.W.: The Sydney coordinated adaptive traffic (SCAT) system philosophy and benefits. IEEE Trans. Veh. Technol. **29**, 130–137 (1980)
15. Zhou, B., Cao, J., Li, J.: An adaptive traffic light control scheme and its implementation in WSN-based ITS. Int. J. Smart Sens. Intell. Syst. **6**, 1559–1581 (2013)
16. Yousef, K.M., Al-Karaki, J.N., Shatnawi, A.M.: Intelligent traffic light flow control system using wireless sensors networks. J. Inf. Sci. Eng. **26**, 753–769 (2010)
17. Zhou, B., Cao, J., Zeng, X., Wu, H.: Adaptive traffic light control in wireless sensor network-based intelligent transportation system. In: 2010 IEEE 72nd Vehicular Technology Conference - Fall, pp. 1–5 (2010)
18. D'Acierno, L., Gallo, M., Montella, B.: An ant colony optimisation algorithm for solving the asymmetric traffic assignment problem. Eur. J. Oper. Res. **217**, 459–469 (2012)
19. Pappis, C.P., Mamdani, E.H.: A fuzzy logic controller for a traffic junction. IEEE Trans. Syst. Man Cybern. **7**, 707–717 (1977)
20. Bingham, E.: Reinforcement learning in neurofuzzy traffic signal control. Eur. J. Oper. Res. **131**, 232–241 (2001)
21. Srinivasan, D., Choy, M.C., Cheu, R.L.: Neural networks for real-time traffic signal control. IEEE Trans. Intell. Transp. Syst. **7**, 261–272 (2006)
22. Faye, S., Chaudet, C., Demeure, I.: Un algorithme distribué de contrôle des feux de circulation sur plusieurs intersections par un réseau de capteurs sans fil. In: Nouvelles Technologies de la

Répartition/Colloque francophone sur l'ingénierie des protocoles (NOTERE/CFIP), Anglet, France, p. 1 (2012). ⟨hal-00781122⟩

23. Faye, S., Chaudet, C., Demeure, I.: A distributed algorithm for adaptive traffic lights control. In: 2012 15th International IEEE Conference on Intelligent Transportation Systems, pp. 1572–1577 (2012)
24. Johri, M., Goel, A., Tiwari, A.K.: Dynamic traffic signal control algorithm in intelligent transportation system through wireless sensor networks. Int. J. Eng. Sci. Res. **2**, 9 (2012)
25. Chakraborty, P.S.: Real time optimized traffic management algorithm. IJCSIT **6**, 119–136 (2014)
26. Rida, N., Hasbi, A.: Dynamic traffic lights control for isolated intersection based wireless sensor network. In: Ben Ahmed, M., Boudhir, A.A., Younes, A. (eds.) Innovations in Smart Cities Applications, 2 edn., pp. 1036–1044 (2019)

Hybrid Heuristic Optimization of an Integrated Production Distribution System with Stock and Transportation Costs

Bensakhria Mohamed(✉) and Samir Abdelhamid

Department of Industrial Engineering, University of Batna 2, Batna, Algeria
m.bensakhria@univ-batna2.dz

Abstract. In this paper we address the integration of two-level supply chain with multiple items, production facility and retailers' demand over a considered discrete time horizon. This two-level production distribution system features capacitated production facility supplying several retailers located in the same region. If production does take place, this process incurs a fixed setup cost as well as unit production costs. In addition, deliveries are made from the plant to the retailers by a limited number of capacitated vehicles and routing costs are incurred. This work aims at implementing a solution to minimize the sum of the costs at the production facility and the retailers. The methodology adopted to tackle this issue is based on a hybrid heuristic, greedy and genetic algorithms that uses strong formulation to provide a good solution of a guaranteed quality that are as good or better than those provided by the MIP optimizer with a considerably larger run time. The results demonstrate that the proposed heuristics are effective and performs impressively in terms of computational efficiency and solution quality.

Keywords: Optimization · Production-distribution systems · Routing decision · Inventory · Genetic algorithms

1 Introduction

Over the recent years modeling and simulation of production routing problems (PRP) has been well developed, given the huge role it plays in the analysis of the discrete behavior of this type of systems, this later combines lot-sizing decisions, inventory management and routing. In the PRP within each period production facility must determine whether or not to make the product and decide the corresponding quantity in which this product can be inventoried at the retailers, Considering production setup cost and variable cost, transportation and routing costs [1, 2].

According to recent studies, integrated supply chain planning problems consider the optimization of stock and transport costs separately which is significant and cannot be neglected, Li et al. [4] a supply chain may be seen as a composed layer. Deliveries are made between each pair of layers. Each pair represents one level of the distribution network; transportation is considered the building block of any supply chain. Many shipment policies have been proposed in literature for this problem. For instance, a

M. R. Senouci et al. (Eds.): CSA 2020, LNNS 199, pp. 118–130, 2021.
https://doi.org/10.1007/978-3-030-69418-0_11

current trend in the application of freight distribution systems involving two or more levels where at first freight travels using ships and then is loaded on trucks to distribute it over the warehouses or retailers [3]. The reminder of this paper organized as follows: Sect. 2 presents the notation used throughout the paper and introduces the literature concept of the PRP. Mathematical formulation presented in Sect. 3. Section 4 a two-phase algorithm is presented to solve the proposed model and. The experimentation and results are provided in Sect. 5. Finally, Sect. 6 provides the conclusions drawn and future research lines.

2 State of the Art

In the research literature these four decisions can be found in Bhatnagar et al. [8], Sarmiento and Nagi [9]. Regarding the large number of studies in this area different tools and techniques have been used in past publication for the optimization of the PRP problem. Mathematical techniques which have been widely used in the past include heuristic procedures. Several metaheuristics, such as Boudia using memetic algorithm with dynamic population management for an integrated production distribution problem [10], GRASP [12], tabu search [13, 14], and ALNS [18]. Ettl et al. [11] they considered a queuing network model with constraint nonlinear optimization to minimize the inventory cost while satisfying customer service level requirements, the main disadvantage of this approach is that it does not consider the production capacity, which is a key constraint in the PRP systems. Chang et al. [15] discussed a serial production system with finite storage space that follows a base-stock policy. They developed a phase-type approximation to estimate the effective arrival rate of customers in which they can determine the optimal base-stock level. Bard et al. [17] they presented a comparative analysis of a series of heuristics for an inventory routing problem (IRP) using branch-and-price framework. Golden et al. [16] were the first to investigate the interrelated problem of inventory allocation and vehicle routing. They developed a heuristic for designing an integrated delivery planning system aimed at comparing the distribution rule used by the company with their approach.

Recently, new forms of relationships in the supply chain have been adopted. One of these is called vendor managed inventory (VMI), Archetti et al. [7] have introduced a particular policy, known as deterministic order up to level policy, the problem is to evaluate for each time period the quantity to ship to each retailer and the vehicle route, they present a mixed integer linear programming model with additional valid inequalities to strengthen the linear relaxation. Chitsaz et al. [5] presented a general model for the assembly routing problem (ARP). The problem formulated as a mixed-integer linear program and they proposed a three-phase decomposition metaheuristic that relies on the iterative solution of different subproblems. Leopoldo et al. [6] developed a heuristic algorithm to solve the selective and periodic inventory routing problem. They proposed a reduce and optimize approach with the introduction of a new inequality.

In this paper, the problem is mathematically formulated by a MILP model. Due to the complexity of the problem which was shown to be NP-hard by Arkin et al. [19], two well-known heuristics available in the literature of PRP (Genetic Algorithm (GA) and Greedy Algorithm (GA)), are first adapted to (LSP) then (VRP). To more effectively solve larger problems, a two-phase algorithm is also presented to solve the proposed model. The algorithm is greedy considering at each step, several alternatives are generated and the best one is selected. To evaluate the model performance, different numerical experiments were performed. The results show the model and the heuristics are effective for the problem.

3 Problem Description

In this research proposal we aim to address the identified points; we will evaluate the inventory cost by integrating production, inventory, and routing decisions that minimize total costs, while ensuring that all demands are met. We consider a set of retailers $\mathbf{r} = \{1 \ldots R\}$ geographically dispersed around the production facility, products delivered to retailers along a planning discrete time horizon $\mathbf{t} = \{1 \ldots T\}$. Each retailer r has a fixed nonnegative deterministic demand $\mathbf{d_{rt}}$ that must be satisfied over a time horizon t. back orders are not permitted a limited amount of the end product can be restocked from the production facility, the storage capacity at the retailers are limited to $\mathbf{C_r}$ with unit holding cost $\mathbf{h_r}$ which depends on the retailer.

A product quantity is produced or made available at the production facility and a quantity is consumed at retailer, no capacity limit is assumed on production site but the amount produced per item in each period is restricted to $\mathbf{P_{max}}$, the storage capacity is limited to $\mathbf{C_p}$ with a holding cost $\mathbf{h_p}$. A fixed production setup cost $\mathbf{F_{pc}}$ and a variable setup cost $\mathbf{V_{pc}}$ is incurred. A fleet of **Z** homogeneous vehicles with capacity **Q** capable of serving several retailers on the same route is responsible for shipping the product from the production facility to retailers. A cost of $\mathbf{C_{ij}}$ is realized when the link among two pair **i** and **j** is traversed. VRP must be solved daily to determine which customers to visit and how much product to deliver to each retailer. From the comparative study of different extensions of the PRP problems and our objectives the problem can be formulated as a mixed integer linear programming (MILP) with a multi-heuristics-based approach.

Indexes

$t = 1 \ldots T$: set of periods

$r = 1 \ldots R$: set of retailers

$z = 1 \ldots Z$: set of vehicles

d_{rt}: demand of retailer r in period t
C_r: storage capacity of retailer r
C_p: storage capacity of production facility
P_{max}: production capacity
h_r: inventory holding cost of retailer r
h_p: inventory holding cost of production facility
Q: vehicle capacity
F_{pc}: fixed setup cost
V_{pc}: variable setup cost
$C_{i,j}$: fixed transportation cost

Decision variables

D_{rzt}: The quantity to be delivered to retailer r using vehicle z in period t
I_p^t: Inventory level at production facility in planning period
I_r^t: Inventory level at retailer r in planning period t
P_t: quantity to be produced in period t
Z_t: Production quantity in planning period t
Y_t: Binary variable equal 1 if there is production on planning period t, otherwise equal 0
X_{rzt}: Binary variable equal 1 if vehicle Z is used in planning period t to serve retailer r in period t, otherwise equal 0
U_{rzt}: Binary variable equal 1 if retailer r is served by vehicle z in planning period t, otherwise equal 0

$$\text{Min} = \sum\nolimits_{t=1}^{T} \left(\begin{array}{c} F_{pc}.y_T + V_{pc}.Z_t + h_p.I_p^t \\ + \sum_r h_r.I_r^t + \sum_r \sum_z C_{i,j}.X_{rzt} \end{array} \right) \tag{1}$$

$$I_p^t = I_p^{t-1} + P_t - \sum\nolimits_r \sum\nolimits_z D_{rzt} \ \forall r \in R,\ z \in Z,\ t \in T \tag{2}$$

$$I_r^t = I_r^{t-1} + \sum\nolimits_z D_{rzt} - d_{rt} \ \forall r \in R,\ z \in Z,\ t \in T \tag{3}$$

$$I_p^t \le C_p \ \forall t \in T \tag{4}$$

$$I_r^t \le C_r \ \forall t \in T \tag{5}$$

$$Pt \le P_{max}.Y_t \ \forall t \in T \tag{6}$$

$$D_{rzt} \le Q.U_{rzt} \ \forall r \in R,\ z \in Z,\ t \in T \tag{7}$$

$$\sum\nolimits_r D_{rzt} \le Q \ \forall r \in R,\ z \in Z,\ t \in T \tag{8}$$

$$\sum\nolimits_z U_{rzt} \le 1 \ \forall z \in Z,\ t \in T \tag{9}$$

$$\sum\nolimits_r U_{rzt} \le 1 \ \forall r \in R,\ t \in T \tag{10}$$

$$\sum\nolimits_r D_{rzt} \le Q.\ X_{rzt} \ \forall z \in Z,\ t \in T \tag{11}$$

$$X_{rzt},\ Y_t,\ U_{rzt} \in \{0, 1\} \ \forall r \in R,\ z \in Z,\ t \in T \tag{12}$$

$$P_t,\ I_p^t,\ I_r^t,\ D_{rzt} :\geq 0 \forall r \in R,\ z \in Z,\ t \in T \tag{13}$$

The mathematical formulation for the proposed problem is presented as follows: The objective function given in (1) used to minimize the overall costs of production setup, transportation, holding costs at the production facility and the retailers. Constraints (2) relates to the inventory balance equations at production facility, where the inventory level at the end of period t can be obtained by adding the inventory level at the end of period t − 1. Similarly, inventory balance equations at the retailer r, as shown in constraints (3), is equal to the inventory level in period t − 1, by adding the total quantity shipped in period t and subtracting the total quantity shipped to the other retailers. Constraints (4) and (5) guarantee the upper bounds on inventory levels at production facility or retailers. Constraints (6) confirm that the inventory level of the production facility does not exceed the maximum level. Constraints (7) link the continuous variable D_{rzt} to the binary variable which means if there is a delivery to a retailer in planning period t, then it must be 1. Constraints (8) define the vehicle capacity. Constraints (9) ensure that each retailer is visited at most once in a given period. Similarly, constraints (10) guarantee the vehicle loading and routing restrictions used mostly once per planning period. Constraints (11) define number of trips during planning horizon. Finally, constraints (12) and (13) are variable definitions.

Initial attempts to solve instances of this size with CPLEX 8.1 were not encouraging. This was not surprising since the model is NP-hard, to deal with such a complex problem is to decompose it into smaller problems of reduced complexity for which optimal or near optimal solutions can be determined.

4 Proposed Method

Optimally solving such a convenient problem is in general difficult due to its combinatorial nature that integrates two well-known problems, lot sizing and vehicle routing problems. We present a solution procedure based on a two-phase algorithm to solve the proposed model, decomposition and integration. The initial phase solves the shortest path constraints for the production facility submodel, which contains all the original constraints. Resulting solutions to this phase is always feasible but not optimal and thus gives an upper bound solution to our problem. Successfully solving this problem determines production quantities, quantities to be delivered to each retailer and each retailer that must be visited at each time period. Fixing the routing decisions in the initial phase define the number of retailers to be visited and the quantity to be delivered at each time period. Consequently, in the next phase we propose a heuristic to solve the routing problem in the distribution submodel. In the next iteration the results of the distribution submodel considered as an input to update the routing decisions for the next iteration. The procedure continues this iterative solve feedback solve until a stopping criterion is achieved or attain a fixed number of iterations. The main steps of the iterative method illustrated in Algorithm 1.

Algorithm 1

Begin

Step 1: Estimate demand and production quantity for each retailer that needs delivery in period t
Step 2: Solve production problem using Greedy algorithm
Step 3: Solve the routing problem using Genetic algorithm
Step 4: Calculate sol and set best sol = sol
Repeat
Repeat
Step 5: Update demand and production quantity for each retailer
Step 6: Run diversification mechanism
Step 7: Solve production problem using retailers' updated costs using Greedy algorithm
Step 8: Solve routing problem using Genetic algorithm
Step 9: Update sol if necessary
Until a stopping criterion in met
Step 10: Calculate the sol
If (sol < best sol)
Set best sol = sol
End if
Until a stopping criterion in met
End.

With the aim to simplify our solution, we set some goals based on the retailer's demand and the vehicles capacity, to be more precise the sum of retailers' demands in planning period t (D_{rt}). The production unit model is demonstrated in relations (14) to (19):

$$\text{Min} = \sum\nolimits_{t=1}^{T} \left(F_{pc} . y_T + V_{pc} . Z_t + h_p . I_p^t \right) \tag{14}$$

$$I_p^t = I_p^{t-1} + P_t - D_{rt} \; \forall r \in R, \; z \in Z, \; t \in T \tag{15}$$

$$I_p^t \leq C_p \; \forall t \in T \tag{16}$$

$$P_t \leq P_{max} . Y_t \; \forall t \in T \tag{17}$$

$$Y_t \in \{0, 1\} \; \forall r \in R, z \in Z, t \in T \tag{18}$$

$$P_t, I_p^t \geq 0 \, t \in T \tag{19}$$

Then the algorithm proceeds to the second phase that require all the routing decisions which include retailers that has served by each vehicle at each time period. Given that the initial phase allocates retailers to vehicles and considering vehicle capacity. Relations (20) to (30) demonstrate the distribution model. The sum of the production quantities Pt work as an update for the next iteration of the second phase. The routing phase model is:

$$\mathrm{Min} = \sum\nolimits_{t=1}^{T} \left(\sum\nolimits_{r} \sum\nolimits_{z} C_{i,j}.X_{rzt} + \sum\nolimits_{r} h_r.I_r^t \right) \tag{20}$$

$$I_p^t = I_p^{t-1} + P_t - \sum\nolimits_{r} \sum\nolimits_{z} D_{rzt} \, \forall r \in R, z \in Z, t \in T \tag{21}$$

$$I_r^t = I_r^{t-1} + \sum_{z} D_{rzt} - drt \, \forall r \in R, z \in Z, t \in T \tag{22}$$

$$I_r^t \leq C_r \, \forall t \in T \tag{23}$$

$$D_{rzt} \leq Q.U_{rzt} \, \forall r \in R, z \in Z, t \in T \tag{24}$$

$$\sum\nolimits_{r} D_{rzt} \leq Q \, \forall r \in R, z \in Z, t \in T \tag{25}$$

$$\sum\nolimits_{z} U_{rzt} \leq 1 \, \forall z \in Z, t \in T \tag{26}$$

$$\sum\nolimits_{r} U_{rzt} \leq 1 \, \forall r \in R, t \in T \tag{27}$$

$$\sum\nolimits_{r} D_{rzt} \leq Q.X_{rzt} \, \forall z \in Z, t \in T \tag{28}$$

$$X_{rzt}, U_{rzt} \in \{0, 1\} \, \forall r \in R, z \in Z, t \in T \tag{29}$$

$$P_t, I_p^t, I_r^t, D_{rzt} :\geq 0 \, \forall r \in R, z \in Z, t \in T \tag{30}$$

Given the fact that during the initial tests that we performed to solve the production submodel problem a greedy search algorithm attempts to generate a feasible solutions, we found that this heuristic solutions are unlikely to perform well and even when they can be adapted, they may not be able to find good optimal feasible solutions in a reasonable time. This implies the alternative of applying extended heuristics, genetic algorithms. Hence the question which needs to be asked is how to adapt two different heuristics in an effective way.

4.1 Proposed Genetic Algorithm

In the first phase chromosomes are created for an optimal size that we describe the routing phase, in which corresponds in calculating vehicle routes at each period. A chromosome is a sequence of nodes, indicating the order in which vehicles must visit all retailers. In this case we consider a set of retailers to be served by a fleet of identical vehicles in planning periods; we set a two-dimensional matrix that each cell contains the retailer's demand in a given time period.

The binary variables U_{rzt} that imply if a vehicle z is used in planning period t to serve a retailer r is the main issue to identify the other decision variables. The above procedure is repeated until a stopping criterion is achieved. The details of the basic genetic algorithm are given below in Algorithm 2.

Algorithm 2

Begin

Step 1: Population size (p)
Step 2: Number of generations (n)
Step 3: Create a delivery with a random order of retailers
Step 4: Do this for p times to create a population
Step 5: Select the best delivery from the population
Step 6: Set the initial feasible solution as local best solution and global best solution

Genetic algorithm

```
Run for n times
Loop1
Run for p times
Loop2
Select a random set of delivery from the population
Get the fittest and return
          //Crossover
Parent1 = roulette wheel selection ()
Parent2 = roulette wheel selection ()
Child = Crossover (Parent1, Parent2)
          //Mutation
Swap random two retailers in the child
endLoop2
          //New population is created
Calculate the fitness value for each individual
Sort the population by their fitness values
Update local best Solution and global best Solution
endLoop1
final local solution = local best Solution
final global solution = global best Solution
```

End

4.2 Genetic Operators

The mechanism for generating the initial population allows the production of individuals' population as the basis for future generations. The choice of the initial population is important because it can make more or less fast the convergence to the global optimum.

Selection and Crossover

The first step in genetic algorithm it starts with selection in which individuals are selected for crossover according to their relative fitness, in this way the selection step implemented in practice by the creation of a mating pool for the next generation. The choice of a selection technique in the suggested genetic algorithm is the roulette wheel operator which used to select the initial population to create new children, the probability of choosing an individual depends directly on its fitness.

In the next step the crossover occurs when two members of a population (chromosomes) are selected for reproduction, and is usually based upon their relative fitness in solving the problem. Two parents P1 and P2 are selected randomly we use the order crossover (OX) with slightly modification, it starts by choosing two crossover points then applying sliding motion to fill up the left out holes by sending the mapped positions. It copies the portion of permutation elements between the crossover points from the cut string directly to the offspring, placing them in the same absolute position. An example of offspring generation is illustrated in Fig. 1.

Mutation

Mutation may be defined as a small perturbation to prevent the genetic algorithm converging to a local minimum; we opted to use the operator left to right flip for the mutation process. An example of mutation consists of choosing randomly two genes from the chromosome and exchanging their values is illustrated in Fig. 2. As noted, to escape such a point we implemented a diversification mechanism. The rule we adopt it captures, for each retailer, the complementarities with the other retailers served in the same route. Our principle of correction is to start by calculate the total sum of the quantity transported by vehicles. After that, we have to browse each group gene by gene and if it is higher than the maximum available vehicle capacity, we must search in the next sequence of group's nodes which succeed this position of excess capacity, a retailer who has not been served and its insertion does not cause overload of vehicle we insert it in the next position.

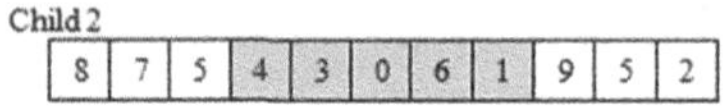

Fig. 1. Order crossover operator

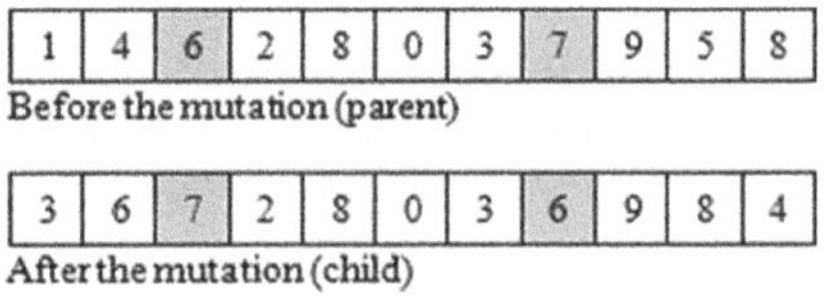

Fig. 2. Mutation operation

5 Experimental Results

Our algorithms are implemented in Matlab environment. We used IBM ILOG CPLEX 12.6 with the default settings to solve the proposed problem. All experiments were performed on an Intel Core 2Duo processor, with 4096Ko of memory cache, 4 GB of RAM and a frequency of 3.0 GHz using the Windows 7 operating system. Problem instances are created by randomly generating parameters as follows:

Number of periods (T), retailers (R), vehicles (V), production capacity (Pmax) and vehicle capacity (Q).

(i) Demand quantities $d_{rt} = U\ [5, 25]$
(ii) Variable production setup costs $V_{pc} = U\ [50,100]$
(iii) Fixed production setup costs $F_{pc} = U\ [500, 1000]$
(iv) Inventory holding costs in production facility and retailers $h = U\ [5, 10]$
(v) Fixed transportation costs $C_{i,j} = U\ [200, 300]$
(vi) Vehicle capacity 2α, 1.5β ($\alpha = \sum_{r,t} d_{r,t}/R \times T$)
(vii) Production capacity 2α, 1.5β ($\beta = \sum_{r,t} d_{r,t}/R \times T$)

5.1 Computational Results on the Small and Large Sets of Instances

Let us first compare the effectiveness of the iterative method proposed in this paper on the small size of instances. In the latter part we present our best solutions and average results on random test problems. A summary of the results for small size test problems is reported in Table 1. In columns 2–4 a comparison of objective functions of the IBM ILOG CPLEX 12.6, Proposed Genetic Algorithm (PGA) then the optimal solution or the best solution is determined by the standard solver IBM ILOG CPLEX 12.6 after one hour.

Table 1. Comparison of objective functions of PGA and IBM ILOG CPLEX on small instances

Problem N T K	CPLEX 1h	PGA	Optimal solutions
4 5 2	23465	23465	6/6
4 5 3	24158	24158	6/6
4 10 2	34028	34101	3/6
4 10 3	42919	43899	2/6
6 5 2	61520	61520	6/6
6 5 3	56676	56676	6/6
6 10 2	55318	55318	6/6
6 10 3	45297	45297	6/6
8 5 2	65754	65890	2/6
8 5 3	63378	66524	2/6
8 10 2	73985	74520	4/6
8 10 3	55550	55550	6/6
10 5 2	74586	77002	3/6
10 5 3	91357	100211	1/6
10 10 2	86542	100254	0/6
10 10 3	86524	99652	0/6
Average	**58816.06**	**61502.31**	**3/6**

In terms of average optimal solutions, the PGA is able to produce better results at the beginning afterward could not further improve the solution quality due to the algorithm sensitivity to the number of periods (T) and the increase number of vehicles. Whereas PGA algorithm could provide better average results for 13 out of 32 test problems and consequently on the overall number of instances solved to optimality, 252 of 648 (39%).

Table 2. Comparison of objective functions of PPGA AND IBM ILOG CPLEX on large instances

Problem N T K	CPLEX 2h	PGA average	PGA worst
30 10 10	163560	179965	182569
30 10 15	215766	236524	245879
30 15 10	310026	356900	365852
30 15 15	383894	425696	440652
40 10 10	523088	536265	575632
40 10 15	179534	193654	199652
40 15 10	257486	280654	290652
40 15 15	335069	340259	391368
50 10 10	466532	500256	520236
50 10 15	559502	625438	643652
50 15 10	307254	350236	364258
50 15 15	284769	312589	343560
60 10 10	402569	480659	499856
60 10 15	501394	576301	601254
60 15 10	665841	758963	796502
Average	**383805**	**428395,06**	**451883,50**

Again, the results for large instances are similar to those observed on the objective function in Table 2. The table illustrates again how PGA determines solutions of good quality compared to the standard solver IBM ILOG CPLEX 12.6, which is the most successful approach on large instances. The standard solver sometimes cannot find feasible solutions whereas both heuristics find one.

6 Conclusions and Future Work

We have proposed a two-phase iterative method based on two heuristics to solve the PRP. In the first phase, a lot-sizing problem is solved using a greedy heuristic. The first phase decides when and how much to produce at each period, when to visit retailers, and how much to deliver at each visit. The second phase takes routing decision into account using genetic algorithm approach. In the GA construction phase, a randomized version of a previously developed construction heuristic is used. Which involves two random neighborhood search mechanisms, the crossover and mutation operations, suitable designs are developed. In order to validate the performance of proposed method, a number of randomly generated test problems are used. We have tested these approaches on small and large size instances. The results show that the proposed method can provide optimal solutions. On average, proposed genetic algorithm outperforms the standard solver IBM ILOG CPLEX 12.6 and generates solutions that are within 2% from the optimal solution.

Our next step is to develop and test different meta-heuristics or other hybrid methods. Then, many more iterations would be possible and more elaborated diversification schemes.

References

1. Geoffrion, A.M., Powers, R.F.: 20 years of strategic distribution system design: an evolutionary perspective. Interfaces **25**(5), 105–127 (1995)
2. Thomas, D.J., Griffin, P.M.: Coordinated supply chain management. Eur. J. Oper. Res. **94**(1), 1–15 (1996)
3. Infante, D., Paletta, G., Vocaturo, F.: A ship-truck intermodal transportation problem. Marit. Econ. Logist. **11**, 247–259 (2009)
4. Li, J., Sava, A., Xie, X.: Performance evaluation and optimization of a two-stage production-distribution system with batch orders and finite transportation time. In: The 16th World Congress of IFAC, Prague (2005)
5. Chitsaz, M., Cordeau, J.-F., Jans, R.: A unified decomposition matheuristic for assembly, production, and inventory routing. INFORMS J. Comput. **31**(1), 134–152 (2019)
6. Cárdenas-Barrón, L.E., González-Velarde, J.L., TreviñoGarza, G., Garza-Nuñez, D.: Heuristic algorithm based on reduce and optimize approach for a selective and periodic inventory routing problem in a waste vegetable oil collection environment. Int. J. Prod. Econ. **211**, 44–59 (2019)
7. Archetti, C., Bertazzi, L., Laporte, G., Speranza, M.G.: A branch-and-cut algorithm for a vendor managed inventory routing problem. Transp. Sci. **41**(3), 382–91 (2007)
8. Bhatnagar, R., Chandra, P.: Goyal SK models for multi-plant coordination. Eur. J. Oper. Res. **67**(2), 141–160 (1993)
9. Sarmiento, A.M., Nagi, R.: A review of integrated analysis of production-distribution systems. IIE Trans. **11**(3), 1061–1074 (1999)
10. Boudia, M., Prins, C.: A memetic algorithm with dynamic population management for an integrated production-distribution problem. Eur. J. Oper. Res. **195**(3), 703–715 (2009)
11. Ettl, M., Feigin, G.E., Lin, G.Y., Yao, D.D.: A supply network model with base-stock control and service requirements. Oper. Res. **48**(2), 216–232 (2000)
12. Boudia, M., Louly, M.A.O., Prins, C.: Areactive GRASP and path relinking for a combined production-distribution problem. Comput. Oper. Res. **34**(11), 3402–3419 (2007)
13. Armentano, V.A., Shiguemoto, A.L., Løkketangen, A.: Tabu search with path relinking for an integrated production-distribution problem. Comput. Oper. Res. **38**(8), 1199–1209 (2011)
14. Bard, J.F., Nananukul, N.: The integrated production-inventory-distribution- routing problem. J. Sched. **12**(3), 257–280 (2009)
15. Chang, K.-H., Lu, Y.-S.: Inventory management in a base-stock controlled serial production system with finite storage space. Math. Comput. Model. **54**, 2750–2759 (2011)
16. Golden, B., Assad, A., Levy, L., Gheysens, F.: The fleet size and mix vehicle routing problem. Comput. Oper. Res. **11**, 49–66 (1984)
17. Bard, J.F., Nananukul, N.: Heuristics for a multiperiod inventory routing problem with production decisions. Comput. Ind. Eng. **57**(3), 713–723 (2009)

18. Adulyasak, Y., Cordeau, J.-F., Jans, R.: Optimization based adaptive large neighborhood search for the production routing problem. Transp. Sci. **48**(1), 20–45 (2014)
19. Arkin, E., Joneja, D., Roundy, R.: Computational complexity of uncapacitated multi echelon production planning problems. Oper. Res. Lett. **8**(2), 61–66 (1989)

Hierarchical Colored Petri Nets for the Verification of SysML Designs-Activity-Based Slicing Approach

Messaoud Rahim[1,2](✉), Malika Boukala-Ioualalen[2], and Ahmed Hammad[3]

[1] Mathematics and Computer Sciences Department, MEDEA University, Médéa, Algeria
messaoud.rahim@univ-medea.dz
[2] MOVEP Laboratory, USTHB, Algiers, Algeria
mioualalen@usthb.dz
[3] FEMTO-ST Institute, UMR CNRS 6174, Besançon, France
ahmed.hammad@femto-st.fr

Abstract. Requirements verification at early phases of the design process is one of the main challenges when developing critical and complex systems. In this paper, we focus on the verification of SysML functional requirements on activity diagrams. Our contribution consists in the proposition of a slicing based verification approach guided by the SysML relationships between requirements, blocks, and activities. The objective is to provide a verification methodology for complex systems with many components. Our approach is based on formalizing activity diagrams using Hierarchical Coloured Petri Nets (HCPNs). The proposed slicing permits to alleviate the verification process. For verifying a given requirement, the slicing consists in extracting a minimized fragment (slice) of the HCPNs, which is sound and sufficient to realize the verification. The approach is illustrated by a case study, where we specify and we verify a fire protection system for data-centers.

Keywords: SysML · Activity diagram · Requirements · Hierarchical Colored Petri Nets · Model slicing · Formal verification

1 Introduction

SysML (System Modeling Language) [9] is an OMG standard modelling language supported by leading organizations from the systems engineering industry, including the International Council On Systems Engineering (INCOSE). SysML covers four perspectives on system modeling: structure, behavior, requirement, and parametric diagrams. The requirement diagram is one of the most important models provided in SysML. Thanks to this diagram, SysML offers an explicit representation of the different kinds of relationships between requirements and other design artefacts [8]. These relationships are of great importance. They can be exploited to guide several aspects related to system design, particularly requirements verification.

M. R. Senouci et al. (Eds.): CSA 2020, LNNS 199, pp. 131–142, 2021.
https://doi.org/10.1007/978-3-030-69418-0_12

The activity diagram allows to specify the dynamic aspects of the system. For describing the full behaviour of complex systems with many components, SysML activity diagram can be composed of several sub-activities. Furthermore, swimlanes can be used to allocate actions to system components (SysML blocks). Despite the various advantages of SysML, it remains a semi-formal language without possibilities of formally verifying models described by it. For this reason, using SysML in conjunction with formal methods is suggested to provide a formal validation for SysML specifications.

Model checking techniques [2] are based on automatic examination of all reachable states to check whether a model of a given system satisfies the desired properties. However, as known, the major flaw of model-checking resides in the state explosion problem. Slicing the model according to the property to verify is one of the approaches aiming to alleviate this problem [6].

This paper concentrates on the verification of SysML functional requirements linked to behavioral models. Other requirements classes such as performance and availability are not considered. It proposes a methodology based on model-checking to verify requirements on activity diagrams.

In contrast to other verification approaches which are based on translating activity diagrams into Petri nets [1,5], our approach consider the translation of activity diagrams with swimlanes and sub-activities. This allows to represent structural and functional decomposition of complex systems as they are captured by the activity diagrams. In addition, the verification in our approach is guided by the SysML relationships between requirements, blocks, and activities, which enables the use of a slicing based verification. For verifying a given requirement, the slicing consists in extracting a minimized fragment (slice) of the HCPN model translated from the activity diagram which is sound and sufficient to realize the verification.

The remainder of this paper is organized as follows: Sect. 2 discusses the related work. Section 3 presents a background about the related concepts. Section 4 introduces our verification methodology. In Sect. 5, we illustrate our approach by a case study. Finally, in Sect. 6, we conclude and we outline directions for future works.

2 Related Work

The idea of translating SysML specifications to get formal semantic models is interesting for verification purposes [4,7]. For activity diagrams, several approaches based on their translation into Petri net models were proposed, each one was based on using specific classes of Petri net depending on the proposed goals to achieve. The work in [7] introduces an approach using SysML for structural specification and Petri net for expressing the system behavior. In [1], a technique was proposed to map SysML activities into Time Petri net for validating the non functional requirements of real-time systems with energy constraints. A bi-directional transformation of UML 2 activities into Petri nets was proposed in [14]. In this translation, only the basic activity constructs was considered. To consider the semantics of data flows, translating activity diagram into coloured Petri nets was proposed in [15]. Activity calling and swimlanes were not considered in these works. Our work considers the semantics of activities with swimlanes, call behavior actions and data flows. These previous approaches intended to verify activity diagrams by model-checking techniques. However, verifying models for complex systems is very challenging task; namely regarding the state space explosion problem. One way to alleviate this problem is model slicing [13]. Several works have proposed model slicing approaches, but very few literatures are reported on slicing of activity diagrams based on requirements and almost all of them address test case generation. As examples, Ray et al. [12] have used conditioned slicing for test case design through activity diagram. The work in [3], proposes a static slicing technique to enhance the comprehension of Use Case Map models. In contrast to these works, we propose model slicing on the formal semantic model translated from the activity diagrams for verification purpose. So, the verification is guided by the SysML specification and the slicing criteria are taken from the SysML requirement diagram.

3 Background

3.1 SysML Requirement and Activity Diagrams

Requirement diagrams are used for specifying requirements and the exiting relationships between them and the other artifact. Our work exploits the "satisfy" and the "Verify" relationships. The "Satisfy" relationship is a dependency between a requirement and a SysML block. It means that the requirement is satisfied by the block. The "Verify" relationship is a dependency between a requirement and a test case. It means that the test case is used to verify the requirement. A test case can be a sequence, a state machine or an activity diagram. Our work considers test cases described by activity diagrams.

The activity diagram is a formalism for describing the system behaviors. The basic constructs of an activity are actions and control nodes. A specific type of action is the call behavior action. A call behavior action permits to invoke an activity when it starts. Activities can be with Swimlanes or activity partitions. Swimlanes are used to allocate actions to SysML blocks. Formally, an activity with swimlanes and call behavior actions is tuple $Act = \langle N, E, Part, Alloc, CAct, call\rangle$ where:

- N is a finite set of nodes partitioned into subsets ($N = N_a \cup N_c \cup N_o$), where N_a is the set of action nodes, N_c is the set of control nodes and N_o is the set of object nodes which represents input and output pins and input and output activity parameters. N_a is partitioned into a set of simple action nodes N_{sa} and a set of call behavior action nodes N_{ca} ($\forall a \in N_{ca}$ and a call $act_i : act_i \in CAct$). N_c is further partitioned into subsets : N_d of decision nodes, N_m of merge nodes, N_f of fork nodes, N_j of join nodes, N_i of initial nodes, N_{ff} of flow final nodes and N_{af} of activity final nodes.

- E is a finite set of edges each connecting either two object nodes with each other or an object node with an action or a control node, i.e., $E \subset (N \times N)$.
- $Part = \{B_1, ...B_n\}$ is a finite set of activity partitions.
- $Alloc$: $N \rightarrow Part$ is a function that allocates activity nodes to activity partitions.
- $CAct$ is the set of the activity invoked by the call behavior actions.
- $call$: $N_{ca} \rightarrow CAct$ is a function that associate a called activity to each call behavior action.

3.2 Hierarchical Coloured Petri Nets

Hierarchical Colored Petri Nets (HCPNs) introduce a facility for building a large CPN model by using a number of small CPNs. In HCPNs, the notion of substitution transition is provided to allow the user to relate a transition to a more complex CPN, called subnet. A substitution transition has input and output places called socket places. More formally, a HCPN is a tuple HCPN = ⟨S, P, T, A, Σ, St, SA, P_{por}, P_{soc}, PS⟩ where,

- S is the set of pages (subnets), P is the set of places and T is the set of transitions.
- $A \subseteq (P \times (T \cup St)) \cup ((T \cup St) \times P)$: is the set of arcs.
- Σ: is the set of non-empty types, called colour sets.
- St: is the set of substitution transitions, SA: St $\rightarrow$ S is a page assignment function, $P_{por} \subseteq$ P is the set of port places, $P_{soc} \subseteq$ P is the set of socket places, and PS: $P_{por} \rightarrow P_{soc}$ is a port-socket relation function that assigns a port places to a socket places to relate the subnet to the main net.

HCPNs are chosen not only due to their expressive power to capture the semantics of activity diagram constructs including data flows, but also because they allow to represent structural and functional decomposition of complex systems.

4 Methodology

Our methodology for verifying SysML system design aims to verify SysML functional requirements. The requirements of the system are described in a SysML requirement diagram and the system behavior is described by an activity diagram. This latter will be structured with activity partititions. Each one represents a SysML block. Also, it can contain call behavior actions that call other activities for modelling complex behaviors. We verify if the SysML functional requirements are verified by the activity diagram.

For that, we define, as detailed in the Fig. 1, a slicing based verification approach to overcome the state space explosion problem. The first step of the proposed verification consists in translating the SysML activity diagram into a HCPN. Then, we extract the blocks and the activities which satisfy (respectively. verify) requirements. This step exploits "satisfy" and "verify" relationships. After that, requirements will be translated into temporal logic formulas and information about blocks and activities will be used to slice the HCPN. To avoid to perform many translations, the slicing will be applied in the HCPN. Finally, we check the temporal logic formulas in the sliced HCPN and we return the verification results.

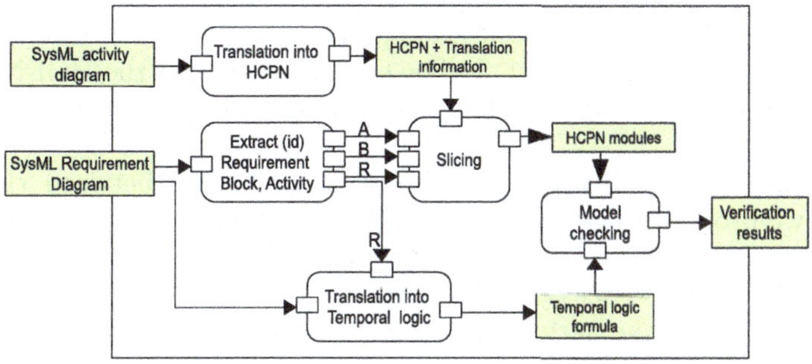

Fig. 1. The verification methodology

4.1 Translation of Activity Diagrams

The translations of basic activity constructs are done as in our previous works [10,11]. So, in this section, we only present the translation of call behavior actions and activity partitions (swimlanes).

The main idea of our translation is as follow: let $AD = \langle setAct \rangle$ be an activity diagram consisting on a set of activities (main activity and sub-activities), where each activity $AName = \langle N, E, Part, Alloc, CAct, call \rangle$ can be partitioned into set of activity partitions $Part$ and can contain call behavior actions. The translation of AD into HCPN = ⟨S, P, T, A, Σ, St, SA, P_{por}, P_{soc}, PS⟩ consists in translating each activity $Act \in setAct$ into a HCPN page $Act\_page \in S$ and each call behavior action $ActionA$ that call the activity Act into a substitution transition $SubTAction\_A \in st$, where $SA(SubTAction\_A) = Act\_page$. If $AName$ is partitioned, each activity partition $B \in Part$ is translated into a substitution transition $TB \in St$ and a HCPN page $B\_page \in S$, where $SA(TB) = B\_page$.

The translation of activity diagrams into HCPNs is fully automated. It is performed based on model-driven approach using the ATL transformation language.

4.2 The Slicing Step

Let R be a requirement to verify in an activity Act partitioned in activity partitions $Part = \{B_1, ...B_n\}$ and R is satisfied by a block $B_i \in Part$. The idea of the proposed slicing consists in determining, for the requirement R, the activity partitions which are needed to its verification. Because R has "satisfy" relationship with B_i, B_i is one of the needed activity partitions. However, in some cases, the actions allocated to B_i are not sufficient to verify R. This is due to their connections with other actions allocated to other activity partitions. Namely, when the actions allocated to B_i require, in their inputs, outputs from actions allocated to other activity partitions. For that, we compute the set of activity partitions from which B_i has a direct or indirect dependency. Let $dep(B_i, Act)$ be the function used to perform this computation. Each activity partition $B_j \notin dep(B_i, Act)$ is not needed for the verification of R. The slicing is done on the HCPN model by deleting, for each $B_j \notin dep(B_i, Act)$, its corresponding substitution transition TB_j.

For computing the direct and indirect dependencies between the activity partitions, we propose to construct a Dependency Graph Between Activity Partitions $DGBAP = \langle Part, A \rangle$, where, $DGBAP$ is a directed graph which has the elements of $Part$, as vertices and the elements of $A \subset Part \times Part$ as directed edges($(B_i, B_j) \in A$ if the activity partition B_j has at least one input link from the activity partition B_i). So, computing $dep(B_i, Act)$ will consists in performing the function $compute\_dep(B_i, DGBAP)$ as listed in the Algorithm 1. Given $DGBAP$ and an activity partition B_i, the algorithm allows to compute a list of activity partitions from which B_i is dependent.

Algorithm 1. $compute\_dep(B_i, DGBAP)$
Inputs : (B_i: activity partition, $DGBAP = \langle Part, A \rangle$)
Output: (DL: a list of activity partitions)

```
DL ← {};Stack P; Push(P, Bi);
while (Not empty(P)) do
   Bj ← Pop(P); DL ← DL ∪ {Bj};
   for all (Bk, Bj) ∈ A) do
      if (Bk ∉ DL) then
         Push(P, Bk);
      end if
   end for
end while
```

4.3 Verification of Requirements

In order to allow the verification of SysML requirements, we propose to express them as temporal properties about the activity diagram elements using the AcTRL language [10,11] and then we translate them into ASK-CTL formulas as in [11]. Finally, the CPNTools is used as model-checking tool for analysing the generated ASK-CTL formulas on the HCPN slices.

5 Case Study: Fire Protection System in Data-Center

We consider a fire protection system in data-center as a case study. A fire protection system includes an automatic fire detection system, fire alarms, and a fire suppression system. Automatic fire detection system is designed to detect fire in its earliest stages. It is based on heat detectors which detect fire and notify information to detection central. To confirm a fire, a cross-zoned system can be adopted. It permits to reduce unwanted and nuisance alarms caused by detectors. In cross-zoned system, a fire is confirmed only if two detectors notify a fire to the detection central which activates the fire suppression system, the internal and the external alarms. A detection by a single detector will only be notified to the system administrator. In data-centers, fire suppression system must be based on gas extinguishing. When a fire is confirmed, the detection central activates an auto actuation device to suppress the fire automatically.

5.1 SysML Specification

The aim of this step is to produce a SysML specification. The SysML Block Definition Diagram in Fig. 2 describes the structure of the system. The requirements of the fire protection system are specified in a SysML requirement diagram. So, in this paper, we consider only a sub-set of requirements. As shown in Fig. 3, the main requirement of the system is to protect the data-center from fires. This requirement is composed of three requirements (R0, R3, R4). The requirement R0 itself is composed of two requirements (R1 and R2). The requirement R1 deals with the detection of fires. It is satisfied by the "Fire Detector" block and verified by the "Detect Fire" activity. The requirement R2 deals with the confirmation of fire detection. It is satisfied by the "Detection System" Block. The requirement R3 is to ensure the sounding of alarms when the detection of a fire is confirmed. It is satisfied by the "Alerting System" block. The requirement R4 deals with the suppression of fires. It is satisfied by the "Suppression System" block. The requirement R2, R3 and R4 are to verify by the "Protect from fire" activity (see Fig. 4).

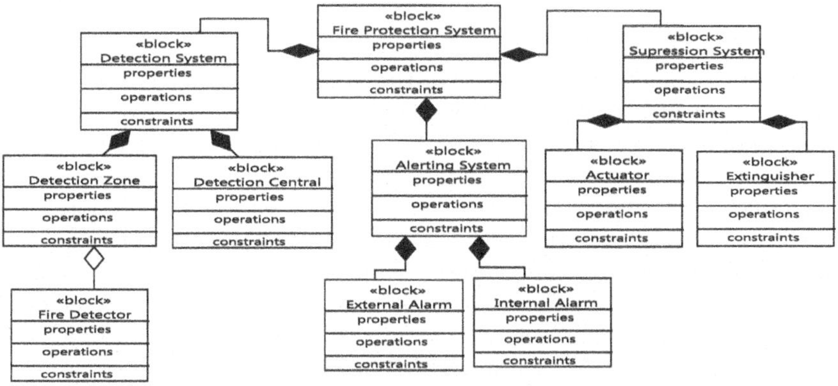

Fig. 2. Block definition diagram for the fire protection system

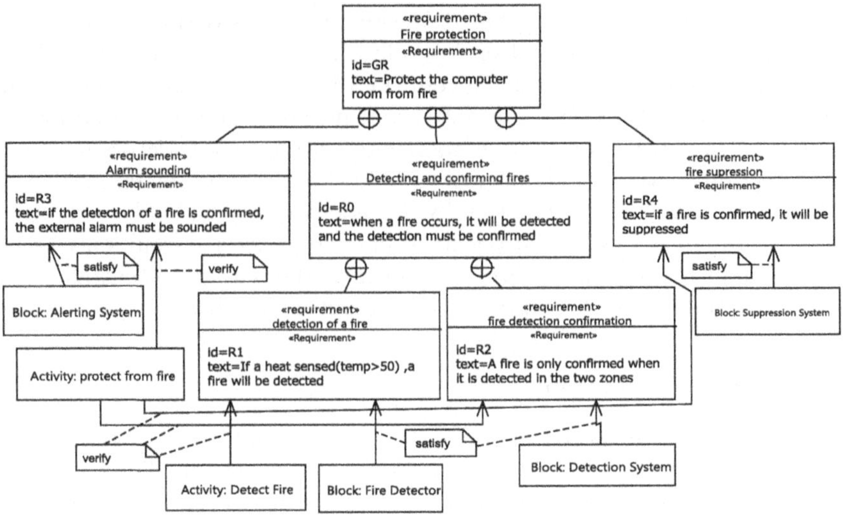

Fig. 3. A requirement diagram for the fire protection system

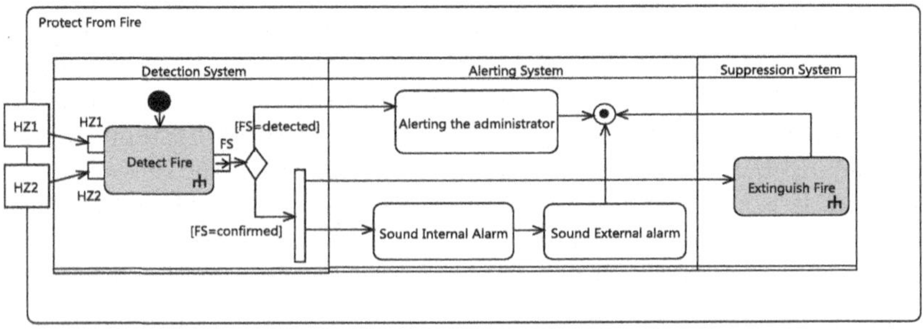

Fig. 4. Activity diagram for the fire protection system

We specify the indented behavior of the fire protection system by the activity diagram presented in the Fig. 4. We create three activity partitions, the first represents the detection system block, the second represents the alerting system block and the third represents the suppression system block. If a fire occurs, it will be detected by the fire detection system. For this, the "Detection System" activity partition contains the "Detect Fire" call behavior action which allows to sense the heat by the fire detectors(in Zone 1 and in Zone 2) and then, determines if a fire is occurred. The activity diagram representing the "Detect Fire" sub-activity is given in Fig. 5. For the actions assigned to the "Alerting System" activity partition, two cases are modelled: fire detected but not confirmed and fire confirmed. In the first case, we only produce an alert for the system administrator. For the second case, we sound the internal and the external alarms. Also, the call behavior action "extinguish fire" assigned to the "suppression System" activity partition is activated in the second case. The models related to the sub-activity called by the "extinguish fire" call behavior action are not presented in this paper.

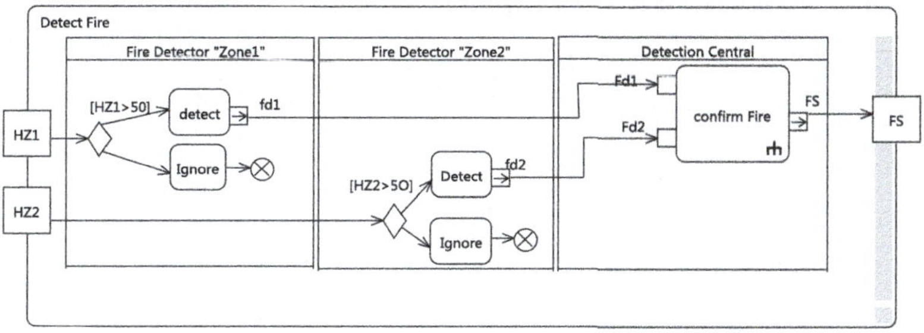

Fig. 5. Activity diagram for the "Detect Fire" sub-activity

5.2 Translation into HCPN

We show in the Fig. 6 the prime page of the HCPN model derived from the activity "Protect from fire". As explained in Sect. 4.1, the HCPN model in its prime page contains three substitution transitions.

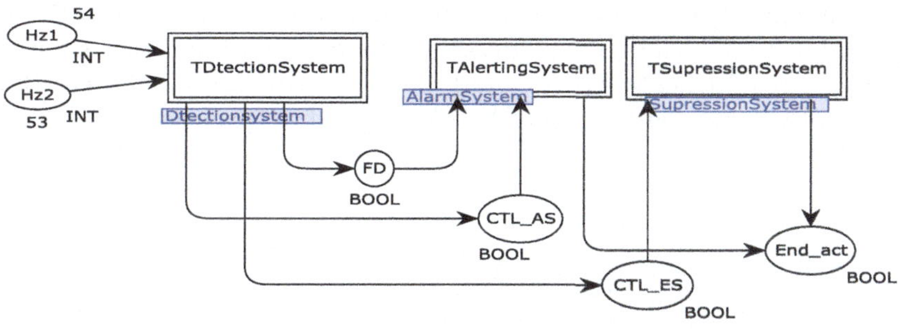

Fig. 6. The HCPN model derived from the "Protect from fire" activity

As example, in the Fig. 7, we show the HCPN sub-page assigned to the "TDetectionSystem" substitution transition. It contains the translation of the fork node, the decision node and the "Detect Fire" call behavior action allocated to "Detection System" activity partition (see Fig. 4). The "Detect Fire" call behavior action is translated into a substitution transition ("Detect_fire"). The HCPN sub-page assigned to this latter is shown in Fig. 8, it is obtained from the translation of the activity "Detect Fire" (see Fig. 5) called by the "Detect Fire" call behavior action (see Fig. 4). In the "Detect Fire" activity, we have three activity partitions. They are translated into substitution transitions.

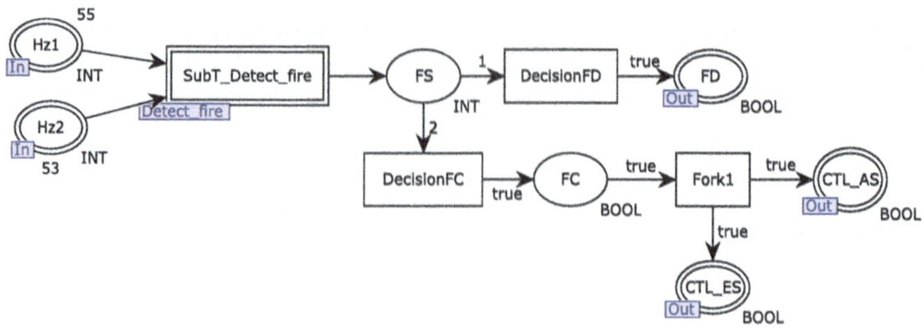

Fig. 7. The HCPN "Detection system" sub-page

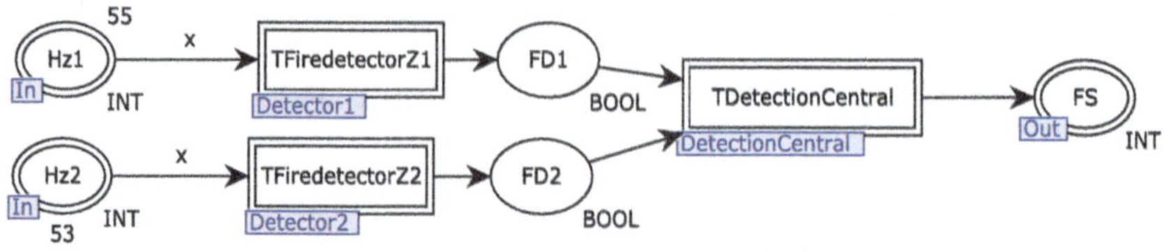

Fig. 8. The HCPN model derived from the "Detect fire" activity

5.3 Slicing Examples and Verification of Requirements

After the translation step, we can begin the verification of requirements using our slicing technique. This section presents and explains some slicing examples when verifying the requirements of the fire protection system. From the requirement diagram, the requirements R2, R3 and R4 are to verify by the activity "Protect from Fire". First, we construct the DGBAP related to this activity (see Sect.4.2 and Fig. 9).

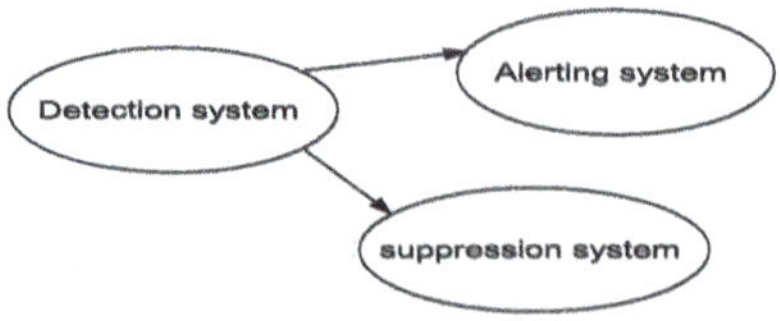

Fig. 9. The DGBAP for the activity "Protect from fire"

As the requirement R2 is satisfied by the block "Detection System", according to the constructed DGBAP, its verification is independent from the other activity partitions. So, the slicing consists in conserving only the substitution transition related to the "Detection System" activity partition.

The requirement R3 is satisfied by the block "Alerting System", according to the constructed DGBAP, its verification is dependent from the "Detection System" and "Alerting System" activity partitions. The slicing consists in deleting the substitution transition related to the "Suppression System" activity partition.

The requirement R4 is satisfied by the block "Suppression System", according to the constructed DGBAP, its verification is dependent from the "Detection System" and "Suppression System" activity partitions. The slicing consists in deleting the substitution transition related the "Alerting System" activity partition.

6 Conclusion

In this paper, we presented a slicing based approach for the verification of SysML requirements on activity diagrams. The verification is based on translating activity diagrams into HCPNs. The proposed slicing approach is guided by the relationships between requirements, activities and blocks. The slicing can be applied, for a given requirement, to verify on an activity diagram with activity partitions. The slicing is done for such requirements on the HCPN translated from the activity diagram. The effectiveness of the proposed technique depends on the activity partitions dependencies and differs from a requirement to another. The approach was illustrated by a case study, where a SysML activity diagram with activity partitions and call behavior actions was translated into HCPN. The HCPN was then sliced to verify SysML requirements.

As future work, it is important to consider the improvement of the proposed slicing technique which can be achieved through the application of our approach on more complicated case studies.

References

1. Andrade, E., Macie, P., Callou, G., Nogueira, B.: A methodology for mapping SysML activity diagram to time petri net for requirement validation of embedded real-time systems with energy constraints. In: Third International Conference on Digital Society, ICDS 2009, pp. 266–271 (2009)
2. Berard, B., et al.: Systems and Software Verification: Model-Checking Techniques and Tools, 1st edn. Springer, Cham (2010). https://doi.org/10.1007/978-3-662-04558-9
3. Binalialhag, T., Hassine, J., Amyot, D.: Static slicing of use case maps requirements models. Softw. Syst. Modeling **18**(4), 2465–2505 (2019)
4. Debbabi, M., Hassaine, F., Jarraya, Y., Soeanu, A., Alawneh, L.: Verification and Validation in Systems Engineering: Assessing UML/SysML Design Models, 1st edn. Springer, New York (2010). https://doi.org/10.1007/978-3-642-15228-3
5. Foures, D., Vincent, A., Pascal, J.: ACTIVITYDIAGRAM2PETRINET : transformation-based model in accordance with the Omg SysML specifications. In: 2011 Proceedings of the Eurosis, European Simulation and Modelling Conference, pp. 429–434 (2011)

6. Ji, W., Wei, D., Zhi-Chang, Q.: Slicing hierarchical automata for model checking UML statecharts. In: George, C., Miao, H. (eds.) Formal Methods and Software Engineering, pp. 435–446. Springer, Heidelberg (2002). https://doi.org/10.1007/3-540-36103-0_45
7. Linhares, M., de Oliveira, R., Farines, J., Vernadat, F.: Introducing the modeling and verification process in SysML. In: IEEE Conference on 2007 Emerging Technologies and Factory Automation, ETFA, pp. 344–351, September 2007
8. Nejati, S., Sabetzadeh, M., Falessi, D., Briand, L., Coq, T.: A SysML-based approach to traceability management and design slicing in support of safety certification: Framework, tool support, and case studies. Inf. Softw. Technol. **54**(6), 569–590 (2012)
9. OMG: OMG Systems Modeling Language (OMG SysML^{TM}) Version 1.6 (2019). Downloadable from https://www.omg.org/spec/SysML
10. Rahim, M., Boukala-Ioualalen, M., Hammad, A.: Petri nets based approach for modular verification of SysML requirements on activity diagrams. In: PNSE 2014, A satellite event of Petri Nets 2014 and ACSD 2014, Tunis, Tunisia, pp. 233–248, June 2014
11. Rahim, M., Hammad, A., Ioualalen, M.: A methodology for verifying SysML requirements using activity diagrams. Innovations Syst. Softw. Eng. **13**(1), 19–33 (2017)
12. Ray, M., Barpanda, S.S., Mohapatra, D.P.: Test case design using conditioned slicing of activity diagram. Int. J. Recent Trends Eng. (IJRTE) **1**, 117–120 (2009)
13. Singh, R., Arora, V.: Literature Analysis on Model based Slicing. Int. J. Comput. Appl. **70**(16), 45–51 (2013)
14. Staines, A.S.: Simplified bi-directional transformation of UML activities into petri nets. In: Proceedings of the 10th WSEAS International Conference on Software Engineering, Parallel and Distributed Systems, SEPADS 2011, Cambridge, UK, pp. 24–29. World Scientific and Engineering Academy and Society (WSEAS) (2011)
15. Störrle, H.: Semantics and verification of data flow in UML 2.0 activities. Electron. Notes Theoret. Comput. Sci. **127**(4), 35–52 (2005)

A Colored Petri Net to Model Message Differences in Energy Harvesting WSNs

Nourredine Oukas[1,3(✉)] and Menouar Boulif[2,3]

[1] Normal Superior School, Kouba, Algiers, Algeria
oukas@ens-kouba.dz

[2] Department of Computer Science, M'Hamed Bougara University of Boumerdes, Boumerdes, Algeria
boumen7@gmail.com

[3] LIMOSE, M'Hamed Bougara University of Boumerdes, Independence Avenue, 35000 Boumerdes, Algeria

Abstract. This paper proposes a modelling for Energy Harvesting Wireless Sensor Networks (EHWSNs) by using Coloured Generalized Stochastic Petri Nets (CGSPN). Given that the transmission process consumes the most part of the energy, we consider it in more detail. Therefore, in contrast to the related works in the literature, the proposed formulation differentiates between messages since energy consumption can significantly differ according to the type of information to send. Furthermore, in order to get a more realistic model, we also consider that the sensor has a limited buffer capacity. We conduct some experiments by feeding the model with varying input parameters to show that it can predict the actual behaviour of the network.

Keywords: Wireless sensor network · Coloured generalized stochastic petri nets · Energy harvesting · Energy quantization

1 Introduction

Wireless Sensor Networks (WSNs) consist of hundreds or thousands of sensor nodes (SNs) that can intercommunicate. Figure 1 depicts a typical WSN where sensors are deployed in a field looking for a given event to happen. When an incident occurs, its surrounding sensors report the event and try to send a message to the sink that aggregates the received messages and send them to the final user for further processing. The sensor-neighbours relationship is the heart of this event reporting system and a key issue in its performance.

Due to its small size, an SN is characterized by having limited capabilities of processing, storage, and battery charge. There are several techniques to improve the power supply of the WSN nodes which is a key element to increase the network lifetime and to enhance its reliability. Among these techniques Energy harvesting (EH) [4,14] consists of getting energy from the environment (sun, wind, heat, pressure, and others) and converting it to electrical energy [17].

M. R. Senouci et al. (Eds.): CSA 2020, LNNS 199, pp. 143–152, 2021.
https://doi.org/10.1007/978-3-030-69418-0_13

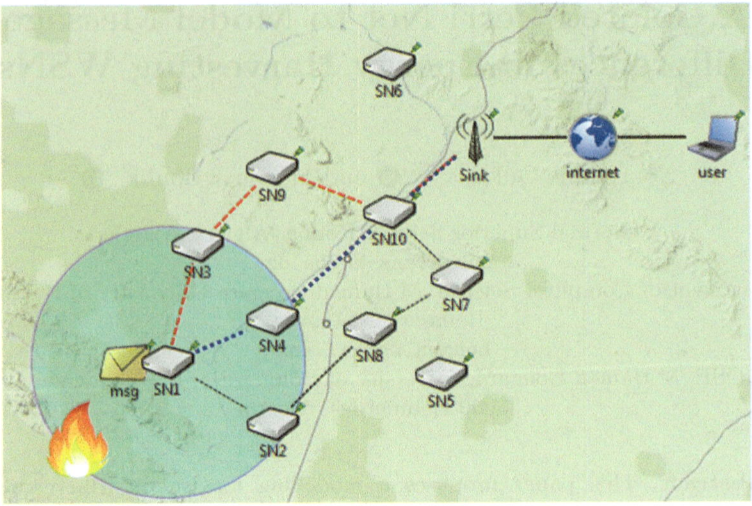

Fig. 1. A wireless sensor network

WSNs formed with SNs equipped with energy harvesting systems are called energy harvesting wireless sensor networks (EHWSNs) [1,13].

For examining the behaviour of a WSN and to predict its performances before its actual deployment, we can use either simulators or mathematical models. Generalized Stochastic Petri Nets (GSPN) and Coloured Petri nets are widely used mediums to model WSN architecture, energy consumption, communication traffic and others. In this area, Hakmi et al. in [10] have proposed a GSPN modeling for studying performances of the radio frequency transmission in wireless sensor networks. The authors studied two models: in the first, the high priority requests have non-preemptive priority over lower ones while, in the second model, they applied the preemptive priority to the high priority requests. According to this study, the authors proved that the preemptive priority is favorable to higher priority customers because they are not at all influenced by lower priority customers. In [5], an energy-saving and latency efficiency techniques are proposed by using GSPN formalism for full-duplex wireless sensor nodes, based on a combination of normal vacation and working vacation policies. They called it the two thresholds working vacation policy. Gharbi and Charabi introduced an algorithmic approach based on GSPN, for modeling and analyzing finite-source wireless networks with retrial phenomenon and two server classes [9]. The particularity of this approach is the direct computing of the infinitesimal generator of the proposed GSPN without generating neither the reachability graph nor the underlying Markov chain. In the proposed modeling, the authors assume that the servers of one class are faster than those of the second one.

In [2], Azgomi and Khalili present a Colored Petri Nets formulation for modeling and performance evaluation of a medium access control protocol in WSNs. Damaso et al. [6] present another approach for evaluating the WSN lifetime by simulating its power consumption. The proposed approach is based on Coloured Stochastic Petri Nets (CSPN) and consists of a fully automated process for evaluating the WSN lifetime, a set of reusable Coloured Petri Net models that express the power consumption of WSN communication protocols, and a strategy for composing CSPN models of WSN applications and protocols. The authors carried out an experimental evaluation that compares their simulation results to other findings taken from the literature in order to assess their power consumption models.

In [15,16], the authors tackle Energy harvesting considerations. In [16] they proposed a GSPN modeling of energy harvesting and energy consumption in an SN. They represented the sensor battery energy by using quantization (a quantum is the amount of energy that is required to send one short message). They conducted an experimental analysis to show the influence of energy harvesting on the mean response time and the mean battery charge.

To the best of our knowledge, the difference between short and long messages has not been taken into account in communication traffic modeling. Therefore, we propose a CGSPNs modeling that takes into account the difference between control and data messages. The choice of the CGSPN formalism is motivated by its straightforwardness to deal with message differences and to its ability to consider with relatively less effort several actual circumstances such as buffer capacity limitation, battery energy quantization, energy harvesting, and energy consumption.

The rest of this paper is organized as follows: In Sect. 2, we give a concise description of the CGSPN formalism. In Sect. 3, we construct a CGSPN model for EHWSNs. In the next section, we show and discuss some experimental analysis. Finally, we draw our conclusion as well as some directions for future works.

2 CGSPN Formalism

CGSPN formalism is an extension of simple Petri nets. The main difference between simple and coloured Petri nets is that tokens may have arbitrarily specified attributes [3,7,11,12,18]. Each token has attached to it a data value called the token colour. The token colours can be selected and modified by the firing of transitions.

Each place in CGSPN serves as container of tokens. It is considered as a set of objects derived from a single class (token type) and each various object is considered as a new colour. If two marks have different attribute values, they are considered as two token with different colours. Structured types are user-defined and may contain any number of base types or other structured types just like a C++ struct statement. Another useful extension that is valuable for many real-life applications is the specification of a place capacity (i.e the maximum number of tokens in the place). This maximum number of tokens that may exist in the place is shown in square brackets near the place, but omitted if the capacity is unlimited.

Directed arcs connect places and transitions like in simple Petri nets. Input arcs of transitions and their inscriptions describe how many tokens are consumed by transition firing. Transition output arcs define what tokens are added to the output place after the transition firing. New tokens of the output place type are created if no input variable is specified at an output arc. The attributes of a new token are set to their default values initially.

Transitions can be fired when all necessary input tokens are available and an optional guard function is satisfied. The firing delay for timed transitions describes the probability distribution of the delay that needs to elapse between the transition enabling and firing. For each transition, a guard function can be assigned to restrict the marks for which it may be enabled. Timed transitions have different service behavior semantics. The server semantic specifies whether the transition might be concurrently enabled with itself (*Infinite server semantic*) or not (*Single server semantic*) [18].

Quantitative analysis of a CGSPN allows several performance measures. They define what is computed during an analysis. We cite for example:

- The number of tokens in a place.
- The mean number of tokens in a place over a time interval, the mean number in steady-state, or the expected number at a certain point in time.
- The number of transition firings.
- The throughput of transitions: It is measured by transition firings and averaging over the time interval of interest.

The interested reader is referred to the rich body of literature on this topic.

3 Proposed Model

The proposed model as shown in Fig. 2, focuses on message exchanges in the sensor-neighbours relationship. In this communication two types of messages are exchanged: control messages and data messages. Due to their length, these two types of messages have not the same impact on the consumed energy. We define three token types as shown in Table 1. The first is *Msg* which represents a class of messages. Several colors can be instanced from this class (we use two colors: *data* message and *hello* message). The second token type is *Neighbor* which represents a container that assembles a sensor with a message. The last token type is *Level.* Each object (color) from this class contains a certain number of quantum. Herein, for simplicity, we suppose that each level contains one quantum of energy.

The communication system is modeled as follows: When a message arrives to a sensor by the firing of the transition *Coming*, it is stocked in the buffer if there is enough space (see the buffer capacity limit *[B]* in the related place). Otherwise, the token must await. The first message in the buffer queue will be immediately transmitted to a neighbor of the next hop. If the immediate transition *Begin_sendC* fires, the selected message from the buffer queue is of type *"hello"*. After that, the firing of the transition *SendHello* means a successful sending of this control message to a next hop neighbor. If instead of *Begin_sendC* the transition *Begin_sendD* fires, the message in the head of the queue is of the data type and hence, the firing of the transition *SendData* simulates the end of

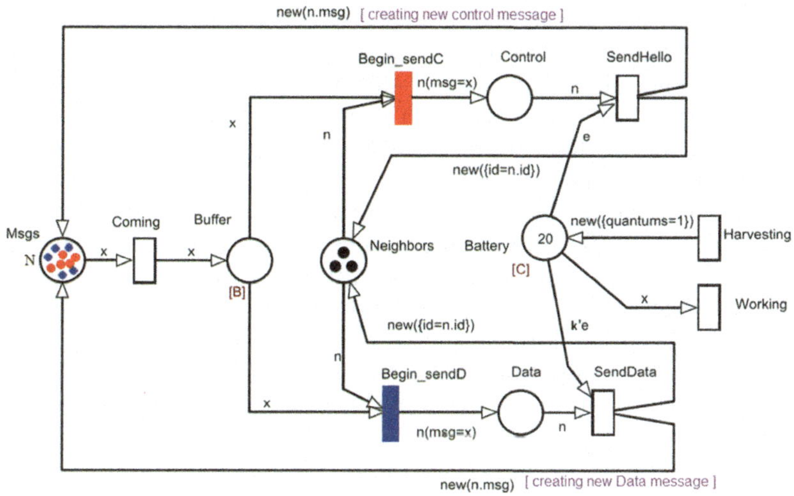

Fig. 2. CGSPN modelling for EH-WSNs

Table 1. Token types description

Token type	Description	*C++* Structure
Msg	Message class. Each instanced object (coloured token) of this class represents a message with a precise type	*struct Msg {string type; int length;}*
Neighbor	A container class for an idle neighbor and a message with a precise type	*struct Neighbor {int id; Msg msg;}*
Level	An instanced object of this class defines the number of quantum of remaining energy in the battery	*struct Level {int quantum;}*

the transmission. The sending of a control message consumes one quantum of energy, whereas, the sending of a data message consumes k quantum, where k can be defined according to the WSN domain of application that governs the proportional length between control and information messages. The firing delays *hello_d* and *data_d* are associated to the *SendHello* and *SendData* transitions respectively. Table 3 summarizes the main proprieties of the timed transitions whereas Table 4 describes the immediate transitions.

Furthermore, all timed transitions are supposed to use a *single server* semantic. Each one of the two immediate transitions has a local guard function as is mentioned in Table 4. Some arcs that have special inscriptions are labeled with their expressions [18]. Some others are labeled with a single character such as x, n and e. To make it more clear, for example, the transition *Begin_sendC* has two input arcs from the *Buffer* and *Neighbors* places labeled with x and n respectively. This transition has also a guard function that selects only control messages. The selected message is directly assigned to x and the selected neighbor is assigned to n. The firing of this immediate transition aggregates the message x with the neighbor n (see *Neighbor* structure in Table 1) by assigning x to the msg member of the container n.

In the model, energy harvesting considerations are modeled with the timed transition *Harvesting*. To meet actual circumstances, inter-firing delays are supposed exponentially distributed. After each firing of the transition *Harvesting*, a new quantum of energy is added to the SN battery. If the battery is full (i.e it contains C quantum), no token can be received. The transition *Working* represents the energy consumption of the remaining tasks an SN should do, such as processing, sensing, etc. Each firing of this transition consumes a quantum of energy.

Table 2. Places description

Place	Signification	Token type	Capacity	Initial marking
$Msgs$	Source of a messages	Msg	/	$N = 10$
$Buffer$	Sensor buffer	Msg	$B = 5$	0
$Neighbors$	Neighbours of next hop	Neighbor	/	$s = 3$
$Control$	Control messages in sending state	Neighbor	/	0
$Data$	Data messages in sending state	Neighbor	/	0
$Battery$	Battery charge level	Quantum	C	$C = 20$

Table 3. Timed transitions description

Timed transition	Signification	Firing delay	Value
$Coming$	Arrival of a message	$come\_d$	500
$SendData$	Data message successful sending	$data\_d$	0.005
$SendHello$	Control message successful sending	$hello\_d$	0.001
$Harvesting$	Energy recovery	$harvest\_d$	60
$Working$	Working energy consumption	$work\_d$	300

Table 4. Immediate transitions description

Immediate transition	Signification	Filter	Priority
Begin_sendD	Sending start of data message	*[x.type== "data"]*	1
Begin_sendC	Sending start of control message	*[x.type== "hello"]*	2

4 Experimental Analysis

We use **TimeNet** tool [19] to conduct our experimental analysis. The input parameter values are mentioned in Tables 2 and 3.

Firstly, the quantitative steady-state analysis requires the verification of the ergodicity condition [8]. Ergodic models are the most interesting because they guaranty the existence of a stationary regime of the modeled system, allowing us to calculate a stationary solution.

Our proposed Petri nets model is *C-bounded* and satisfies the *vivacity* property. Therefore, it meets the conditions of the ergodicity theorem that guarantees the existence of a stationary state.

Resolving the network in the steady-state by TimeNet with the highlighted input values gives the graphs of Fig. 3 and 4. Figure 3(a) represents the evolution of the mean battery charge. It is clear that the charge of the battery converges to the value of 19.7 quantum. The mean number of messages in the buffer converges to the value of 4.89 (see Fig. 3). Figure 4(a) and Fig 4(b) represent the evolution of the throughput of the transition SendData and SendHello respectively.

4.1 Influence of EH Delay on the Battery and the Buffer

Figure 5 shows the influence of the energy harvesting delay on the mean battery load. As expected, when the energy harvesting delay increases, the mean battery charge increases too. Notice that when the time to accumulate one quantum of energy is less than 300 s, the mean battery charge is always greater than 50%.

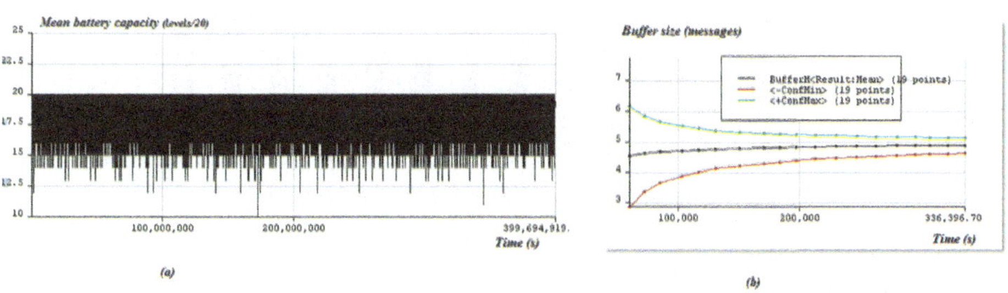

Fig. 3. (a) Battery charge level (b) Number of messages in the buffer

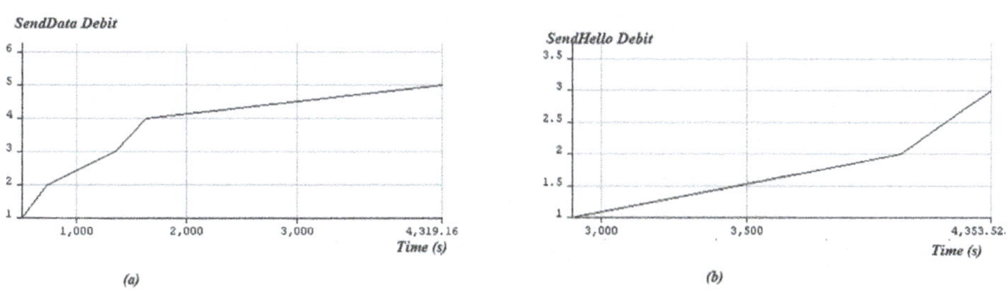

Fig. 4. (a) SendData Debit (b): SendHello Debit

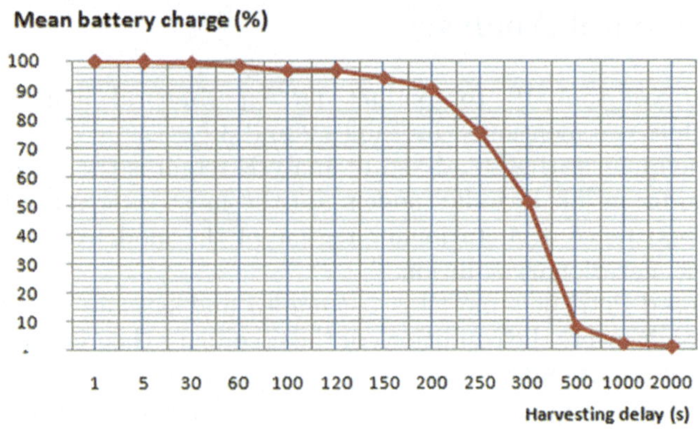

Fig. 5. Mean battery charge versus energy harvesting delay

Figure 6 shows the influence of the energy harvesting delay on the mean number of messages in the buffer. Given that the buffer length is 5 messages in this scenario, with the input configuration we defined earlier, the buffer is full in most of the time.

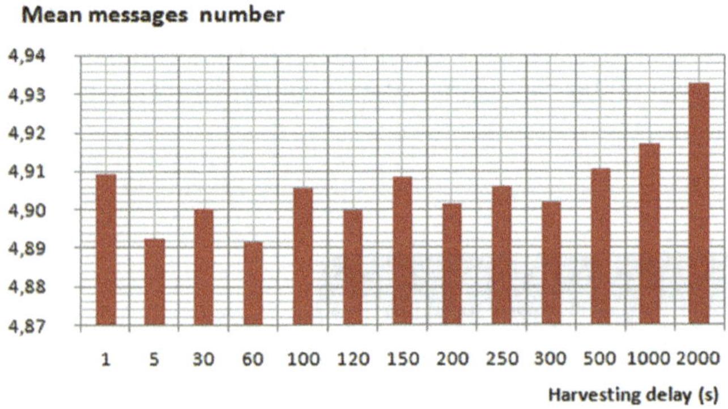

Fig. 6. Mean number of messages in the buffer versus EH delay

Summarizing, the obtained results prove that our model can simulate the actual behavior of this kind of networks and hence, by varying the values of the input parameters, we can evaluate their performances at a very low cost.

5 Conclusion and Future Works

In this paper, we proposed a Coloured Generalised Stochastic Petri Net (CGSPN) model for energy harvesting WSNs. By considered two types of messages: control and data messages we analyzed the energy consumption in such networks when they are equipped with an energy harvesting system. CGSPN allowed us to perform some experimental analyses to show the ability of our model to predict the behavior of the WSN before its actual deployment at a very low cost.

In our future works, we will try to enhance our CGSPN model by incorporating other actual circumstances such as repeated calls, sleeping mode, breakdowns and repairs with different message types.

References

1. Adu-Manu, K.S., Adam, N., Tapparello, C., Ayatollahi, H., Heinzelman, W.: Energy-harvesting wireless sensor networks (EH-WSNs): a review. ACM Trans. Sensor Netw. (TOSN) **14**(2), 10 (2018)
2. Azgomi, M.A., Khalili, A.: Performance evaluation of sensor medium access control protocol using coloured petri nets. Electron. Notes Theoret. Comput. Sci. **242**(2), 31–42 (2009). https://doi.org/10.1016/j.entcs.2009.06.021, http://www.sciencedirect.com/science/article/pii/S157106610900200X. Proceedings of the First Workshop on Formal Methods for Wireless Systems (FMWS 2008)
3. Ballman, K.: Coloured petri nets: basic concepts, analysis methods and practical use, volume 3. Am. Math. Monthly **105**(5), 493 (1998)
4. Basagni, S., Naderi, M.Y., Petrioli, C., Spenza, D.: Wireless sensor networks with energy harvesting. In: Mobile Ad Hoc Networking: Cutting Edge Directions, pp. 701–736 (2013)
5. Boutoumi, B., Gharbi, N.: Two thresholds working vacation policy for improving energy consumption and latency in WSNs. In: International Conference on Queueing Theory and Network Applications, pp. 168–181. Springer, Cham (2018)
6. Dâmaso, A., Rosa, N., Maciel, P.: Using coloured petri nets for evaluating the power consumption of wireless sensor networks. Int. J. Distrib. Sensor Netw. **10**(6), 423537 (2014)
7. Ezpeleta, J., Couvreur, J.M., Silva, M.: A new technique for finding a generating family of siphons, traps and st-components. application to colored petri nets. In: International Conference on Application and Theory of Petri Nets, pp. 126–147. Springer, Heidelberg (1991)
8. Florin, G., Fraize, C., Natkin, S.: Stochastic petri nets: properties, applications and tools. Microelectron. Reliab. **31**(4), 669–697 (1991)
9. Gharbi, N., Charabi, L.: Wireless networks with retrials and heterogeneous servers: comparing random server and fastest free server disciplines. Int. J. Adv. Netw. Serv. **5**(1 & 2), 2012 (2012)
10. Hakmi, S., Lekadir, O., Aïssani, D.: Application of generalized stochastic petri nets to performance modeling of the RF communication in sensor networks. In: International Conference on Verification and Evaluation of Computer and Communication Systems, pp. 33–47. Springer, Cham (2017)
11. Jensen, K.: Coloured Petri Nets: Basic Concepts, Analysis Methods and Practical Use, vol. 1. Springer, Heidelberg (2013)
12. Jensen, K., Kristensen, L.M.: Coloured Petri Nets: Modelling and Validation of Concurrent Systems. Springer, Heidelberg (2009)
13. Kausar, A.Z., Reza, A.W., Saleh, M.U., Ramiah, H.: Energizing wireless sensor networks by energy harvesting systems: scopes, challenges and approaches. Renew. Sustain. Energy Rev. **38**, 973–989 (2014). https://doi.org/10.1016/j.rser.2014.07.035, http://www.sciencedirect.com/science/article/pii/S1364032114004870

14. Onishi, T., Ogose, S.: Lifetime extension of wireless sensor networks with energy harvesting. J. Signal Process. **22**(2), 77–86 (2018)
15. Oukas, N., Boulif, M.: Energy-consumption-aware modelling and performance evaluation for eh-wsns. In: Proceedings of the Second Conference on Informatics and Applied Mathematics, IAM 2019, pp. 57–62 (2019). http://labstic.univ-guelma.dz/sites/labstic.univ-guelma.dz/files/
16. Oukas, N., Boulif, M.: A petri net modeling for WSN sensors with renewable energy harvesting capability. In: Hatti, M. (ed.) Smart Energy Empowerment in Smart and Resilient Cities, pp. 524–534. Springer, Cham (2020)
17. Tan, Y.K., Panda, S.K.: Review of energy harvesting technologies for sustainable wireless sensor network. In: Sustainable Wireless Sensor Networks, pp. 15–43 (2010)
18. Zimmermann, A.: Stochastic Discrete Event Systems. Springer, Heidelberg (2007)
19. Zimmermann, A.: Modelling and performance evaluation with TimeNET 4.4. In: International Conference on Quantitative Evaluation of Systems, pp. 300–303. Springer, Cham (2017)

Chaotic Binary Particle Swarm with Anti Stagnation Strategy on Feature Selection

Hayet Djellali[1,2(✉)] and Nadjette Dendani[1,3]

[1] Computer Science Department, Badji Mokhtar University, Annaba, Algeria
{hayetdjellali,n_dendani}@yahoo.fr
[2] LRS Laboratory, Badji Mokhtar University, Annaba, Algeria
[3] Labged Laboratory, Badji Mokhtar University, Annaba, Algeria

Abstract. This paper investigates Chaotic Restart Binary Particle Swarm PSO (ChResBPSO) algorithm applied to feature selection. In this study, to escape local optima of particle swarm PSO and solve the stagnation problem, we add new particles using prior information about current global best and its neighborhood. The solution adopted is to update the particles using N-Best previous particles, their neighbors and new random particles. The information on worst features is incorporated to direct the novel solutions to avoid them. Various chaotic systems replace the main parameters of PSO to find the best chaotic map. Experiments conducted on UCI data: hepatitis, breast cancer, colon cancer, DLBCL validate that chaotic PSO with anti-stagnation criterion outperforms the state of the art methods (BPSO), chaotic BPSO, artificial bee colony (ABC). The novel ChRes-BPSO enhances the final solution in term of accuracy and minimal number of features.

Keyword: Anti-stagnation strategy · Worst features · Chaotic restart particle swarm PSO · Chaotic map · Feature selection

1 Introduction

Feature selection is an important technique in pattern recognition, artificial intelligence and data mining applications. Classification tasks succeed if the process of feature extraction is efficiently carried out and optimized by feature selection (FS). Feature Selection techniques search a minimal subset of relevant features and discard noisy and redundant of them. Various approaches of feature selection were applied [1], such as, filters [2], wrappers [3–5] and embedded methods are the main categories.

Particle Swarm Optimization [5, 6] is a well-known optimization algorithm, a swarm consists of N particles moving in search space. Firstly, the population of particles is initialized randomly to provide diversity in solutions. Secondly, the PSO algorithm searches optimal solutions called global best by updating the position and velocity of each particle including its personal best at each generation. The quality of solution is evaluated using fitness function.

Feature selection as many optimization problems occur in discrete space. Consequently, Kennedy and Eberhart introduced binary PSO (BPSO) [7], which can be applied

M. R. Senouci et al. (Eds.): CSA 2020, LNNS 199, pp. 153–162, 2021.
https://doi.org/10.1007/978-3-030-69418-0_14

to discrete binary features. To improve PSO algorithm, the chaos was introduced [8]. The chaotic optimization maintains population diversity. According to previous studies [9], the PSO convergence is influenced by the parameters inertia weight w, learning coefficients c1, c2, and random numbers r1 and r2.

Different chaotic PSO (CPSO) variants have also been proposed by researchers [8, 10, 11] and the results have indicated good performance of the chaotic PSO variants. This paper [12] presents a novel optimization algorithm called catfish binary particle swarm optimization (CatfishBPSO). The introduction of new particles (Catfish) into the search space which replace particles having the worst fitness by a new particle if the fitness of the global best particle has not improved for a consecutive iteration. Experimental results demonstrate that Catfish BPSO outperforms PSO for classification accuracy. Authors [13] describe and evaluate a new version of particle swarm algorithm of flexible process plans optimization. To explore search space and gain diversity, proposed algorithm is implemented to prevent the early convergence of PSO by embedding ten chaotic maps and outperform experimentally genetic algorithm, simulated annealing and particle swarm.

This study focuses on searching a tradeoff between previous good particles to keep and new particles generated taking profit from both best and worst particles. Standard and Chaotic PSO can be trapped in local optima, to boost PSO exploration and find the optimal solution, we reuse the previous best particles including global best and nearest best particles. Moreover, the solution proposes to remove the worst particles but use this prior information and benefit from it. We discard N% of the worst population and replaced it with two subpopulations:

1. N-Best neighbor particles.
2. We generate new random particles without worst features extracted from current bad particles discarded.

The idea is to generate random particles but forbid the use of features (set to zero) that are statistically highly used in bad particles. Consequently, even removed particles guide the search process by rejecting bad features.

The paper develops different sections. Section 2 describes features selection methods: binary and chaotic Particle Swarm (BPSO, ChPSO) and related work. Section 3 gives a detailed description of the proposed FS methods. Experimental results and discussions are presented in Sect. 4 and 5. finally conclusion in Sect. 6.

2 Methods and Related Works

2.1 Binary PSO

In a binary space, a particle can move to corners of a hypercube by swapping several bits; consequently, the overall particle velocity may be described by the number of bits changed per iteration [7].

$$V(I, t+1) = w * V(I, t) + c1 * r1 * (PBest(I, t) - X(I, t)) \\ + c2 * r2 * (GBest(I, t) - X(I, t)); \tag{1}$$

C1, c2 learning coefficient; r1 and r2 random parameters; PBest: personal best GBest: global best.

Position

$$X(I, t+1) = X(I, t) + V(I, t+1) \tag{2}$$

Binary PSO

$$\text{Position} = \begin{cases} 1 \text{ if sigmoid } (V(I, t) > \text{rand} \\ 0 \text{ otherwise} \end{cases}$$

2.2 Chaotic Binary PSO

Chaotic PSO [10] aims to enhance the performance of PSO. In PSO, the chaos helps to maintain population diversity. In ChPSO, sequences generated by the chaotic map substitute the random parameters r1 and r2 of PSO. In literature, various chaotic maps were proposed, we present below 4 among them:

Iterative map [14]

$$Y(k+1) = \sin((s * pi)/y(k)) \tag{3}$$

Where s ϵ [0,1].

Sine [15] **map**

$$Y(k+1) = s/4 * \sin(\pi * Y(k)) \tag{4}$$

Where 0 < s < = 4.

Sinusoidal [16],

$$Y(k+1) = s * (Y(k))^2 + \sin(\pi * Y(k)) \tag{5}$$

s = 2.3;

Logistic [9] **map**

$$Y_{k+1} = s * Y_k * (1 - Y_k) \tag{6}$$

Y_k is the kth chaotic number, with k: iteration number, y0 ϵ (0,1).

s = 4 is used for experiments.

The velocity V with chaos is given by:

$$\begin{aligned} V(I, t+1) &= w * V(I, t) + \text{Chaos}(ch, a) * (\text{PBest}(I, t) - X(I, t)) \\ &+ c2 * (1 - \text{Chaos}(ch, a))(\text{GBest}(I, t) - X(I, t)); \end{aligned} \tag{7}$$

Feature selection employs sigmoid function for velocity to discretize position (Eq. 2); the position might be zero which means feature not selected and 1 if selected.

2.3 Related Work

Various papers considered chaotic particle swarm [17], the chaotic accelerated PSO [9] and Chaotic differential evolution Algorithm [18], chaotic firefly algorithm [19] outperform the original version of PSO and other approaches. The results of these studies validate the use of chaos in evolutionary and swarm algorithms. Alatas et al. [20], combine the strengths of PSO and chaos by inserting 12 chaotic maps into PSO. The chaotic PSO enhances the performance compared to other algorithms. Recently, Djellali and Ghoualmi [21] investigated chaos in both the initializations and the bodies of particle swarms in feature selection and demonstrate the success of that approach.

3 Proposed Modified Binary Particle Swarm

In this study, to escape local optima of particle swarm PSO and solve the stagnation problem. We incorporate new particles using prior information about current global best and its neighborhood. In fact, it is observed empirically that the local best might stay unchanged during several iterations leading to fail in finding the best solution. The solution is to count the number of iterations where the error does not decrease and discard the bad particles after computing statistical information about irrelevant features and keep the best particles and continue to benefit from them.

Binary Restart Particle swarm BResPSO and chaotic Restart BPSO (ChResBPSO) are subject to this modification to evaluate the impact of the updating population (strategy Table 1) on their performances. We remove the N % of the worst particles and replaced it with N-Best Neighbors and new random particles according to these scenarios (Table 1). Due to the stagnation (no improvement of solutions in several iterations), the first scenario replaced 40% of particles and the second 50% of worst particles.

Table 1. Updating particles

First scenario	Second scenario
60% N-Best	**50%** N-Best
20% N-Best neighbors **20%** Random Particles with features removed from WorstFeatList	**20%** N-Best Neighbors **30%** Random Particles with features removed from WorstFeatList

3.1 Binary Restart PSO

Binary Restart Particle swarm BResPSO is enhanced version of BPSO by updating its particles if the global best does not change for several iterations. We use prior information from worst particles, consequently, we calculate the frequency of each features in worst particles and build a red list of features used to forbid the selection of features belonging to that set. The modified particles are selected randomly but highly influenced by this WorstFeat list of features. The code is given below:

Pseudo Code of BResPSO

```
Input CountLimit=6;
BEGIN
Initialization of particles;
   For it=1: Maxiterations
      For i=1: NParticles
        Compute Velocity (equation1) and
        Position (equation 2) with its fitness ;
        Compute Personnal best and globalBest;
      End
     IF  GlobalBest  does not change CounLimit iterations
      Compute NewParticles = N-BestNeighbor() ;
       Calculate Worst Features List (WorstFeat) ;
       Generate new particles (NPartWF) without features
        in WorstFeat ;
       New particles = merge old 60% of previous best
       particles,  N-BestNeighbors  and NPartWF;
     EndIF
   END
END
```

3.2 Chaotic Restart BPSO

This paper introduces a new chaotic Restart Binary particles swarm optimization (ChResBPSO). Thus different chaotic maps used as efficient alternatives to pseudo random sequences. To assess the proposed algorithm, we focus on using different chaotic systems to find the best chaotic map for ChResBPSO. Chaotic Restart PSO (ChResBPSO) uses updating mechanism of particles with chaos in the body of PSO (replace random parameters r1 and r2 with chaotic parameters ch1 and ch2). The novelty introduced in chaotic Binary PSO is the particles restart strategy consisting of updating the population of particles if the global best did not improve for several iterations. The solution investigates the replacement of bad particles with new one but forbidding worst features extracted from previous bad solutions.

```
Pseudo Code of Chaotic ChResBPSO
Input CountLimit=5; oldGlobBest of iteration #1;
BEGIN
Initialization of particles; it=1 ;
     while (it<=maxiter) {
      For i=1: NParticles
        Calculate chaotic ch1 and ch2 ;
        For d = 1 to D Update the Chaotic Ch value ; end
           Compute New Velocity (equation7) and
        New Position (equation 2) with its fitness ;
        Calculate Personnal best and globalBest;
      End
    IF  GlobalBest  does not change CountLimit iterations
         Compute NewParticles = N-BestNeighbor();
         Calculate Worst Features List (WorstFeat);
         Generate new particles (NPartWF) without
           features in WorstFeat ;
        New particles = merge old 60% of previous best
           particles,  N-BestNeighbors  and NPartWF;
    EndIF   it=it+1 ;}
END
NewParticles = N-BestNeighbor() {
    Sort particles with asending order of error
    Keep 60% of well ranked Particles according to
     fitness value.
   Calculate the N-Best(N=10)Neighbors particles ;we
      disturb  the M=3 bit of each Best particles
        randomly from(1 to 0) or from (0 to 1).
   Generate X+Y% of  new particles according to table1
      senarios1 and 2.
    X is the proportions of N-BestNeighbors and Y the
      propottion of Random particles without bad
      features.
The output will be the new population of particles.}
```

4 Experimental Results

In this section, we first explain the settings parameters, then, we evaluate the performance of the proposed feature selection approach chaotic Restart BPSO ChResBPSO with chaotic Binary PSO (ChBPSO), Binary PSO (BPSO), and Artificial Bee colony ABC [22]. Support Vector machine is an efficient machine learning technique that uses kernel to construct linear classification boundaries in higher dimensional space [23]. The fitness is the accuracy of support vector machine classifier.

4.1 Feature Selection Algorithms Setting Parameters and Database

We tune the parameters setting of Chaotic Restart PSO: ChResBPSO ABC, chaotic BPSO and BPSO. The standard 10 fold cross-validation techniques measured the classification accuracies of algorithms on training data. The database is divided into training set (70%) and test set (30%). The average accuracy of the algorithms is computed on 20 independent runs. MATLAB 7.0 is used to implement our algorithms. The parameters setting of algorithms are presented in Table 2.

Table 2. Setting parameters

Algorithms	Setting parameters
ABC	Maximal iteration = 100; Population size = 80; Fitness function: Accuracy of SVM classifier;
ChResBPSO	MaxIteration = 50; W = 0.72; c1 = c2 = 2.05
Chaotic BPSO	MaxIteration = 50; W = 0.72;chaotic map: sinusoidal, logistic
BPSO	Population size = 100; inertia weight = 0.7; personal learning coefficient c1 = 2.05; global learning coefficient c2 = 2.05; max iteration = 100; fitness function: Accuracy of SVM classifier

The machine learning UCI [24] dataset allow us to test the performance of the proposed feature selection algorithms. The dataset used are Wisconsin Breast Cancer Dataset WBCD, hepatitis, colon cancer, DLBCL. The data properties are presented in Table 3.

Table 3. Dataset tested in the experimental study

Dataset	Number of features	Class	Number of instances
Hepatitis	19	2	80
Colon tumor	2000	2	62
DLBCL	4026	2	45
WDBC	30	2	569

5 Discussion

The experiments carried out on Hepatitis data in Table 4 indicate that Chaotic Restart PSO (ChResBPSO) with logistic map outperform all chaotic variants of BPSO, ABC on mean train accuracy and number of features but ABC is better for test accuracy (86.10%). It is shown from experimental results on WDBC in Table 5 that chaotic Restart BPSO

Table 4. Chaotic restart PSO performance on hepatitis

FS methods	Train SVM accuracy %	Test SVM accuracy	# features	Chaotic map
ChResBPSO	99.06 ± 0.015	62.50 ± 0.56	**8**	Sinusoidal
ChBPSO	93.58 ± 0.020	66.67 ± 0.052	4	
ChResBPSO	**99.69 ± 0.0098**	**63.40 ± 0.0614**	**9**	Logistic
ChBPSO	92.11 ± 0.031	62.50 ± 0.022	5	
ABC	87.20 ± 0.013	**86.10** ± 0.018	12	
BPSO	84.30 ± 0.026	81.19 ± 0.009	9	

Table 5. Chaotic restart PSO performance on WDBC

FS methods	Train SVM accuracy %	Test SVM accuracy	# features	Chaotic map
ChResBPSO	**99.16 ± 0.0039**	**97.26 ± 0.0099**	**13**	Sinusoidal
ChBPSO	95.90 ± 0.003	96.93 ± 0.0014	8	Sinusoidal
ABC	**99.35** ± 0.041	95.32 ± 0.025	15	
BPSO	96.80 ± 0.036	94.60 ± 0.014	17	
ChResBPSO	**99.03 ± 0.0052**	**97.60 ± 0.0076**	**12**	Logistic
ChBPSO	91.63 ± 0.078	93.77 ± 0.007	7	Logistic

achieves mean accuracy (train 99.03%, test 97.60% and 12 features) with logistic map and outperforms all methods show in Table 5.

It can be learned from Table 6 on Colon cancer that chaotic Restart BPSO realizes (100.00%, 85.27% and 992 features) with logistic map better than all variant of PSO and ABC algorithm.

Table 6. Chaotic restart PSO performance on colon

FS methods	Train SVM accuracy %	Test SVM accuracy	# feature	Chaotic map
ChResBPSO	100.00 ± 0.000	73.70 ± 0.035	1002	Sinusoidal
ChBPSO	86.15 ± 0.0	63.16 ± 0.00	910	Sinusoidal
ChResBPSO	**100.00 ± 0.00**	**85.27 ± 0.0355**	**992**	Logistic
ChBPSO	86.27 ± 0.0	63.16 ± 0.0	915	Logistic
BPSO	82.44 ± 0.0	81.78 ± 0.0	1023	
ABC	**88.35** ± 0.0	**81.24** ± 0.0	**980**	

The results on DLBCL data in Table 7 demonstrate that chaotic ChBPSO with sinusoidal map realizes the best test accuracy 89.71% and 1917 features where ABC achieve less features 1532.

Experiments show that chaotic Restart BPSO outperforms with logistic map all methods on WDBC and colon in term of accuracy and minimal number of features. However, ChaoticBPSO is better on DLBCL and ABC on hepatitis. Adding the information on worst features extracted from bad particles with low accuracy enhances the quality of solution in term of higher accuracy and less features.

Table 7. Chaotic restart PSO performance on DLBCL

FS methods	Train SVM accuracy %	Test SVM accuracy	# features	Chaotic map
ChResBPSO	100.00 ± 0.011	60.00 ± 0.05	2015	Sinusoidal
ChBPSO	87.25 ± 0.021	89.71 ± 0.071	1917	
ChResBPSO	100.00 ± 0.00	50.00 ± 0.061	1999	Logistic
ChBPSO	85.72 ± 0.021	62.50 ± 0.092	1992	Logistic
BPSO	90.12 ± 0.036	88.98 ± 0.041	2065	
ABC	90.08 ± 0.071	81.19 ± 0.052	1532	

6 Conclusion

This paper demonstrates how to enhance the quality of solutions by disturbing populations of particles. Prior information on best solutions and worst features list provide a rigorous way to improve the global best. Chaotic restart binary BPSO is compared to chaotic BPSO, binary PSO and ABC. We found that the proposed chaotic restart binary PSO outperforms in term of accuracy and minimal number of features all variant of PSO and competitive with ABC method.

Statistical information on best previous particles and worst particles deliver information about good and bad features and allow the algorithm to guide the search process through better solutions without exploring wide search space and keep diversity.

References

1. Bolon-Canedo, V., Sánchez-Marono, N., Alonso-Betanzos, A., Benítez, J.M., Herrera, F.: A review of microarray datasets and applied feature selection methods. Inf. Sci. **282**, 111–135 (2014)
2. Djellali, H. Zine, N.G., Azizi, N.: Two stages feature selection based on filter ranking methods and SVMRFE on medical applications. In: Modelling and Implementation of Complex Systems. Lecture Notes in Networks and Systems, vol. 1. Springer, Cham (2016)

3. Djellali, H., Guessoum, S., Layachi, S., Ghoualmi-Zine, N.: Fast correlation based filter combined with genetic algorithm and particle swarm on feature selection. In: The 5th IEEE International Conference on Electrical Engineering, Boumerdes (ICEE-B) (2017)
4. Djellali, H., Djebbar, A., Ghoualmi-Zine, N., Azizi, N.: Hybrid artificial bees colony and particle swarm on feature selection. In: IFIP International Conference on Computational Intelligence CIIA, pp 93–105 (2018)
5. Kennedy, J. Eberhart, R.C.: Particle swarm optimization. In: IEEE International Conference on Neural Networks, vol. 4, pp. 1942–1948 (1995)
6. Xue, B., Zhang, M., Browne, W.N., Yao, X.: A survey on evolutionary computation approaches to feature selection. IEEE Trans. Evol. Comput. **20**(4), 606–626 (2016)
7. Kennedy, J., Eberhart, R.C.: A discrete binary version of the particle swarm algorithm. In: Proceedings of the IEEE International Conference on Systems, Man and Cybernetics, vol. 5, Orlando, FL, USA. pp. 4104–4108 (1997)
8. Meng, H.J., Zheng, P., Wu, R.Y., Hao, X.J., Xie, Z.: A hybrid particle swarm algorithm with embedded chaotic search. In: IEEE Conference on Cybernetics and Intelligent Systems, vol. 1, no. 2, pp. 367–71 (2004)
9. Gandomi, A.H., Jin Yun, G., Yang, X.S., Talatahari, S.: Chaos-enhanced accelerated particle swarm optimization. Commun. Nonlinear Sci. Numer. Simul. **18**, 327–340 (2013)
10. Xiang, T., Liao, X., Wong, K.W.: An improved particle swarm optimization algorithm combined with piecewise linear chaotic map. Appl. Math. Comput. **190**, 1637–1645 (2007)
11. Liu, B., Wang, L., Jin, Y.H., Tang, F., Huang, D.X.: Improved particle swarm optimization combined with chaos. Chaos Solitons Fract. **25**, 1261–1271 (2005)
12. Chuang, L.-Y., Tsai, S.-W., Yang, C.-H.: Improved binary particle swarm optimization using catfish effect for feature selection. Int. J. Expert Syst. Appl. **38**, 12699–12707 (2011)
13. Petrovic, M., Mitic, M., Vukovic, N., Mmiljkovic, Z.: Chaotic particle swarm optimization algorithm for flexible process planning. Int. J. Adv. Manuf. Technol. **85**(9–12), 2535–2555 (2016)
14. May, R.M.: Simple mathematical-models with very complicated dynamics. Nature **261**, 459–467 (1976)
15. Devaney, R.L.: An Introduction to Chaotic Dynamical Systems. Addison-Wesley, Boston (1987)
16. Li, Y.T., Deng, S.J., Xiao, D.: A novel hash algorithm construction based on chaotic neural network. Neural Comput Appl. **20**, 133–141 (2011)
17. Alatas, B., Akin, E., Ozer, A.B.: Chaos embedded particle swarm optimization algorithms. Chaos Solitons Fract. **40**, 1715–1734 (2009)
18. Gandomi, A.H. Yang, X.S., Talatahari, S., Alavi. A.H.: Firefly algorithm with chaos. Commun. Nonlinear Sci. Numer. Simul. https://doi.org/10.1016/j.cnsns.2012.06.009. (2016, in press)
19. Zhenyu, G., Bo, C., Min, Y., Binggang, C.: Self adaptive chaos differential evolution. Adv. Nat. Comput. 972–975 (2006)
20. Alatas, B.: Chaotic bee colony algorithms for global numerical optimization. Expert Syst. Appl. **37**, 5682–5687 (2010)
21. Djellali, H., Ghoualmi, N.: Improved chaotic initialization of particle swarm applied to feature selection. In: IEEE International Conference on Networking, ICNAS, Algeria, pp. 1–5 (2019)
22. Karaboga. D.: An idea based on honey bee swarm for numerical optimization. Technical report TR06, Erciyes University, Engineering Faculty (2005)
23. Vapnick, V.: Statistical Learning Theory. Wiley, New York (1998)
24. Machine Learning Repository UCI. https://archive.ics.uci.edu/ml/datasets.html

IoT, Networks, and Web Technologies

Advanced Sensorless Weather Station Implementation Using ESP32

Jugurtha Chettouh[1] and Samia Mezzah[2(✉)]

[1] Faculty of Technology, Bejaia University, 06000 Bejaia, Algeria
[2] Laboratory of Medical Informatics (LIMED), Bejaia University, 06000 Bejaia, Algeria
samia.mezzah@univ-bejaia.dz

Abstract. The Internet-of-Things (IoT) concept has been largely integrated into heterogeneous systems and it has recently become more relevant to practical implementations because of the growth of ubiquitous communication and remote data access techniques. This paper presents an IoT system that basically gathers weather-related data information from a forecast API (Application Programming Interface) for a specific configurable location and displays them on TFT touch screen. The implemented system uses a low-cost ESP32 microcontroller and exploits its integrated Wi-Fi to achieve transmitting and receiving communications. Moreover, we develop an embedded web server on ESP32 using SPIFFS (Serial Peripheral Interface Flash File System) to monitor and control power consumption mode over a local network and a secured internet connection via a laptop, cellphone, or tablet. These functionalities are implemented by using FreeRTOS that provides multitasking support for the ESP32 dual-core microcontroller.

Keywords: IoT system · ESP32 FreeRTOS · Sensorless weather station

1 Introduction

Progress in electronic and telecommunication fields are currently boosting an important technological shift that changes human world interaction [1]. These new technologies enhance greatly data availability and enable access to both useful services, like information aggregation for well-being, comfort and productivity, and vital services such as health care and education. IoT, which is the most important emerging technology, consists of connecting physical objects to the Internet [2]. The size, power consumption, and cost of electronic components that are needed to support Internet use capabilities play a critical role in the widespread adoption of IoT [3]. In this sense, System-on-Chip (SoC) solutions, like ESP microcontroller chips, offer low cost and efficient components that integrate all the necessary connectivity modules of the complete system on a single chip.

Initially, the ESP chip, with its built-in IEEE 802.11 b/g/n networking protocol support module, was mainly used as a second low-cost controller to provide Wi-Fi capabilities and network access for another controller via a serial port [4]. However, more and more projects began using the spare processing power of these chips to also provide

M. R. Senouci et al. (Eds.): CSA 2020, LNNS 199, pp. 165–174, 2021.
https://doi.org/10.1007/978-3-030-69418-0_15

general I/O control so that in many cases no additional control processor was required [5, 6]. This prompted Espressif, the chip's manufacturer, to bring out the next generation of chips, ESP32, that contain additional functional modules and more powerful processing cores. ESP32 chips work with a clock rate of up to 240 MHz and integrate [7]: a Tensilica Xtensa LX6 32-bit CPU in both dual-core and single-core versions, an additional power-efficient coprocessor, 36 GPIO pins, dual-mode BLE (Bluetooth Low Energy), 520 KB SRAM, 448 KB ROM, in-built antenna switches, power amplifier, low-noise receive amplifier, hardware accelerators for AES/SSL/TLS and many Input/Output peripherals that include capacitive touch, ADCs, DACs, I^2C, UART, SPI, I^2S, RMII, …etc.

The ESP32 Chip is highly complex and cannot easily be used on its own especially for prototyping. Therefore, several development boards are available that incorporate the ESP32 chip and associated hardware to simplify the task of project development. Some of the most popular ESP32 development boards are: Adafruit Huzzah32, NodeMCU-32 s and ESP32 DevKitC [5]. In this work, we used an Adafruit Huzzah32 board that contains the ESP32 WROOM32 module, which, in turn, contains the dual-core ESP32 D0WDQ6 chip and a 4 MB SPI flash memory (see Fig. 1).

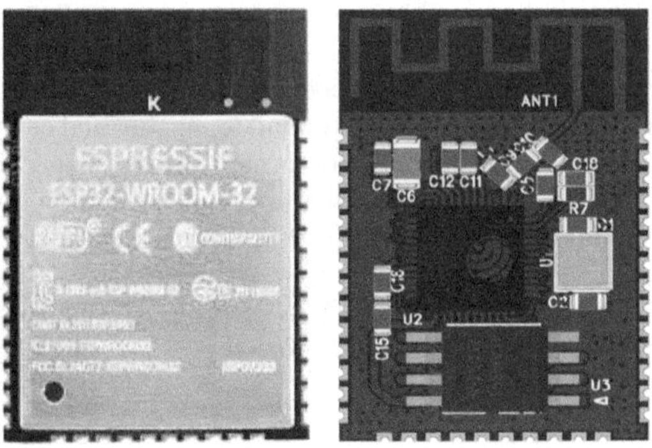

Fig. 1. The ESP-WROOM-32 module

2 System Description

Traditional weather station systems require many sensors, data monitoring and collection systems as well as regular maintenance equipment and skilled operators [8]. Alternatively, sensorless weather stations offer a good solution for low-cost application with minimum effort. The realized prototype is deployed to get and display weather information of a specified location in real-time [9]. Both the transmitting and receiving communications are managed using integrated WiFi. This system pulls weather data using the OpenWeatherMap API. Beside this information gathering functionality, the ESP32 assure 3 other functions which are (see Fig. 2):

1. Display control: to display date, time, Wi-Fi strength, current weather conditionals, 4-day forecast and moon phases.

2. Web server: to allow remote control of power mode (normal, screen off or deep sleep) through a Web portal or a mobile application.
3. Touch detection: to manage activation/deactivation of deep-sleep mode using a touch sensor.

3 Hardware Components

In this work, we use the HUZZAH 32 board [10] and a TFT FeatherWing touch screen [11]. The TFT FeatherWing is a display shield with built-in microSD card socket. This TFT display is 2.4 inches diagonal with a bright four white-LED backlight. It displays 16-bit color images of 240 × 320 pixels. This display comes with an attached resistive touchscreen. It also includes a reset button and an on/off switch connected to the Feather Enable pin. The TFT display, SD card and touch screen use the SPI interface to communicate with the controller; That means MISO, MOSI and SCK ports are used whenever either are accessed.

Great addition to the used board is the incorporated Lipoly battery charger. This will let the feather run on a rechargeable battery. Hence, to supply this embedded system with power, we used 3.7 V Lithium Polymer (Lipo) battery. To economize battery energy, we

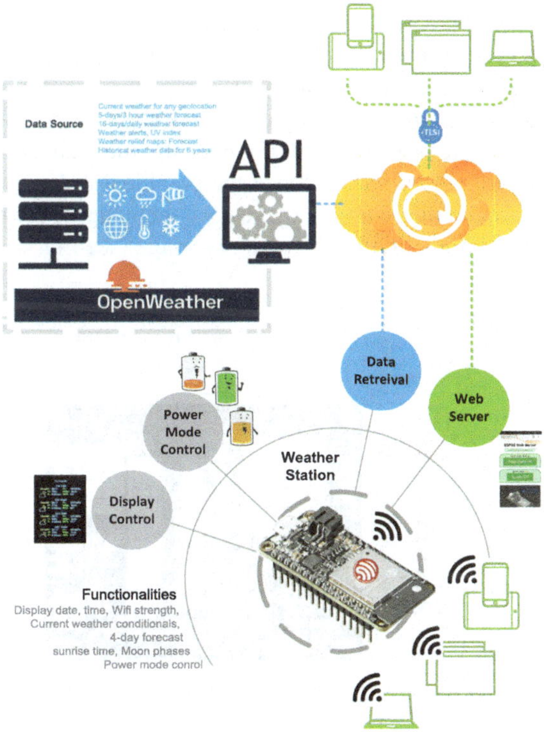

Fig. 2. System functionalities

added power mode control using a touch sensor capability of pin A4. Details of hardware components and pin assignment are given in Fig. 3.

4 Software Implementation

Since we use compatible hardware components, the prototype assembly is simple without any extra wire or need for electronics components. Hence, a major effort is devoted to software development.

The ESP32 application code was developed using Arduino 1.8.7 IDE, FreeRTOS and many hardware libraries. To program the ESP32 development board using Arduino IDE, the ESP32 toolchain has to be installed which uses "xtensa-esp32-elf-gcc" & "xtensa-esp32-elf-g++ " compilers. The most important extension feature of this library is the FreeRTOS kernel support. The implementation of FreeRTOS support for the ESP32 microcontroller is very important because it provides support for multitasking.

4.1 Multitasking Programming

The used ESP-IDF (IoT Development Framework) FreeRTOS is a modified version of vanilla FreeRTOS which supports symmetric multiprocessing (SMP) [12]. ESP32 is dual-core containing a Protocol CPU (known as Core 0) and an Application CPU (known as Core 1). The two cores are identical in practice and share the same memory. This allows the two cores to run tasks interchangeably between them.

In order to optimize the code execution, it is recommended to run modem tasks (Connecting to Wi-Fi, transmitting and receiving packets …etc.) independently from the other tasks; So that those processes will not interrupt the main core which may have heavy tasks (display tasks in our case). Since the graphical tasks should be initialized before modem tasks, we put them on core1, and gave them higher priority and stack size because those tasks have higher CPU and memory usage (see Fig. 4).

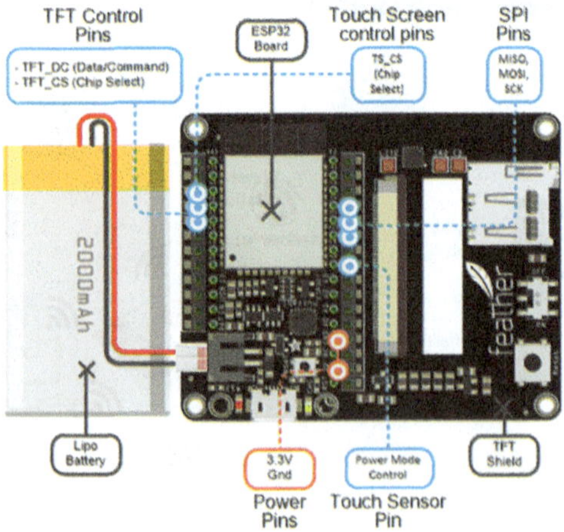

Fig. 3. Weather Station Hardware components.

4.2 Display Tasks

For display code, we use Mini_Grafx library which is a graphics library for embedded devices containing a framebuffer to avoid flickering: all drawing operations are made on buffer stored in memory. When drawing writing is finished, the whole content of the frame buffer is displayed. The library contains drawing routines and the driver for ILI9341 based displays. This driver considers 565 coded color (5bit for red, 6bit for green and 5bit for blue). The available configuration function allows the setting of bit depth and palettes which determine the buffer size.

4.3 Weather Data Collection

ESP32 can operate in three different modes: Station mode, Soft Access Point mode, and both at the same time. In this work, we configured the ESP-32 to act in station mode since it needs internet access to send and receive requests. Modem tasks are based on not blocking WiFi library function calls. For data retrieval, ESP32 sends HTTP requests to OpenWeatherMap API which will respond with a JSON file containing weather information (forecasts, Current Conditions…etc.) of the given location as shown by Fig. 5. The GET HTTP request used contains the location (City ID), an identification key (API key), and additional configuration parameters.

The Json_Streaming_Parser library is used for (byte by byte) parsing json streams because it is well adapted to embedded systems with limited memory by using Json-Listner module that generate notification when interesting parts in the feed are received. The esp8266-weather-station library is also used. It contains client modules to fetch data from given number of weather servers, Time client module to set the clock and functions to calculate lunar phase and illumination, sun and moon properties for a given date, time and location.

```
//graphical task definition with priority 3 to be executed on core1
xTaskCreatePinnedToCore(
  Graphical_tasks,      /*Task function*/
  "Task1",              /*Task name*/
  12000,                /*Task Stack Size*/
  NULL,                 /*Task parameter pointer*/
  3,                    /*Task priority level = 3*/
  &Task1,               /*Task handle*/
  1);                   /*Execution core1 selected*/
Delay(500);
//Modem task definition with priority 1 to be executed on core0
xTaskCreatePinnedToCore(
  Modem_tasks,
  "Task2",
  10000,
  NULL,
  1,                    /*Task priority level = 1*/
  &Task2,
  0);                   /*Execution core1 selected*/
Delay(500);
```

Fig. 4. Multitasking parameters

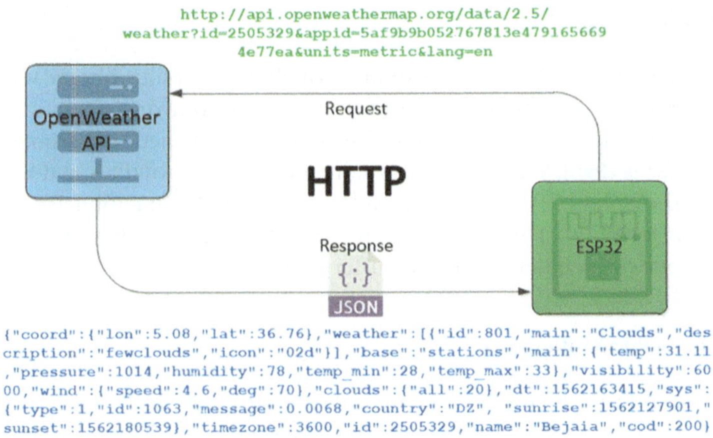

Fig. 5. Data retrieval protocol

4.4 ESP-32 Web Server

The web server is used to control power consumption mode through the web. Hence, Users can control the Screen and Deep-Sleep status using the web server page. Figure 6 explains This process. The developed web server, which serves HTML and CSS files stored on the ESP32 filesystem, is based on SPIFFS (Serial Peripheral Interface Flash File System) library. SPIFFS is a file system intended for SPI flash devices on embedded targets.

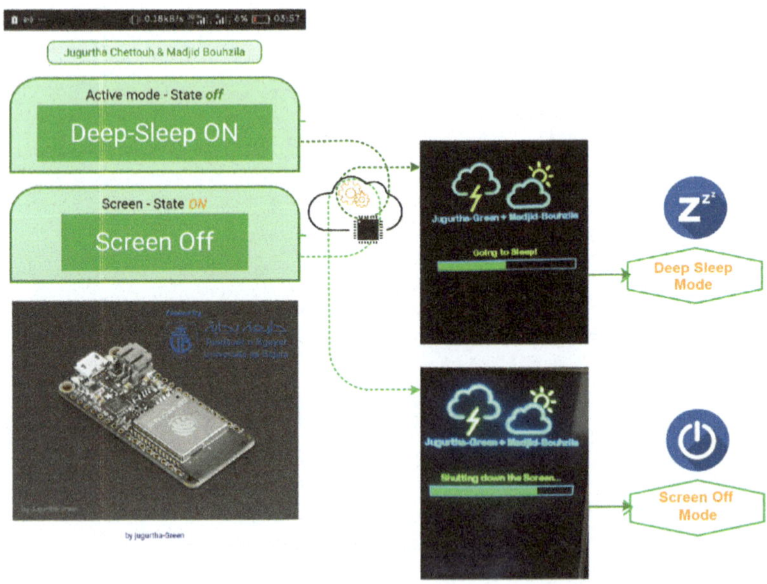

Fig. 6. Web Server page and Power Control functions.

mDNS Web Server implementation. One of the greatest features ESP32 provides is that it cannot only connect to an existing WiFi network and act as a Web Server, but it can also set up a network of its own, allowing other devices to connect directly to it and access web pages. mDNS stands for multicast Domain Name System. It is a network protocol for resolving hostnames to IP addresses in small local networks. Hence, we can easily access to the ESP-32 Web Server using local name server (in our case it is esp32.local) instead of IP address. Users can also access the ESP-32 Web Server from outside (over Internet) without the need for Port Forwarding, this is done by an intermediate server like serveo.net or ngrok.io which are a combination of SSH, HTTP & TCP server that create a sub-domain (ours is "esp32-jugu.serveo.net") that forward traffics to the local network from the internet. In addition, we created an android application, which allows access to the ESP Web Server in one click.

Security Layer implementation. We added an extra layer of security for control access capabilities to deny access for unauthorized users when a user tries to access the ESP-32 over internet. The authorized user has to authenticate using "username" and "password" in order to control the device using http requests over the ESP-32 Web Server. Authentication operations are done with an external server using PHP as a back-end programming language. Figure 7 shows the process of Authentication. The login page has been created using HTML5, CSS3, and JavaScript. The web server code uses WiFiClintSecure library that implement support for secure connections using TLS/SSL (Transport Layer security/Secure Socket Layer). There are three ways to establish a secure connection using WiFiClintSecure: using a root Certificate Authority (CA) cert, a root CA cert plus a client cert and key or a Pre-Shared Key (PSK).

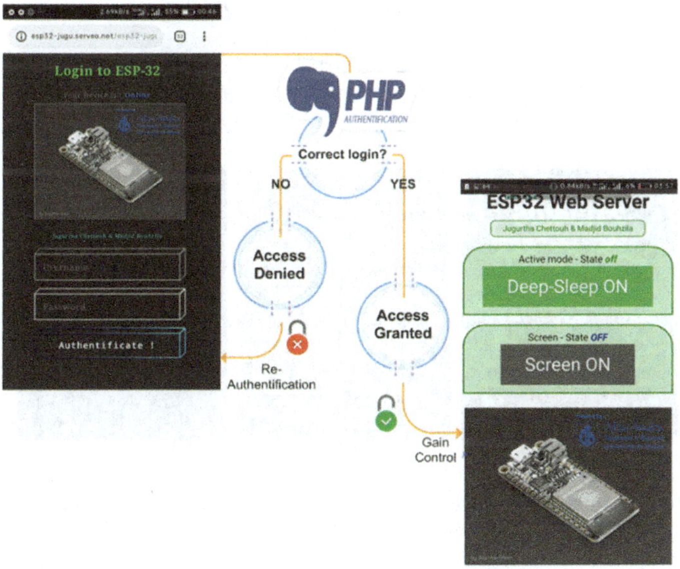

Fig. 7. Authentication process

5 Implementation Results

For the implementation, we chose the Arduino platform as a development environment, into which we have installed ESP32 development support. We proceeded to implement each functionality separately and then test the components. After completing the entire assembly, we performed the testing of the entire embedded system.

Figure 8 shows the prototype of our embedded system. As shown in Fig. 9 and Fig. 10, the developed code controls the TFT screen to display weather information, according to the data retrieval configuration, and to show the status of different functional processes (initialization, updating data, activation of power modes …etc.).

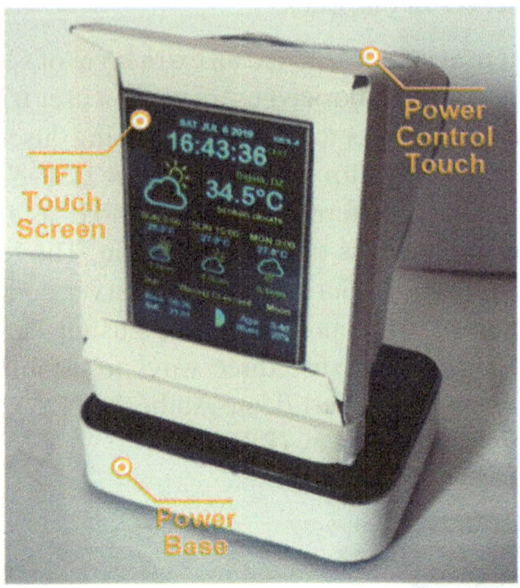

Fig. 8. The final prototype

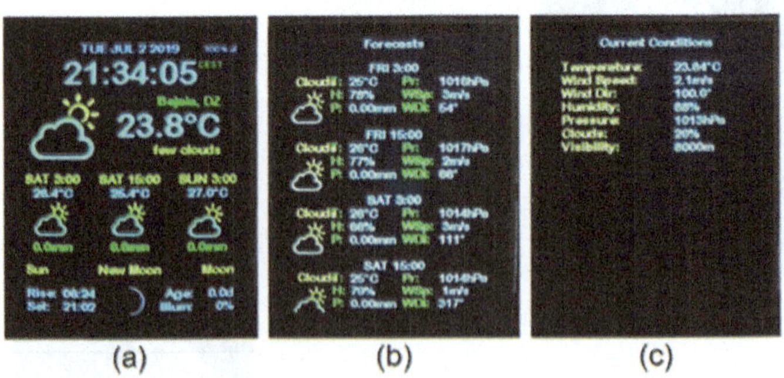

Fig. 9. Weather information screens: (a) main screen (b) forecasts (c) current conditions

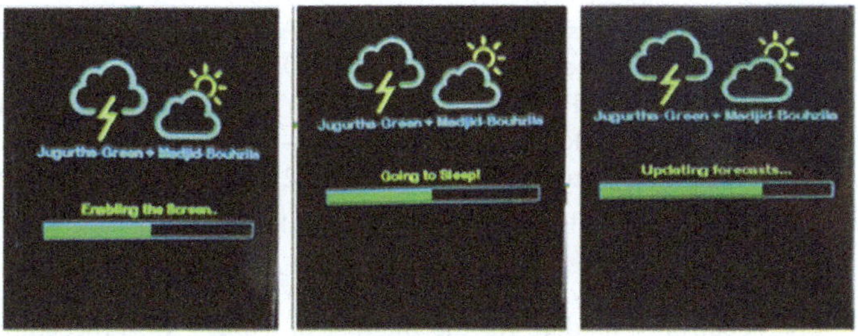

Fig. 10. Some Processing screens

6 Conclusion

The contribution describes the realization of an Internet of Things device with an ESP32 microcontroller with the dual-core implementation of data retrieval and processing beside wireless communication tasks. The main objective was to design an application for an ESP32 chip that will use most of its capabilities.

Hardware components used are the adafruit HUZZAH 32 board and the 2.4" TFT FeatherWing. The main part of this realization is software components development. We used accessible hardware and communication libraries for the development of different routines for data collection, processing and display as well as web server and security layer implementation.

The first leading idea was to use low-end hardware and software optimization to build an IoT embedded system, however the developed system is generic and could be reused and easily adapted to deploy other embedded systems because it includes main components that characterize an IoT embedded system which are network connectivity (Wi-Fi), HTTP based communication, data collection and remote system control (internal web server).

The presented work can be extended in two directions; First, it will be useful to include OTA (Over-The-Air) Programming feature which allows updating microcontroller code to the ESP32 using a browser, without the need to make a physical connection between the ESP32 and the computer. Second, the software design of this embedded system may be refined to maximize the use of ESP32 capabilities to create architectural design patterns and methods that are independent and reusable to implement any IoT embedded systems. This general framework will be characterized by: Device configurability, more functionalities for remote access (Control, reporting, monitoring … etc.) as well as self-test and diagnostic capability.

References

1. Elazhary, H.: Internet of Things (IoT), mobile cloud, cloudlet, mobile IoT, IoT cloud, fog, mobile edge, and edge emerging computing paradigms: Disambiguation and research directions. J. Netw. Comput. Appl. **128**, 105–14 (2019)
2. Nord, J.H., Koohang, A., Paliszkiewicz, J.: The Internet of Things: review and theoretical framework. Exp. Syst. Appl. **133**, 97–108 (2019). ISSN 0957–4174

3. Ojo, M.O., Giordano, S., Procissi, G., Seitanidis, I.N.: A review of low-end, middle-end, and high-end Iot devices. IEEE Access **6**, 70528–70554 (2018)
4. Ibrahim, D.: The Complete ESP32 Projects Guide, 59 Experiments with Arduino IDE and Python. Elektor International Media (2019). ISBN 978-1-907920-75-2
5. Babiuch, M., Foltýnek, P., Smutný, P.: Using the ESP32 microcontroller for data processing. In: 20th International Carpathian Control Conference (ICCC), Krakow-Wieliczka, Poland, pp. 1–6 (2019)
6. Maier, A., Sharp, A., Vagapov, Y.: Comparative analysis and practical implementation of the ESP32 microcontroller module for the Internet of Things. In: Proceedings of Internet Technology Applications (ITA), pp. 143–148 (2017)
7. Espressif Systems. ESP32 Technical Reference Manual, 2018 Version 4
8. Morón, C., Diaz, J.P., Ferrández, D., Saiz, P.: Design, development and implementation of a weather station prototype for renewable energy systems. Energies **11**, 2234 (2018)
9. Adafruit Learning System. ESP8266 WiFi Weather Station with Color TFT Display (2018)
10. Adafruit Learning System. Adafruit HUZZAH32 - ESP32 Feather (2019)
11. Adafruit Learning System. Adafruit 2.4 TFT FeatherWing (2019)
12. Espressif Systems. FreeRTOS ESP-IDF Programming Guide Version 4.1

Perceptual Masking Model Using Lifting Wavelet Transform for Fingerprint Images

Mohamed Lamine Benseddik(✉), Khalil Zebbiche, Mohamed Salah Azzaz, and Said Sadoudi

Ecole Militaire Polytechnique, BP 17 Bordj El Bahri, 16111 Algiers, Algeria
benseddik.mohamedlamine@gmail.com

Abstract. In this work, an efficient pixel-wise perceptual masking model that operates in the Lifting Wavelet Transform (LWT) is proposed for fingerprint images. The proposed model simulates the Human Visual System (HVS) behavior in a better way by exploiting four of its characteristics, namely: the frequency band sensitivity, the local brightness, the texture masking and the edge masking. It allows estimating the maximum unnoticeable noise for each coefficient. Experimental results show that the proposed model has high-noise tolerance when compared to a number of related state-of-the-art models.

Keywords: Perceptual masking model · Lifting Wavelet Transform · Human Visual System · Fingerprint image

1 Introduction

Over many decades, understanding and modeling the human visual system (HVS) have attracted many researchers to design novel perceptual models that exploit its characteristics [1]. Particularly, in the fields of image processing and machine vision, HVS modeling plays a major role, in the sense that it allows the design of applications with better performance from a visual quality point of view. For instance, in image compression, it reduces the visibility of introduced artifacts, and in watermarking, it allows to hide information more robustly in images [2]. However, in most cases, perceptual models are only simplifications derived from psycho-physical experiments and even though exhibiting some differences, they are all based on modeling and combining common visual phenomena, such as: background luminance, edge proximity, texture activity and spatial frequency [3].

In the literature, many perceptual models have been proposed for both spatial and transform domains [4]. Particularly, due to its temporal-spatial characteristics which simulate the HVS behavior in a better way, the wavelet transform received much attention, resulting in the development of several perceptual models for different wavelet-based decompositions, such as: Discrete Wavelet Transform (DWT) [4, 5], Dual Tree Complex Wavelet Transform (DTCWT) [6, 7] and Lifting Wavelet Transform (LWT) [8, 9]. With regard to the LWT domain, only few works have been published that address perceptual masking. Taha *et al.* [8] have proposed an Accumulative Lifting Difference (ALD)

M. R. Senouci et al. (Eds.): CSA 2020, LNNS 199, pp. 175–184, 2021.
https://doi.org/10.1007/978-3-030-69418-0_16

function as texture mapping model based on the LWT. This function was combined with luminance masking to build a block-based perceptual mapping model to estimate Just Noticeable Distortion (JND) for digital images. The idea was later improved in [9], where authors have proposed to combine the ALD function with a simplified edge detection model and luminance masking model to determine the maximum tolerated modifications for each 5×5 block of the LWT approximation band. In the case of fingerprint images, a wavelet-based perceptual model, called Just Perceptual Weighting (JPW) has been proposed in [3]. This model exploits three HVS characteristics, namely: spatial frequency sensitivity, local brightness masking and texture masking, to compute a weight for each DWT coefficient.

In this paper, a new perceptual masking model for fingerprint images that operates in the LWT domain is proposed. This model, which is a pixel-wise model, computes a weight for each LWT coefficient that quantifies the amount of unnoticeable changes. It has been designed by combining common HVS characteristics, namely: the frequency band sensitivity, the local brightness, the texture masking and the edges masking. To the best of our knowledge, there has not been any pixel-wise masking model reported in the literature that exploits the abovementioned characteristics and operates in the LWT domain. Experiments have been conducted to evaluate the performance of the proposed model and the obtained results show a clear superiority over other related wavelet-based perceptual models in the literature.

The paper is structured as follows. Section 2 provides a description of the proposed model. Experimental results are reported and discussed in Sect. 3. Conclusions are drawn in Sect. 4.

2 The Proposed Scheme

In this work, we propose a new pixel-wise perceptual masking model in the LWT domain for fingerprint images. Compared to the conventional DWT, the LWT uses lower number of mathematical operations, resulting in a smaller implementation area, less power consumption and lower design complexity [9]. In the case of digital images (i.e. 2-D signals), the LWT decomposition produces at level l: (i) a low-frequency sub-band (LL), (ii) l high-frequency horizontal sub-bands $\{HL_i : i = 0, \ldots, l-1\}$, (iii) l high-frequency vertical sub-bands $\{LH_i : i = 0, \ldots, l-1\}$ and (iv) l high-frequency diagonal sub-bands $\{HH_i : i = 0, \ldots, l-1\}$. For the sake of simplicity, each sub-band will be represented by a pair (λ, θ), where $\lambda = \{1, \ldots, l\}$ is the transform level and $\theta = \{LL, HL, LH, HH\}$ is the sub-band orientation.

The design of the proposed perceptual masking model relies on common considerations stating that the HVS is less sensitive to modifications in [3, 5]:

- High-frequency bands, particularly those bands having orientation of 45°;
- Regions and areas of the image where brightness is high or low;
- Highly textured areas, except near edges where the eye is more sensitive.

Consequently, the proposed perceptual model combines a spatial frequency sensitivity function, and two masking components based on local brightness and texture masking and an edge detection technique, to compute a weight for each LWT coefficient of fingerprint images, as illustrated in Fig. 1.

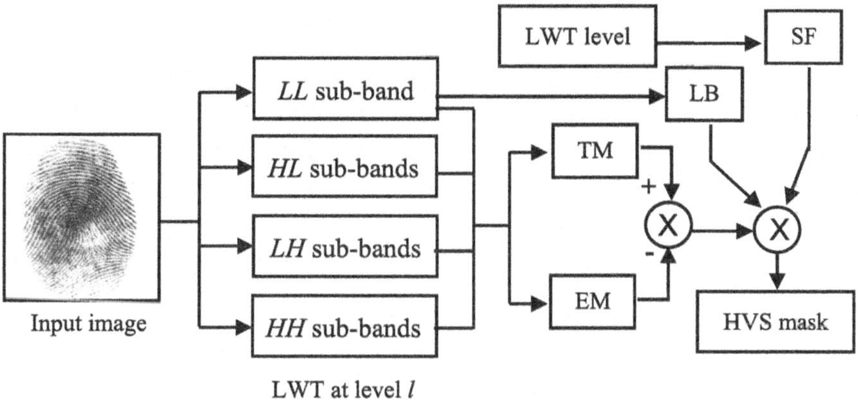

Fig. 1. General framework of the proposed perceptual masking

The mathematical formulation of the proposed perceptual masking model is given as follows

$$vm(\lambda, \theta, i, j) = [\alpha SF(\lambda, \theta)]\big[\beta LB(\lambda, \theta, i, j)\big]\big[\gamma TM(\lambda, \theta, i, j) - \delta EM(\lambda, \theta, i, j)\big] \quad (1)$$

where SF is the spatial frequency sensitivity for a sub-band (λ, θ). $LB(\lambda, \theta, i, j)$ is the local brightness for a coefficient that belongs to sub-band (λ, θ) at location (i, j). $TM(\lambda, \theta, i, j)$ represents the texture masking adjustment for a coefficient in position (i, j) in the sub-band (λ, θ). $EM(\lambda, \theta, i, j)$ is the edge detection value of a coefficient located in position (i, j) in the sub-band (λ, θ). The parameters α, β, γ and δ are weighting factors that control the intensity of each mask and the optimal values obtained through extensive experiments are $\alpha = \gamma = 2.5$, $\beta = 1.5$ and $\delta = 0.01$.

2.1 Spatial Frequency Sensitivity

As stated in [10], the human eye has more sensitivity to patterns and textures, which can be perceived as spatial frequencies and in the case of 2-D patterns, they can be characterized by their magnitude and orientation. Also, it has been shown that the HVS sensitivity depends on the orientation of patterns. Particularly, the human eye has less sensitivity to lines and edges with a 45° orientation than those with vertical or horizontal orientations. Usually, the spatial frequency response is determined by the sensitivity to luminance contrast as a function of spatial frequency, and this is known as contrast sensitivity function (CSF).

In general, a simple approach for implementing CSF in the wavelet domain consists on assigning a single frequency weighting factor to each sub-band, which represents the average sensitivity of the HVS for the covered frequency range [11].

In this work, the spatial frequency sensitivity model proposed in [3, 12] is adopted. It is formulated as

$$SF(\lambda, \theta) = \begin{pmatrix} \sqrt{2}, \text{ if } \theta = HH \\ 1, \quad \text{otherwise} \end{pmatrix} \frac{1}{H(f)(\lambda, \theta)} \tag{2}$$

where $H(f)(\lambda, \theta)$ is the CSF weight and the term $1/H(f)(\lambda, \theta)$ describes the minimal noticeable sensitivity depending on the spatial frequency. The values of $H(f)(\lambda, \theta)$, representing the average of the CSF curve in its corresponding frequency band, are reported in Table 1 as given by Beegan [13].

Table 1. Band-average CSF values for a five-level wavelet decomposition of an image

λ	θ	$H(f)(\lambda, \theta)$
1	HL, LH, HH	1
2	HL, LH, HH	2.56
3	HL, LH, HH	3.16
4	HL, LH, HH	2.87
5	HL, LH, HH	2.35

2.2 Local Brightness Masking

As stated by [5, 9, 14], alterations are most likely not perceptible by the human eye in very dark and very bright areas. This characteristic has been particularly exploited by researchers to design perceptual models for different types of wavelets, such as [4, 5] for the DWT, [6, 7] for the DTCWT and [8, 9] for LWT. In this work, we propose to use the approximation sub-band of a given level to compute the local brightness of all sub-bands at that level, as suggested by [3]. A mathematical formulation is given by

$$LB(\lambda, \theta, i, j) = 1 + L'(\lambda, \theta, i, j) \tag{3}$$

where

$$L'(\lambda, \theta, i, j) = \begin{cases} 1 - x(\lambda, LL, i, j), & \text{if } x(\lambda, \theta, i, j) < 0.5 \\ x(\lambda, LL, i, j), & \text{otherwise} \end{cases} \tag{4}$$

where $x(\lambda, LL, i, j)$ is the value of the LWT coefficient located in (i, j), belonging to the approximation sub-band LL at level λ. It is worth noting that before computing the local brightness LB, the values of the approximation sub-bands are all normalized into the rang [0, 1].

From Eq. 4, one can observe that the local brightness term takes its largest values when the coefficients corresponds to very dark or very bright areas, which is very appropriate for fingerprint images, since such images mainly comprise bright background and dark foreground, where the HVS is less sensitive to modifications.

2.3 Texture Masking

It is well known that the HVS has less sensitivity to modifications in highly textured regions than in smooth and homogeneous areas. Texture masking is a perceptual effect in which some patterns can be hidden or masked in textured areas without being perceived by the human eye. According to [15], by relying on the statistical approach, the texture concentration in a certain area can be calculated by using the differences of the intensities in that area. This fact has been used by Taha *et al.* [8] to propose a simple texture mapping model, which is called Accumulative Lifting Differences (ALD). The computation of the ALD begins by dividing the high-frequency diagonal sub-band *HH* into 5 × 5 blocks. Then, for each block, adjacent coefficients in each row of the block are subtracted from each other and the absolute values of the obtained differences are computed. Finally, the absolute values are summed together to obtain one single value that represents the texture activity of that block, as illustrated in Fig. 2.

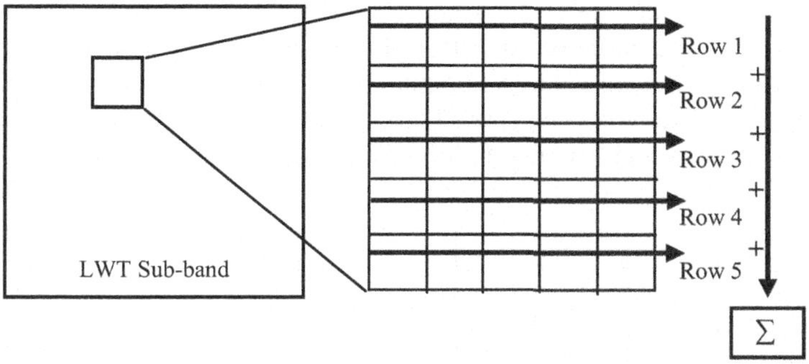

Fig. 2. Accumulative lifting difference computation [9]

In this work, the ALD-based model has been adapted in such a way that a texture weight is computed for each LWT coefficient. Also, instead of using the *HH* sub-band to compute the *TM* of the other sub-bands, each sub-band is used to compute its own texture mask. The perceptual weight describing sensitivity to changes in textured areas is given as

$$TM(\lambda, \theta, i, j) = \sum\nolimits_{i=i-k}^{i=i+k} \sum\nolimits_{j=j-k-1}^{j=j+k} |x(\lambda, \theta, i, j-1) - x(\lambda, \theta, i, j)| \quad (5)$$

where $x(\lambda, LL, i, j)$ is the value of the LWT coefficient located in (i, j), belonging to the sub-band θ at level λ. k is the block size, which has been set to 5 in this work.

2.4 Edge Detection

Although the HVS is less sensitive to changes and noise in highly textured regions, it is more sensitive near edges, which represent the boundaries of these textures. In other words, modifications near edges are more likely to be detected by human eye than other areas [5]. Thus, it is important to extract edges from textured areas in order to build a more accurate perceptual model.

Edge detection is a process of locating boundaries of objects or textures present in an image. In the literature, many edge detection techniques have been proposed and used, ranging from simple detectors to more complex ones [15–17]. However, most of these detectors, including simpler ones such as Robert, Sobel and Prewitt operators, require convolution process that involves the operation of scanning all pixels of the image, hence, resulting in a high computational complexity. To cope with that, Taha *et al.* [9] have proposed to eliminate the convolution task by applying four 5×5 extended Sobel kernels once to each 5×5 block of the LWT approximation sub-band instead of moving kernels over the entire sub-band. The edge value of that block corresponds to the largest value among the absolute values obtained from the application of the four kernels.

In this work, we propose to adopt the same approach as in [9], with three major modifications. Firstly, four 3×3 extended Sobel kernels, shown in Fig. 3, are used instead of the four 5×5 Sobel kernels used in [9]. This modification is justified by the fact that smaller edge detection kernels are more suitable for fingerprint images, which are in fact an alternation of narrow black and white patterns (i.e. ridges and valleys). Secondly, an edge value is computed for each coefficient of the block rather than computing one value for the entire block. Thirdly, each sub-band is used in place of the approximation band to compute its corresponding edge detection mask.

1	2	1
0	0	0
-1	-2	-1

$kernel_1$

1	0	-1
2	0	-2
1	0	-1

$kernel_2$

0	-1	-2
1	0	-1
2	1	0

$kernel_3$

2	1	0
1	0	-1
0	-1	-2

$kernel_4$

Fig. 3. Sobel kernels.

Mathematically, the edge mask term *Em* can be formulated as

$$Em(\lambda, \theta, i, j) = \max_{k=1\cdots4}\left(\left|\sum_{m=-1}^{+1}\sum_{n=-1}^{+1}\left(x(\lambda, \theta, i+m, J+n)kernel_k(m, n)\right)\right|\right) \quad (6)$$

where $kernel_k(m, n)$ is the value of Sobel kernel k (see Fig. 2), at position (m, n).

3 Experimental Results

Experiments have been conducted to evaluate the performance of the proposed perceptual mask on fingerprint images. In all experiments, real fingerprint images of size 480×640 have been used from the Fingerprint Verification Competition (FVC 2004, DB 1_a) database [18]. In this paper, only results of the four sample fingerprint images shown in Fig. 4 are reported since results obtained on other test images are very similar. Note that these images have been chosen for their different visual quality. The performance of the proposed perceptual model has been compared with three models: (i) the block-wise model in the LWT domain proposed by Taha *et al.* [9], (ii) the pixel-wise model in the DWT proposed by Zebbiche *et al.* [3] for fingerprint images, and (iii) the reference pixel-wise DWT model proposed by Barni *et al.* [5].

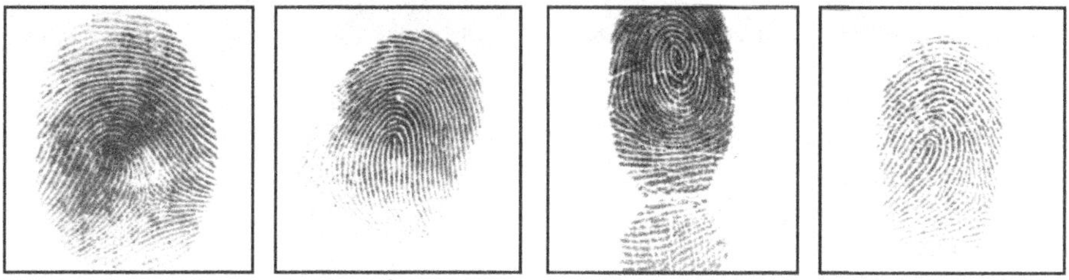

a *b* *c* *d*

Fig. 4. Test images with different visual quality from FVC 2004, DB 1_a: *a* Image 105_2, *b* Image 106_5, *c* Image 109_1 and *d* Image 103_6

For the sake of illustration, the grey adjustment display of the proposed perceptual model for 3-level LWT decomposition of Image 105_2 is shown in Fig. 5. Furthermore, the resulting spatial profiles of the perceptual models are displayed in Fig. 6.

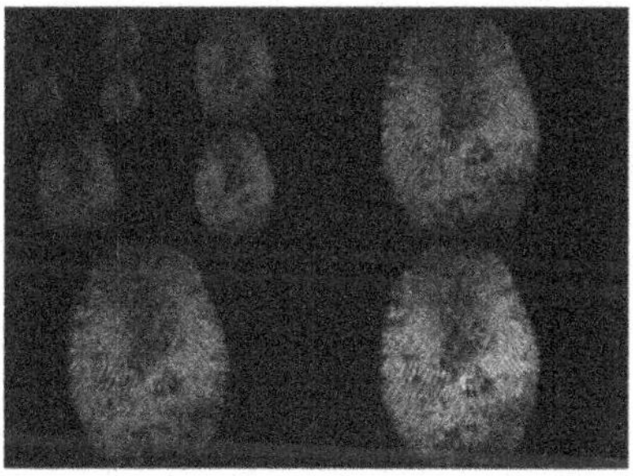

Fig. 5. Grey adjustment display of the proposed perceptual model.

The performance of perceptual models can be assessed by their effectiveness in noise shaping [19]. In most cases, a more efficient model should allow the addition of higher noise energy without causing perceptual alterations. In other words, it should yield a better perceptual quality when the same noise is injected [14].

In our experiments, a random noise is injected into the test images by adding it to the three high frequency sub-bands, by using the equation

$$nx(\lambda, \theta, i, j) = x(\lambda, \theta, i, j) + n(i, j)\ r\ vm(\lambda, \theta, i, j) \tag{7}$$

where $x(\lambda, \theta, i, j)$ is the original sub-band, $nx(\lambda, \theta, i, j)$ is the sub-band after noise injection, n is a random noise with values $\{-1, +1\}$, and r is a factor that controls the noise energy. The Structural SIMilarity (SSIM) index is used as an image quality metric while the noise energy is measured by the Mean Square Error (MSE).

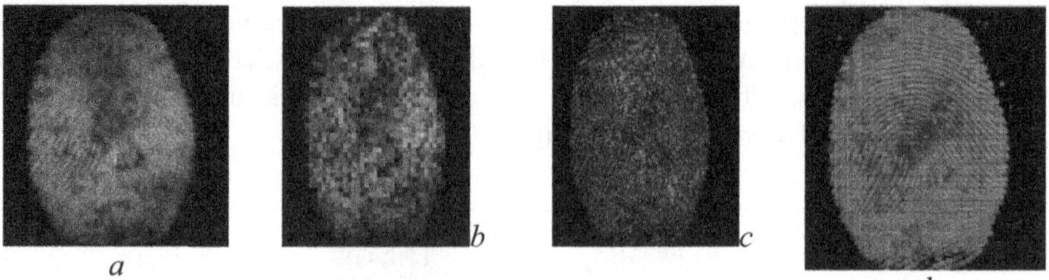

Fig. 6. Resulting spatial profile of: *a* Proposed model, *b* Taha *et al.* model [9], *c* Zebbiche *et al.* model [3] and *d* Barni *et al.* model [5].

In the first experiment, for all test images and for all perceptual models, the value of *r* has been adjusted to obtain the same SSIM value of 0.9750 and then the energy of the injected noise is compared. The obtained results are listed in Table 2.

Table 2. MSE values obtained after noise adding. SSIM = 0.9750

Image	Proposed model	Taha *et al.* model [9]	Zebbiche *et al.* model [3]	Barni *et al.* model [5]
Image 105_2	**71.482**	67.984	4.951	57.114
Image 106_5	**145.898**	144.515	4.758	73.337
Image 109_1	**90.501**	83.976	4.993	80.285
Image 103_6	**103.083**	81.462	3.201	78.54

As it can be noticed in Table 2, for all test images, the proposed model outperforms the other models, by allowing the injection of higher noise energy without causing perceptual alterations. Overall, both proposed model and the one of Taha *et al.* [9] perform well compared to the other models. These results were expected since the models proposed in [3] and [5] are originally designed for DWT and, thus, are not directly applicable in the LWT domain.

In the second experiment, the same noise energy is injected into test images with each perceptual model, by adjusting the parameter *r*, and then the SSIM values are computed and compared. The noise energy is fixed to obtain an MSE value of 80. The obtained results are reported in Table 3.

Similarly, as it can be noticed in Table 3, for all test images, the proposed model offers higher SSIM values when the same noise energy is injected, followed by Taha's model which exhibits close performance.

Moreover, in Fig. 7, are displayed the noise mask of each model, which allow to appreciate the noise distribution, and thus, the mask behavior in the spatial domain. As it can be seen, for all models, except for the one of Taha *et al.*, no noise is injected into the background region. This is due to the fact that Taha's model, unlike the other

Table 3. SSIM values for MSE = 80

Image	Proposed model	Taha *et al.* model [9]	Zebbiche *et al.* model [3]	Barni *et al.* model [5]
Image 105_2	**0.9723**	0.9714	0.7383	0.9660
Image 106_5	**0.9853**	0.9841	0.6838	0.9735
Image 109_1	**0.9775**	0.9756	0.6956	0.9756
Image 103_6	**0.9801**	0.9779	0.6650	0.9705

models which are multiplicative, is an additive model that allows noise adding in non-textured areas where modifications are perceptible. Furthermore, it can be seen that in the proposed model, less noise is injected into the edges and flat regions, resulting in a better visual quality image. One interesting application that highlights these capabilities is image watermarking, where the proposed model can be used to strengthen the presence of the watermark while preserving its imperceptibility. Especially, its ability to avoid adding watermark into the background area, which, in the case of fingerprint images, represents the region of non interest (RONI) that can be removed with any information hidden in it after applying image processing operations such as segmentation or cropping. Note that this advantage is not possible with Taha's model.

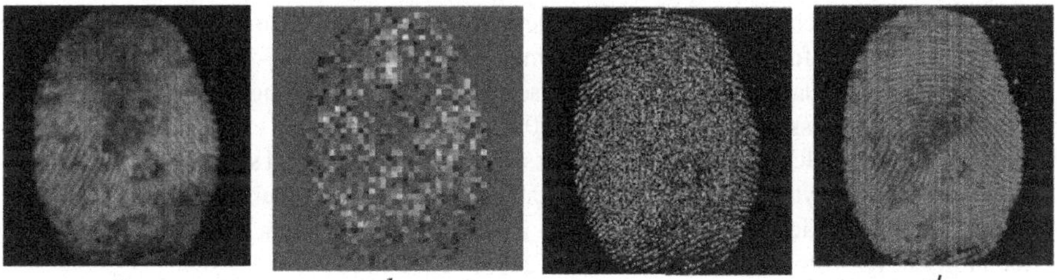

a *b* *c* *d*

Fig. 7. Resulting noise mask (with MSE = 80) of: *a* Proposed model, *b* Taha *et al.* model [9], *c* Zebbiche *et al.* model [3] and *d* Barni *et al.* model [5].

4 Conclusion

In this paper, an efficient pixel-wise LWT-based perceptual masking model for fingerprint images has been proposed. This model exploits four HVS characteristics, namely: the frequency band sensitivity, the local brightness, the texture masking and the edge masking. Experimental results show that the proposed model allows more tolerance to noise injection when compared to other HVS-based masking models. Furthermore, it is the most adapted model to be used for fingerprint images, since it avoids using the background region, besides presenting less distortion around edges and flat areas where modifications are most likely to be perceptible, hence, resulting in a better visual quality image.

References

1. Beghdadi, A., Larabi, M.-C., Bouzerdoum, A., Iftekharuddin, K.: A survey of perceptual image processing methods. Signal Process. Image Commun. **28**(8), 811–831 (2013)
2. Nadenau, M.J., Winkler, S., Alleysson, D., Kunt, M.: Human vision models for perceptually optimized image processing-a review. Proc. IEEE **32** (2000)
3. Zebbiche, K., Khelifi, F.: Efficient wavelet-based perceptual watermark masking for robust fingerprint image watermarking. IET Image Process. **8**(1), 23–32 (2014)
4. Wu, J., Qi, F., Shi, G.: An improved model of pixel adaptive just noticeable difference estimation. In: Proceedings of IEEE International Conference on Acoustics, Speech and Signal Processing (ICASSP), pp. 2454–2457 (2010)
5. Barni, M., Bartolini, F., Piva, A.: Improved wavelet-based watermarking through pixel-wise masking. IEEE Trans. Image Process. **10**(5), 783–791 (2001)
6. Liu, J., She, K.: Robust image watermarking using dual tree complex wavelet transform based on human visual system. In: International Conference on Image Analysis and Signal Processing (IASP), pp. 675–679 (2010)
7. Zebbiche, K., Khelifi, F., Loukhaoukha, K.: Robust additive watermarking in the DTCWT domain based on perceptual masking. Multimedia Tools Appl. **77**(16), 21281–21304 (2018)
8. Taha, T.B., Ehkan, P., Ngadiran, R.: A new perceptual mapping model using lifting wavelet transform. In: MATEC Web of Conferences, vol. 140 (2017)
9. Taha, T.B., Ngadiran, R., Ehkan, P.: Adaptive image watermarking algorithm based on an efficient perceptual mapping model. IEEE Access **6**, 66254–66267 (2018)
10. Cox, I.J., Miller, M.L., Bloom, J.A., Fridrich, J., Kalker, T.: Digital Watermarking and Steganography, 2nd edn. Morgan Kaufmann Publishers, San Francisco (2008)
11. Nadenau, M.J., Reichel, J., Kunt, M.: Wavelet-based color image compression: exploiting the contrast sensitivity function. IEEE Trans. Image Process. **12**(1), 58–70 (2003)
12. Cui, L., Li, W.: Adaptive multiwavelet-based watermarking through JPW masking. IEEE Trans. Image Process. **20**(4), 1047–1060 (2011)
13. Beegan, A.P.: Wavelet-based image compression using human visual system models. Master's thesis, Virginia Polytechnic Institute and State University, Blacksburg, VA (2001)
14. Wan, W., Wu, J., Xie, X., Shi, G.: A novel just noticeable difference model via orientation regularity in DCT domain. IEEE Access **5**, 22953–22964 (2017)
15. Bhardwaj, S., Mittal, A.: A survey on various edge detector techniques. Procedia Technol. **4**, 220–226 (2012)
16. Maini, R., Aggarwal, H.: Study and comparison of various image edge detection techniques. Int. J. Image Process. **3**(1), 1–11 (2009)
17. Ruksar, F., Mirajkar, F.D., Hashmathunissa, B., Ashwini, S.M.: An extensive survey on edge detection techniques. Int. J. Eng. Techn. **3**(6), 63–69 (2017)
18. Fingerprint Verification Competition. http://bias.csr.unibo.it/fvc2004/download.asp
19. Uzair, M., Dony, R.D.: Estimating just-noticeable distortion for images/videos in pixel domain. IET Image Process. **11**(8), 559–567 (2017). https://doi.org/10.1049/iet-ipr.2016.1120

Dynamic Web Service Selection Based on Score Voting

Zeyneb Yasmina Remaci(✉), Fethallah Hadjila, Abdelhak Echialli, and Mohammed Merzoug

Computer Science Department, LRIT Laboratory, Faculty of Sciences, Abou Bakr Belkaid University of Tlemcen, B.P 119, 13000 Tlemcen, Algeria
{Zeynebyasmina.remaci,fethallah.hadjila,Mohamed.merzoug}@univ-tlemcen.dz, echialli@gmail.com

Abstract. The QoS aware service selection is one of the major issues of the service-oriented computing paradigm. Most of the existing works handle the QoS attributes as static variables. In real scenarios, the QoS criteria are dynamic and change according to time and space. Consequently, the solutions provided by the existing approaches are error prone and less appropriate to the real context. In this paper, we tackle the above mentioned issue by proposing a two-stage approach. The first stage reduces the space by adopting the score voting heuristic; the second one finds the TopK service compositions that ensure the global constraints. The experiments show that the proposed heuristic is very promising.

Keywords: Web service · Web service selection · Uncertain quality of service · Backtracking search · Score voting

1 Introduction

The purpose of selecting an appropriate web service based on the quality of service (QoS) attributes is widely handled in recent years [1,4,7]. The QoS attributes are primordial for comparing the services that offer the same functionality. Additionally, they also play a crucial role in selecting the services that compose the users' workflows. According to the literature review, we observe that the majority of the existing methods do not take into account the impact of the dynamic environment on the QoS (such as network load, rush hours). The variance of QoS overtime does not only influence service selection but also affects the service composition. Therefore, the challenge is to propose a selection/aggregation model that handles these alterations as well as the global optimization aspects (i.e., global constraints, aggregated QoS).

1.1 Motivation

In the example cited in Table 1, we assume a user's need that consists of invoking two types of services: an IaaS application that provides virtual machines

M. R. Senouci et al. (Eds.): CSA 2020, LNNS 199, pp. 185–195, 2021.
https://doi.org/10.1007/978-3-030-69418-0_17

with customized configurations (such as RAM/Hard disk) and an image processing application (in the form of software as a service, i.e., SaaS). Each service is characterized by the reputation, and the best practices (denoted R, BP in Table 1). The criteria R and BP are positive and aggregated with additive functions (should be maximized). The selected combination must respect the global constraints,(i.e., the user's requirements), as a result, the global reputation (the QoS sum of the service components) is greater than or equal to 1.2 and the global (aggregated) BP is greater than or equal to 0.7. Additionally, Table 1 shows the QoS variation (see the instances lines t_1, t_2, t_3, t_4) of each service. By considering the elements of the IaaS class (i.e., the services S_{11} and S_{12} that have the same functionality), we notice that the utilization of the mean QoS as a comparison mechanism can create a confusing result. This situation is explained as follows: mean QoS of S_{11} is (0.38, 0.40), likewise the mean QoS of S_{12} is (0.46, 0.44), and thus S_{12} is better than S_{11}, but when we use the median QoS as comparison mechanism, we note that the median QoS of S_{11} is (0.5, 0.4), and the median QoS of S_{12} is (0.3, 0.3). This means that S_{11} is better than S_{12}. This problem is delicate because the initial ordering could generate a nonfeasible composition. For example, if we consider the median QoS of each component service as the representative value, and by employing the aggregation's methods [8] to compute the QoS values of a composite service, we notice that composition $c = <S_{12}, S_{21}>$ is not feasible: MedianR (S_{12}) + MedianR (S_{21}) $= 0.3 + 0.7 < 1.2$ (The first global constraint is not respected). On the other hand, the composition $c' = < S_{11}, S_{21} >$ is feasible since: MedianR (S_{11}) + MedianR (S_{21}) $= 0.5 + 0.7 \geqslant 1.2$. Median BP ($S_{11}$) + MedianBP ($S_{21}$) $= 0.4 + 0.3 \geqslant 0.7$. Actually, three aspects that should be handled in the service selection:

Table 1. Normalized QoS realizations.

	IaaS application				Image processing (SaaS)			
	S_{11}		S_{12}		S_{21}		S_{22}	
	R	BP	R	BP	R	BP	R	BP
t1	0.2	0.1	0.1	0.4	0.5	0.2	0.3	0.5
t2	0.5	0.3	0.9	0.8	0.8	0.3	0.2	0.5
t3	0.1	0.6	0.3	0.1	0.7	0.3	0.4	0.7
t4	0.5	0.6	0.7	0.4	0.5	0.6	0.6	0.4
t5	0.6	0.4	0.3	0.9	0.9	0.5	0.5	0.1

- How to represent the user's needs (global constraints, the objective function, the number of services classes, the control flow).
- How to manage the nondeterministic QoS.
- How to reduce the space search. In fact this problem is NP-Hard [1,2].

1.2 Contributions

To achieve the above requirements, we propose in this paper a general framework for web services selection. This framework is based on a heuristic termed score voting. This principle selects the best services of each class, and it reduces the search space of the compositions (which is exponential in general). Our overall contribution is summarized as follows:

1. Firstly, we rank the services of each class, according to the score voting [3]. Then, we retain the Top-K services having the highest scores concerning the majority score.
2. To rank the service compositions, we leverage an objective function based on the median QoS of the service composition (see Eq. 1), as well as the ratio of satisfied global constraints.
3. We explore the search space constituted of the first K services of each class, by implementing a backtracking search (inspired from the constraint satisfaction problems). Then, we return the Top-K compositions.

The remainder of the paper is organized as follows: Sect. 2 shows an overview of the related works. The third section presents the problem formalization, in the fourth Section we describe the proposed framework as well as the selection algorithms, and finally, we present in the last section our conclusions and perspectives.

2 State of the Art

A lot of approaches were proposed in the literature [1,2,4,10] to deal with service selection and/or composition. In general, these approaches neglect the fluctuation of the environment and manipulate the QoS as deterministic variables. In [14], the authors leverage the biogeography-based optimization (BBO)algorithm to solve the web service composition problem. In [18], the authors propose a new approach to handle the service composition problem by considering user preference. This approach combines the triangular fuzzy with a genetic algorithm. A harmony search is used to handle the web service compositions in [4]. The proposed framework takes into account both the functional facets and nonfunctional facets. The system proposed in [10] uses the harmony search meta-heuristic to select the near optimal compositions. [6], defines the new concept of Averaged-Fuzzy-Dominated-score (denoted AFDetS()) to rank services based on QoS. A new approach called SEFAP is presented in [11] to rank the skyline services. This approach exploits the Multi-Criteria Decision Making methods (fuzzy AHP and PROMETHEE) to perform the ranking. In [12], the authors present their proposition termed ROMETHEE plus. The latter method leverages the classification to reduce the search space. In [15], the authors leverage the concept of skylines as well as the BWM (Best Worst Method) method to sort the existing services. Many current works [5,7,17] handle the service selection depending on nondeterministic QoS. In [17], the authors introduce a new concept termed "dominance ability" (based on the dominance probability). This concept allows the computation of the top k dominant compositions with handling the global constraints. In [7], the authors suppose that each task of the abstract workflow is governed by a Gaussian probability distribution. The initiative of the method consists of decomposing the global constraints into local constraints. The work proposed in [9], addresses the web service selection with the presence of outliers. The authors propose suitable heuristics for decomposing the global constraints into local constraints. In contrast to [7], the actual approach is less sensitive to noise. In [16], the authors employ the entropy and variance concepts to filter out the high QoS uncertainty services.

3 Problem Statements

In this section, we will formalize the problem of Top-k dominant compositions under uncertain QoS. In what follows, we will assume a set of hypotheses and notations in order to simplify the problem specification:

- All the QoS attributes are positive (i.e., the positive attributes need to be maximized).
- The composition model is sequential.
- The QoS criteria of composition are aggregated according to the sum function (such as reputation), if there are multiplicative criteria, then we replace them with their log value and we treat them as additive criteria.
- n: is the number of abstract classes.
- $Cl_1, Cl_2, ..., Cl_n$: are the set of abstract classes, each class contains atomic web services with the same functionality and different QoS.
- m: is the number of services per class.
- r: is the number of QoS attributes.
- l: is the number of service instances (i.e., the number of QoS realizations).
- QoS_{piju} : is the value of the pth QoS attribute related to the uth instance of the service $S_i \in Cl_j$.
- $b_1, b_2, ..., b_r$: are the user's global constraints (i.e., the limits which need to be met by the QoS of the composition).
- $w_1, ...w_r$: are the weight of the QoS, the default value of each w_p is $1/r$.
- k: the size of the returned list of compositions.
- The overall utility of a service composition c is computed as follows:

$$U^{'}(c) = \sum_{p=1}^{r} w_p * ((MedianQ^{'}_p(c) - Qmin^{'}(p))/(Qmax^{'}(p) - Qmin^{'}(p)) \quad (1)$$

- x_1 (resp $x_2, .., x_n$): represents the id of the selected service related to Cl_1 (resp $Cl_2, .., Cl_n$).

$$Qmin^{'}(p) = \sum_{j=1}^{n} Qmin(j,p)\ (the\ first\ normalization\ constant), were \quad (2)$$

$$Qmin(j,p) = MIN_{S_i \in Cl_j}, u \in 1, ..l(QoS_{piju}) \quad (3)$$

$$Qmax'(p) = \sum_{j=1}^{n} Qmax(j,p)\ (the\ second\ normalization\ constant), were \quad (4)$$

$$Qmax(j,p) = MAX_{S_i \in Cl_j, u \in 1,..l}(QoS_{piju}) \quad (5)$$

$$MedianQ'_p(c) = \sum_{j=1}^{n} Median_{u \in 1,..,l}(QoS_{piju}) \quad (6)$$

In this work, we adopt the median QoS Value to evaluate the performance of each candidate service S_{xi}, and the performance of the composition c (see Eq. 1). Because the median values are less sensitive to the variations and the outliers of the data set. In addition, we use Eq. 7 to compare the compositions according to their degree of satisfying the global constraints. Broadly speaking, we say that a composition c is better than another composition c' if the score of c is higher than the score of c' with respect to Eq. 7. If c ties with c', then we rank them according to Eq. 1, the higher the score of U' the better the rank. Formally, c is ranked above c' iff:

$$\begin{cases} 1/r * \sum_{p=1}^{r} CC(p,c) >= 1/r \sum_{p=1}^{r} CC(p,c') \\ 1/r * \sum_{p=1}^{r} CC(p,c) = 1/r \sum_{p=1}^{r} CC(p,c') \\ and\, U'(c) \geqslant U'(c') \end{cases} \quad (7)$$

Where CC (the Constraint Conformance) is defined as:

$$CC(p,c) = \begin{cases} 1, if MedianQ'_p(c) \geqslant b_p \\ 0\ otherwise \end{cases} \quad (8)$$

The function CC returns 1, if the pth global constraint is preserved with respect to the median QoS of the composition c, otherwise it returns 0. In summary, our main objective is to select the Top-K compositions, $C_1, \ldots C_k$ which:

- Maximize the chance of satisfying the global constraints: $1/r * \sum_{p=1}^{r} CC(p, c_y)$, where $y \in 1,..,k$ and $p \in 1,..,r$.
- Maximize the function $U'(.)$.

4 Proposed Approach

In this section, we present our selection framework (see Fig. 1). It is composed of three parts: **The class management module:** its goal is to assign each service to a predefined abstract class, this module also updates the classes by removing outdated services. **The QoS management and integration module:** it stores the dynamic QoS of each service. **The selection module:** The selection module leverages two algorithms (Algorithm 2: Ranking based on score voting and Algorithm 3: backtracking search). The Algorithm 2 aims to downsize the search, while the backtracking search is performed in order to provide Top K dominants service compositions.

4.1 Ranking Based on Score Voting

The first algorithm sorts the services of each abstract task using a score voting procedure. This procedure is a modified version of the model proposed in [3]. The basic idea of our solution is based on the concept of majority-score (or median score). Simply speaking, we compare the QoS of each pair of services $S_i, S_{i'}$ with respect to the score value, if the number of instances is odd, then the returned value coincide with middle of the list, but if the number is even then the returned score is equal to the lower of the two middle values. The tie-breaking is applied iteratively when two or more service's majority-scores are equal: the scores are dropped from each service, then we compute the new score, the service with higher majority score is the winner. Thereafter, we increment the ranking score of the wining service. The more the score is high the better the rank. The pseudo code of Algorithm 1 (also abbreviated as MS-SR) is given below. The elucidation, of Algorithm 2 is given as follows:
In line 2, we initialize the ranked Class $RankedCl_i$ (with an empty arrangement).
In lines 5–7, we initialize the sorting score of each service of the current class i.
In lines 8 up to 11, we compare each pair of services $(S_i, S_{i'})$ of the same class i, through the use of the MajorityScoreComparison function.
In line 12 we update the score of S_i.
In line 16 we rank the elements of $RankedCl_i$ according to the scores updated in 12. We return the ranked classes in line 18. Since the cost of Algorithm 1 is O(l^2), we infer that the cost of Algorithm 2 is $O(nm^2.r.l^2 + n.mlogm)$.

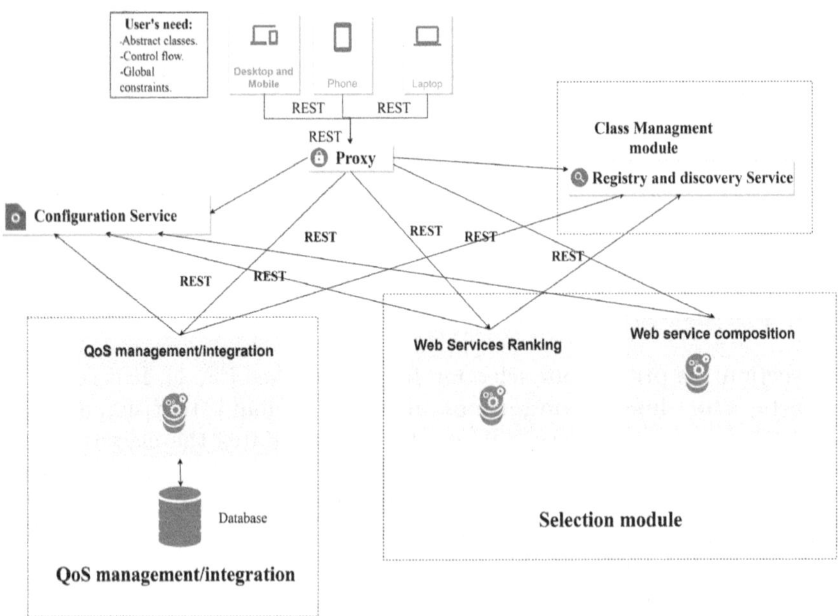

Fig. 1. Service selection framework.

4.2 Backtracking Search

The aim of Backtracking Search is to provide the possible compositions. Then, it selects Top-K optimal compositions with the best ratio of satisfied global constrained (see Eq. 7) and objective function based on the median QoS (see Eq. 1). The pseudo-code of Algorithm 3 is given below. The elucidation of Algorithm 3 is given as follows: In line 2, we investigate all the possible compositions. In line 3, we get the current composition c. In line 4, we compute the ratio of satisfied global constrained. In line 5, we check the fact that the ratio of the preserved global constraints is above the threshold. In line 7, we compare c with the existing TopKComposition elements through the use of Eq. 7 and 1. In line 8, we update the result TopKCompositions, if c is better than an existing composition. We return the final answers in line 3. Lastly, we notice that the computational cost of Algorithm 3 is O($k^n(k.r.n + n + klogk)$).

Algorithm 1. MajorityScoreComparison

Input: $H =< h_1, .., h_l >, G =< g_1, .., g_l >$.
Output: $Score \in \{0, 0.5, 1\}$.
Begin

i=L;
while i >0 **do**
    if (odd(i)) **then**
        $middle = i \div 2 + 1$;
        if ($i == 1$ **and** $h_i == g_i$) **then return** 0.5;
    else if (even (i)) **then**
        $middle = i \div 2$;
    if ($h_{middle} \geq g_{middle}$ **and** $H.type == positive$) **then return** 1;
    else if ($h_{middle} \neq g_{middle}$ **and** $H.type == negative$) **then return** 1;
    else if ($h_{middle} == g_{middle}$) **then return** 0;
    else if ($h_{middle}! = g_{middle}$) **then**
        remove h_{middle} and g_{middle} from both H and G (respectively); $i - -$;
end while
end

---

5 Experimental Studies

In the sequel, we evaluate the performance for our framework in terms of execution time and optimality. We conduct a set of experiments, with several configurations of parameters (see Table 2). The experiments are conducted on a machine having an Intel I3 core 2.53 GHz processor, 4 GB RAM, and running Windows 7. Figure 1 up to 6 are related to Algorithm 1. However, Table 3 is related to Algorithm 1 and Algorithm 3.

Table 2. Parameters and examined values.

Parameters	Values
Number of tasks (n)	5, 15, 25, 35
Number of services (m)	100 to 1100
Number of QOS criteria (r)	2 to 10
Instances (l)	100 to 400
Size of the result (k)	2, 6

Algorithm 2. MajorityScore-ServiceRanking

```
Input: < CL_1, ..., CL_n >,
Output: RankedCl_1, ..., RankedCl_n
Begin
for i=1 to n do
    RankedCl_i =<>;
end for
for i=1 to n do
    for y=1 to m do
        score(y)=0;
    end for
    for j=1 to m do
        for j' = j + 1 to m do
            for p=1 to r  do
                Score(j) = score(j) + 1/r * MajorityScoreComparison(Qp(S_j), Qp(S_j'));
            end for
        end for
    end for
    RankedCl_i=decreasing-sort (Cl_i)
end for
return < RankedCl_1, ..., RankedCl_n >
end
```

As depicted in Fig. 2, we observe that the execution time of the score voting ranking is almost linear with respect to the number of task n. Likewise, the previous experiment, we notice in Fig. 3, that the execution time of the score voting ranking is also linear with respect to the number of criteria r. The Fig. 4 demonstrates the relationship between the execution time and the number of instances. It is clearly shown that the increase of time is polynomial with respect to the variable l. Figure 5 shows the relationship between the execution time and the number of services m. Despite the inherent noise depicted around $m = 300$, the overall behavior of the curve is almost polynomial with respect to the number of services.

According to Table 3, we observe that the respected global constraint ratio of (MS-SR+Algo3) is higher than the ratio relative to (DSR+Algo3). This observation is valid for both $k = 2$ and $k = 6$. These findings confirm the effectiveness

Algorithm 3. BacktrackingSearch

Input: $RankedCl_1, ..., RankedCl_n$,$b_1, b_2, ..b_r$: global constraints, k: size of the results set; t: the minimum

Output: TopKCompositions **Begin**

```
TopKCompositions=<>// the result is initially empty.
for i=1 to k^n do
    c = GetNextComposition(RankedCl_1, ..., RankedCl_n);
    degree = 1/r * Σ_{p=1}^{r} CC(p, c)
    if (degree ⩾ t) then
        if better (c, TopKCompositions) then
            Update (c, TopKCompositions)
end for
return (TopKCompositions)
end
```

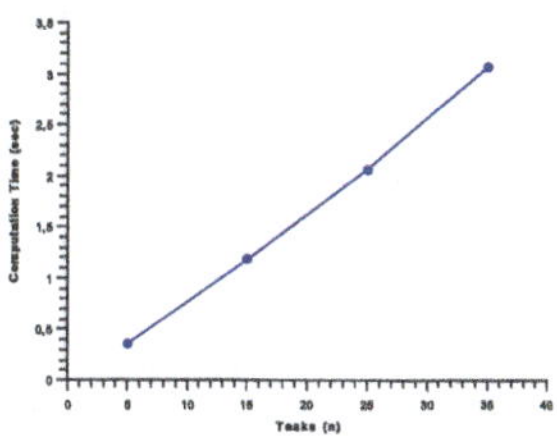

Fig. 2. CPU time versus n (r = 3, m = 200, l = 100).

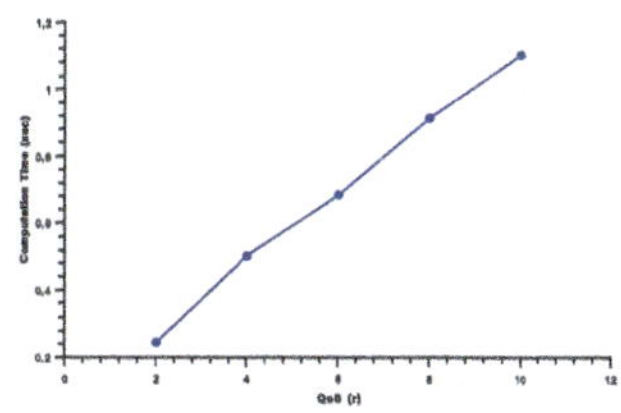

Fig. 3. CPU time versus r (n = 5, m = 200, l = 100).

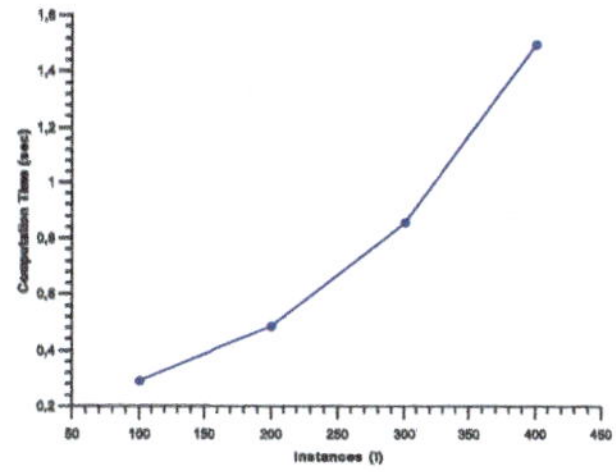

Fig. 4. CPU time versus l (n = 5, r = 3, m = 200).

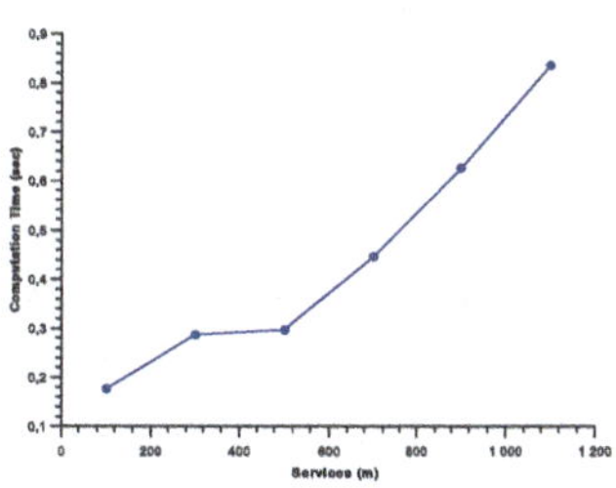

Fig. 5. CPU time versus m (n = 5, r = 3, l = 100).

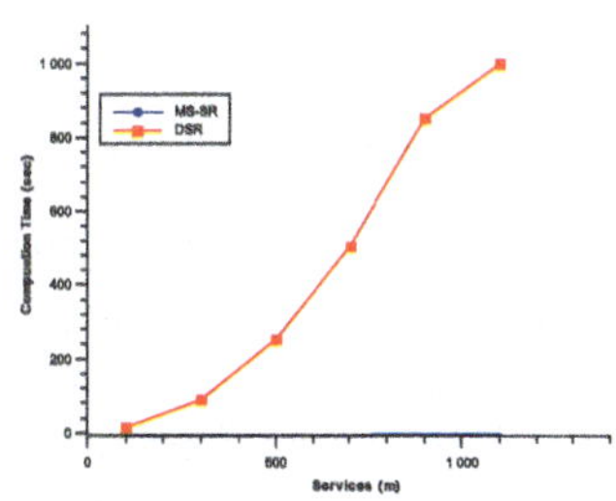

Fig. 6. CPU time comparison: MS-SR Vs DSR.

as well as the efficiency of the score voting heuristic in the area of uncertain service selection. Despite the fact that the Dominance Service Ranking (DSR) [13], and the ranking based on score voting (MS-SR), have the same theoretical complexity, we observe that the gap between their respective curves is increasing with the augmentation of m (see Fig. 6). This observation is justified by the fact that the average execution time of Algorithm 1 is smaller than

Table 3. Performance comparison between MS-SR and DSR.

Parameters	Solutions	Algorithms	Utility $U'(.)$	Respected GC (%)
n = 10, r = 3, m = 200, l = 100	Top2 solutions	MS-SR+ Algo3	0.4310	100
			0.430.3	100
		DSR Algo3	0.5742	33.33
			0.5696	33.33
	Top6 solutions	MS-SR+ Algo3	0.4512	100
			0.4505	100
			0.4502	100
			0.4489	100
			0.4489	100
			0.4484	100
		DSR+ Algo3	0.5015	66
			0.5010	66
			0.4994	66
			0.4984	66
			0.4957	66
			0.4952	66

the average execution time of the dominance probability used in algorithm DSR [13]. In practice, the invocation of the tie breaking rule is not frequent. Thus the average execution time of the function MajorityScoreComparison is largely smaller than the theoretical execution time.

6 Conclusion

In this work, we have addressed the service selection with non deterministic QoS. The proposed solution proceeds in two steps: first, it selects the promising web services of each class according to the score voting principle. Second, it performs a backtracking search to select the TOP K optimal compositions. We have confirmed the effectiveness/efficiency of our approach with a set of experiments. For future work, we will leverage additional ranking heuristics (which can be based on probability theory, possibility theory, fuzzy set theory).

References

1. Alrifai, M., Risse, T., Nejdl, W.: A hybrid approach for efficient web service composition with end-to-end QoS constraints. TWEB **6**(2), 7:1–7:31 (2012)
2. Alrifai, M., Skoutas, D., Risse, T.: Selecting skyline services for QoS-based web service composition. In: Proceedings of the 19th International Conference on World Wide Web, Raleigh, North Carolina, USA, pp. 11–20 (Apr 2010)

3. Balinski, M., Laraki, R.: A theory of measuring, electing and ranking. Proc. Nat. Acad. Sci. **104**(21), 8720–8725 (2007)
4. Bekkouche, A., Benslimane, S.M., Huchard, M., Tibermacine, C., Hadjila, F., Merzoug, M.: QoS-aware optimal and automated semantic web service composition with user's constraints. SOCA **11**(2), 183–201 (2017)
5. Benouaret, K., Benslimane, D., HadjAli, A.: Selecting skyline web services from uncertain QoS. In: 2012 IEEE Ninth International Conference on Services Computing, Honolulu, HI, USA, pp. 523–530 (June 2012)
6. Halfaoui, A., Hadjila, F., Didi, F.: QoS-aware web services selection based on fuzzy dominance. In: Computer Science and Its Applications - 5th IFIP TC 5 International Conference, CIIA, Saida, Proceedings, pp. 291–300 (May 2015)
7. Hwang, S.-Y., Hsu, C.-C., Lee, C.-H.: Service selection for web services with probabilistic QoS. IEEE Trans. Serv. Comput. **8**(3), 467–480 (2015)
8. Rojec-Goldmann, G., Jaeger, M.C., Muhl, G.: QoS aggregation for web service composition using workflow patterns. In: Eighth IEEE International Enterprise Distributed Object Computing Conference (EDOC'04), pp. 149–159 (2004)
9. Kim, M., Oh, B., Jung, J., Lee, K.: Outlier-robust web service selection based on a probabilistic QoS model. IJWGS **12**(2), 162–181 (2016)
10. Merzoug, M., Chikh, M.A., Hadjila, F.: QoS-aware web service selection based on harmony search. In: 2014 4th International Symposium ISKO-Maghreb: Concepts and Tools for knowledge Management (ISKO-Maghreb), pp. 1–6 (2014)
11. Ouadah, A., Hadjali, A., Nader, F., Benouaret, K.: SEFAP: an efficient approach for ranking skyline web services. J. Ambient Intell. Hum. Comput. **10**(2), 709–725 (2018)
12. Purohit, L., Kumar, S.: A classification based web service selection approach. IEEE Trans. Serv. Comput., p. 1 (2018)
13. Remaci, Z.Y., Fethallah, H., Didi, F.: Selecting web service compositions under uncertain QoS. In: 6th IFIP TC 5 International Conference, Computational Intelligence and Its Applications, pp. 622–634. Springer, Cham (May 2018)
14. Sangaiah, A.K., Bian, G.-B., Bozorgi, S.M., Suraki, M.Y., Hosseinabadi, A.A.R., Shareh, M.B.: A novel quality-of-service-aware web services composition using biogeography-based optimization algorithm. Soft Comput. **24**, 1–13 (2019)
15. Serrai, W., Abdelkrim, A., Mokdad, L., Hammal, Y.: An efficient approach for web service selection. In: IEEE Symposium on Computers and Communication, ISCC, Messina, Italy, pp. 167–172 (June 2016)
16. Sun, L., Wang, S., Li, J., Sun, Q., Yang, F.: QoS uncertainty filtering for fast and reliable web service selection. In: 2014 IEEE International Conference on Web Services, ICWS, Anchorage, AK, USA, pp. 550–557 (June 2014)
17. Wen, S., Tang, C., Li, Q., Chiu, D.K.W., Liu, A., Han, X.: Probabilistic top-k dominating services composition with uncertain QoS. SOCA **8**(1), 91–103 (2014)
18. Xu, J., Guo, L., Zhang, R., Hu, H., Wang, F., Pei, Z.: QoS-aware service composition using fuzzy set theory and genetic algorithm. Wireless Pers. Commun. **102**(2), 1009–1028 (2018)

A User-Side POIs Mobile Recommender System

Mohamed Boubenia(✉), Fayçal M'hamed Bouyakoub, and Abdelkader Belkhir

LSI Laboratory, USTHB University, Algiers, Algeria
mboubenia@live.fr, fbouyakoub@usthb.dz, belkhir@lsi-usthb.dz

Abstract. Recommending pertinent Place Of Interests (POIs) is a desirable feature for mobile users, and which is generally served by for-profit proprietary platforms, such as Yelp, TripAdvisor, etc. However, the siloed design of these platforms raises today several issues about privacy, user data portability, and algorithm transparency. To address these issues, we propose a decoupled recommender system (RS) architecture. The idea consists of externalizing the sensitive features, such as the users' preferences and the underlying RS algorithm, from the service's application. Hence, the proposed RS operates as an interchangeable third-party service. We conducted several experiments to evaluate the impact of the decentralization, and we were able to improve the performances by relying on Linked Open Data (LOD) and on an appropriate similarity measure.

Keywords: Mobile recommender system · POIs · Decentralization · Linked Open Data · Privacy

1 Introduction

Nowadays, mobile users tend to access registers to search for relevant Place Of Interests (POIs) (e.g. hotels, restaurants, parks, etc.). These POIs registers are generally well served by for-profit location-based services (LBSs), such as Yelp[1], Foursquare[2], etc., which provide a high user experience. Through the time, these commercial companies have accumulated a huge amount of valuable data about POIs and user personal data, and started to build an entire business model around these datasets. Consequently, the competitive advantage of these companies lies in their back-end data, which leads unfortunately to much of POIs datasets and user personal data being siloed within these proprietary databases with restricted APIs. The other fact is that existing registers differ in several aspects [8]. Thus, a user would tend to perform search on different registers. However, since LBSs are designed as silos, the user has no digital ownership on his personal data (bookmarks, check-ins, etc.), nor the possibility to transport

[1] https://www.yelp.com.
[2] https://www.foursquare.com.

M. R. Senouci et al. (Eds.): CSA 2020, LNNS 199, pp. 196–207, 2021.
https://doi.org/10.1007/978-3-030-69418-0_18

them on other registers. Consequently, his preferences become scattered across registers. In addition to these issues, privacy is becoming another hot concern in such centralized hubs [11]. However, since the controversies of Cambridge Data Analytica [4], the general public realizes the danger of this situation, where a few platforms are gaining a powerful control of data [1].

To tackle such problems, recent efforts have suggested a decentralized and decoupled design [16], in which the service and data would no more be located on the service provider side, letting the user be free to choose where to store his data and allowing him to transport them. However, beyond these aspects, we believe that problems related to the lack of algorithm transparency could persist. In fact, the user will still depends on service's algorithms, such as profiling and personalization, which are sensitive due to their privacy intrusion [14]; their high impact on the user choices (i.e. anchoring effect) [17]; and neutrality issue (i.e. the system benefits the user or the provider?). Therefore, we propose a user side recommender system of POIs that attempts to guarantee: data-ownership and privacy, portability and interoperability, and algorithm transparency.

The rest of the paper is organized as follows: Sect. 2 recalls the principle of RSs. Section 3 exhibits some relevant related works. Section 4 presents our proposed approach. Section 5 details the different conducted experiments, followed by results. Finally, Sect. 6 discusses results and concludes the paper.

2 Recommendation Systems

RSs are a technology that guides the user to the resources which might interest him. The Content-Based (CB) approach is one among the most popular recommendation approaches [10], and relies on three components: *Item profile*, *User profile*, and *Similarity measure.*

The Item profile can be constructed by extracting a set of terms from the description of the item. Next, a weighting function, such as the frequency/inverse document frequency (TF-IDF) [15], is generally used to determine the importance of each extracted term, where the weight w_{ij} of a term t_i in the item I_j is defined as $w_{ij} = TF_{ij} * IDF_i$, such as TF_{ij} is the number of times t_i appears in the item I_j; and IDF_i is used to penalize terms that appear in many items, and is computed as follows: $IDF_i = \log \frac{M}{m_i}$, where M is the total number of items; and m_i is the number of items containing t_i. Next, the item profile is defined as a vector $(w_{1j}, w_{2j}, \ldots, w_{kj})$.

The User profile can be constructed by analyzing the content of the items previously liked by the user, and represented as a vector of weights $(w_{1u}, w_{2u}, \ldots, w_{ku})$, where each weight w_{iu} represents the importance of the term t_i to the user u. TF-IDF can also be applied here, such as $w_{iu} = TF_{iu} * IDF_i$, where TF_{iu} is the number of times t_i appears in the set of items liked by u; and $IDF_i = \log \frac{N}{n_i}$ (i.e. N is the total number of users, and n_i is the number of users having the term t_i in their liked items).

Finally, a Similarity measure is used to compare profiles, and the pertinent items are provided as recommendation list. Here, the *cosine* similarity measure [15] is generally used, and is defined as follows:

$$\cos\left(\overrightarrow{w_j}, \overrightarrow{w_u}\right) = \frac{\overrightarrow{w_j} \cdot \overrightarrow{w_u}}{\|\overrightarrow{w_j}\| \|\overrightarrow{w_u}\|} = \frac{\sum_{i=1}^{k} w_{ij} w_{ic}}{\sqrt{\sum_{i=1}^{k} w_{ij}^2} \sqrt{\sum_{i=1}^{k} w_{iu}^2}} \tag{1}$$

3 Related Work

In order to share user profile across siloed platforms, two different strategies can be distinguished in the literature: *Standardization strategy*, consists of finding common vocabularies and representation of user profile [2,7]; *Mediation strategy*, consists of relying on specialized mediator components which translate the data between different models to a generic model [12]. The main obstacle of the standardization strategy is the lack of a unique common standard. Thus, platforms that want to exchange the user models have to agree on the adopted vocabularies, whereas the main challenge of the mediation strategy is to collect user preferences from the different platforms he use and aggregate them, which also can be challenging. The other fact about these general solutions efforts is that they don't guarantee for the user the digital ownership on his personal data.

Elsewhere, recent researches supported by Tim Berners-Lee [16] proposed a new paradigm of building Web applications. This paradigm consists of two major parts: (i) introducing the Web user identity; and (ii) building applications in high degree of decentralization. For this, they proposed an experimental ecosystem called *Solid*[3] (Social Linked Data), where users can have their own data storage, called *data pod*, in a *Solid*-enabled servers (e.g. Inrupt[4]). The *pod* represents their identity on the Web; contains their data (e.g. preferences) represented in LOD vocabularies; and is equipped with highly granular access control for third-party applications. On the other side, applications would no more have to hold user data, but rather act as clients for these users *data pod*. The motivation behind such design is to decouple data and service, and thus, applications become interchangeable views that user may change without losing his data. However, such decentralized platforms are in their infancy, and lack a lot in terms of functionalities in order to compete centralized solutions and convince users.

4 Proposed Approach

We propose a user side RS of POIs that attempts to guarantee:

- *Data privacy and ownership*: as in [16], we externalize user data from LBSs. Thus, users' preferences are stored on the user personal storage, allowing him to preserve control on his data.
- *User profile portability and interoperability*: as user data are stored on the user personal storage, he will be able to transport them across different platforms. Moreover, to guarantee interoperability between profiles, we semantically represent profiles in a generic model using DBpedia[5] LOD entities. Finally, to

[3] https://solid.mit.edu.
[4] https://solid.inrupt.com.
[5] http://dbpedia.org.

compare entities we use *LODS* [5], an early proposed LOD-based semantic similarity measure.

- *Algorithm transparency*: we also externalize the RS from the LBS application. We believe that a RS operating as third-party service could guarantee neutrality of recommendations (i.e. not biased by the provider's business interests).

The proposed mobile RS architecture is composed of three parties: the user, the target POIs register, and the third-party RS. The third-party RS is triggered on demand by the user when he explicitly asks for recommendations from the third-party RS on a target POIs register. We describe the process as follows (see Fig. 1):

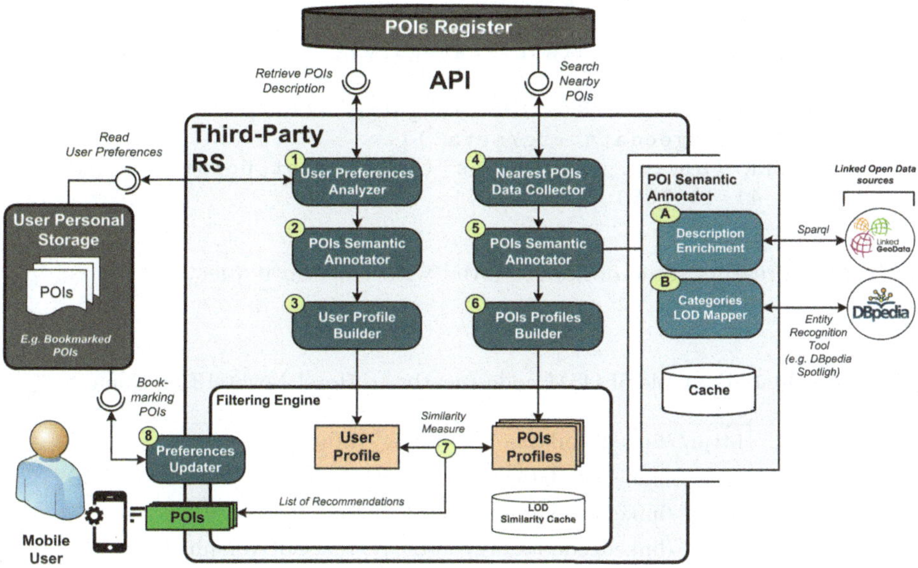

Fig. 1. Overview of the proposed system architecture

1) The *User Preferences Analyzer* module reads the user preferences from his User Personal Storage (e.g. mobile device or the user *data pod* [16]).
2) The *POIs Semantic Annotator* module attaches LOD concepts to the descriptions of POIs. In the step (A), the *Description Enrichment* module enriches the description of each POI with details describing its environment using LinkedGeoData[6] (LGD). For this, we directly start from the GPS coordinates (i.e. Latitude, Longitude) of the POI, and use the SPARQL endpoint[7] of LGD to retrieve nearby nodes to this geographic point (e.g. within a radius of 100 m around). We choose to fetch only the corresponding categories of these collected environmental nodes. For instance, to obtain the

[6] http://linkedgeodata.org/About.
[7] http://linkedgeodata.org/sparql.

environment nodes of Hotel Novotel Barcelona City (latitude = 41.403758, longitude = 2.191363), we can easily use the following SPARQL query:

```
PREFIX rdfs: <http://www.w3.org/2000/01/rdf-schema#>
PREFIX ogc:  <http://www.opengis.net/ont/geosparql#>
PREFIX geom: <http://geovocab.org/geometry#>
PREFIX lgdo: <http://linkedgeodata.org/ontology/>

SELECT DISTINCT ?category
FROM <http://linkedgeodata.org> {
    ?x a lgdo:Amenity.
    ?x rdfs:label ?l.
    ?x geom:geometry [ogc:asWKT ?g].
    ?x rdf:type ?category.

    FILTER(!STRSTARTS(str(?category), 'http://geovocab.org
       /spatial')).
    FILTER(!STRSTARTS(str(?category), 'http://
       linkedgeodata.org/meta')).
    FILTER (bif:st_intersects (?g, bif:st_point (2.191363,
        41.403758), 0.1)) .
}
```

Table 1 presents some nodes retrieved with the above query.

Table 1. Example of OSM nodes nearby to Hotel Novotel Barcelona

http://linkedgeodata.org/ontology/Museum
http://linkedgeodata.org/ontology/Theatre
http://linkedgeodata.org/ontology/Library
http://linkedgeodata.org/ontology/PlaceOfWorship
http://linkedgeodata.org/ontology/Restaurant
http://linkedgeodata.org/ontology/Mall
http://linkedgeodata.org/ontology/Attraction
http://linkedgeodata.org/ontology/Leisure
http://linkedgeodata.org/ontology/Cinema
http://linkedgeodata.org/ontology/BicycleRental
http://linkedgeodata.org/ontology/Parking

In the step (B), the *Categories LOD Mapper* module maps each category to its corresponding LOD entity (e.g. the category *Museum* has the LOD entity http://dbpedia.org/resource/Museum). For this task, a LOD entity recognition tool, such as DBpedia Spotlight[8], can be used. Finally, each POI

[8] https://www.dbpedia-spotlight.org/demo/.

is represented with a set of LOD entities. Categories mapping results can be cached for future needs.

3) The *User Profile Builder* represents the user profile in a generic LOD entity-based model as follows: $P_u = \left\{ \left(e, w\left(u, e\right)\right) \mid e \in E \right\}$; where E is the set of LOD entities extracted in the precedent mapping step (without environmental details), and $w(u, e)$ is a weight that determine the importance of the entity e for u. Since the proposed RS is not intended to store users preferences, it would not be possible to apply IDF. Thus, we apply the simple occurrence frequency TF (e.g. $w(u, dbpedia : Museum) = 7$ means that u has seven times the category *Museum* among his preferences). We also represent environmental preferences in an auxiliary profile as follows: $P_{u\_env} = \left\{ \left(e_{env}, w\left(u, e_{env}\right)\right) \mid e_{env} \in E_{env} \right\}$; where E_{env} denotes the set of LOD entities extracted from neighborhood nodes $N_{env} \in Env(P_{liked})$ of all liked POIs P_{liked}. To weight an environmental entity e_{env}, we choose to use TF and the average physical proximity of entity e_{env} to the liked POI P_{liked} (since e_{env} can appear in multiple N_{env}).
4) The *Nearest POIs Data Collector* module uses the API of the target POIs register, and asks for a list of POIs located nearby the user position. Such functionality is often made available by POIs registers through their APIs. For example, Yelp API provides the RESTful method *search*[9], which takes Latitude and Longitude as parameters to perform a search nearby a point.
5) The *POIs Semantic Annotator* module enriches the description of the collected POIs using LGD, and attaches LOD entities to categories.
6) The *POIs Profile Builder* represents the collected POIs in a LOD entity-based model as follows: $P_o = \left\{ \left(o, w\left(o, e\right)\right) \mid e \in E \right\}$. To weight the entity e in the POI o, we can apply the occurrence frequency TF, and next, combine it with IDF. However, the scope of IDF will be limited to the selected candidates, since the proposed RS is not intended to store all POIs profiles. Also, environmental details are represented in an auxiliary profile as follows: $P_{o\_env} = \left\{ \left(e_{env}, w\left(o, e_{env}\right)\right) \mid e_{env} \in E_{env} \right\}$. To weight environment entities, we use TF-IDF (here the scope of IDF is limited to the set of the collected POIs), as well as the physical proximity to the POI.
7) The *Filtering Engine* module computes the similarity between the user profile and candidates POIs profiles, and keeps the most similar ones. We propose to use *LODS* [5], which is a LOD-based similarity measure that exploits the semantic description of the compared LOD entities. Thus, the system can discover, for instance, that the category *Restaurants* is more similar to *Fast Food* than to *Museums*. This ensures the interoperability and portability of the user profile. To compare profiles, we propose the following formula:

$$Sim\left(P_u, P_o\right) = \frac{Sim_{Profiles}(P_u, P_o) + Sim_{Profiles}(P_{u\_env}, P_{o\_env})}{2} \quad (2)$$

[9] https://www.yelp.com/developers/documentation/v3/business_search.

Where $Sim_{Profiles}$ is our proposed similarity measure that we used to compare two entity-based profiles, as follows:

$$\begin{aligned} &Sim_{Profiles}(P_1, P_2) = \\ &Jaccard(P_1, P_2) + \Big(SemComp(P_1, P_2) * (1 - Jaccard(P_1, P_2))\Big) \end{aligned} \tag{3}$$

The principle of $Sim_{Profiles}$ consists of, first, using $Jaccard(P_1, P_2)$ [9] to measure the syntactical relatedness between P_1 and P_2, then measuring the semantic relatedness with $SemComp$. The contribution of the semantic comparison is weighted, such as, the more P_1 and P_2 are syntactically related (i.e. have common entities), the less is the impact of $SemComp$. In our proposed $SemComp$, we perform a partial matching and not a full matching in order to avoid comparing the shared sub-sets, as follows:

$$SemComp(P_1, P_2) = \begin{cases} Match_{LODS}(M_{10}, M_{01}) & if\ (|M_{11}| = 0) \\ \dfrac{Match_{LODS}(M_{10}, P_2) + Match_{LODS}(P_1, M_{01})}{2} & if\ ((|M_{11}| > 0) \wedge (|M_{10}| > 0) \wedge (|M_{01}| > 0)) \\ Match_{LODS}(P_1, M_{01}) & if\ ((|M_{11}| > 0) \wedge (|M_{10}| = 0) \wedge (|M_{01}| > 0)) \\ Match_{LODS}(M_{10}, P_2) & if\ ((|M_{11}| > 0) \wedge (|M_{10}| > 0) \wedge (|M_{01}| = 0)) \end{cases} \tag{4}$$

Where $SemComp(P_1, P_2) \in [0, 1]$ (1 if the profiles are semantically highly related, 0 otherwise), M_{11} is the set of common entities between P_1 and P_2, M_{10} is the set of entities in P_1 but not in P_2, M_{01} is the set of entities in P_2 but not in P_1, and $Match_{LODS}$ is our proposed function that matches two given sets of LOD entities E_1 and E_2, using $LODS$ to compare each pair (e_1, e_2), such as $e_1 \in E_1$ and $e_2 \in E_2$. $Match_{LODS}$ is defined as follows:

$$Match_{LODS}(E_1, E_2) = \frac{\sum_{e_i \in E_1} \sum_{e_j \in E_2} LODS(e_i, e_j) * w_i * w_j}{|E_1| * |E_2|} \tag{5}$$

Such as, w_i and w_j are respectively the weights associated to entities e_i and e_j. Here, we choose to normalize weights, such as: $\sum_{i \in Profile} w_i = 1$.

8) Finally, the *Preferences Updater* module updates the user's preferences repository by saving the liked POIs on his User Personal Storage.

5 Evaluation

5.1 Experiments

The objective of the evaluation is to measure the effectiveness of the RS in the proposed decentralized architecture. To this end, we choose to conduct several offline experiments on the Yelp dataset[10], containing users' POIs ratings.

[10] https://www.yelp.com/dataset/challenge.

We particularly experiment several variations when comparing profiles, and compare the obtained results with those of the centralized architecture.

The original dataset contains 1543 categories with 22 parent categories. From these parent categories, we selected six categories related to POIs (see Table 2). We also kept only cities having at least 50 POIs, and only users having at least 10 positive ratings (i.e. $rating > 3$) on different POIs. We partitioned the dataset to three sub datasets D1, D2 and D3 (categories were grouped two by two). Finally, we enriched POIs with environmental details (radius of 100 m) using the SPARQL endpoint of LGD. Next, we performed a LOD mapping of categories using DBpedia Spotlight, a LOD entity recognition tool.

Table 2. Statistics on the obtained datasets

N°	Datasets	#Users	#POIs	#Ratings	#Categories
D1	Active life/arts & entertainment	134	971	2239	577
D2	Food/restaurants	101	1369	1425	
D3	Hotels & travel/nightlife	355	1929	11843	

We conducted two kinds of experiments. The first one is a ***Local registers experiment*** (on D1 and D2) and aims to simulate POIs registers specialized in one kind of POIs. Thus, *learning* and *testing* phases were conducted on each dataset separately. The second one is a ***Cross-registers experiments*** (on D3) and aims to simulate the transportation of user preferences from different POIs registers. Here, we simulated users with no preferences in the target category of POIs, and learned user preferences on other categories of POIs (D1 and D2). Moreover, for each experiment, we simulated the two cases: ***Centralized RS*** (i.e. the RS accesses to the whole ratings matrix), and ***Decentralized RS*** (i.e. the RS accesses only the ratings of the user). The objective was to have the results of the centralized RS as baseline, in order to observe the impact of decentralization. Moreover, to study the impact of the proposed solution, we used different approaches to compare profiles, as follows:

- *Cent_LODS*: (centralized) the *LODS*-based approach (Eq. 2) was used.
- *Cent_Cosine_syntactic*: (centralized) the *cosine* function (Eq. 1) was used with the categories of POIs.
- *Cent_Cosine_res*: (centralized) same as *Cent_Cosine_syntactic* but by replacing categories with their LOD resources.
- *Dec_LODS*: as *Cent_LODS* but in decentralized environment.
- *Dec_Cosine_syntactic*: as *Cent_Cosine_syntactic* but in decentralization.
- *Dec_Cosine_res*: as *Cent_Cosine_res* but in decentralization.

The Top-N most similar POIs were provided as recommendation list, ranked according to their similarity score (from the most similar to the less similar). The focus was put on N = 20 items [6].

The performances were measured using the $F1$ metric; the harmonic mean of *precision* and *recall*. *Precision* is the fraction of the recommended items that are relevant to the user; *Recall* is the fraction of the relevant items that are successfully retrieved in a dataset. The *precision* and *recall* metrics are usually formulated as follows [13]:

$$precision = \frac{1}{|Users|} \sum_{u \in Users} \frac{Total\ number\ of\ relevant\ items\ recommended\ to\ u}{Total\ number\ of\ items\ recommended\ to\ u} \tag{6}$$

$$recall = \frac{1}{|Users|} \sum_{u \in Users} \frac{Total\ number\ of\ relevant\ items\ recommended\ to\ u}{Total\ number\ of\ relevant\ items\ to\ u\ present\ in\ the\ system} \tag{7}$$

$F1 \in [0, 1]$ is usually formulated as follows [13]:

$$F1 = \frac{1}{|Users|} \sum_{u \in Users} 2 \times \frac{precision \times recall}{precision + recall} \tag{8}$$

5.2 Results

Results are presented in Fig. 2. The paired *t-test*[11] is used for testing the significance of the difference between results, with a standard[12] significance level set to 0.05.

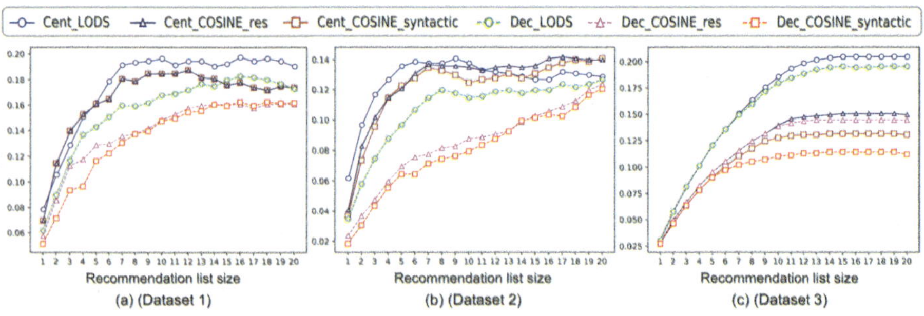

Fig. 2. Performances of the different approaches in the different datasets

Results in Centralized RS. In D1, *Cent_Cosine_syntactic* and *Cent_Cosine_res* had the same results, while with *Cent_LODS*, an average improvement of 5% is observed. In D2, *Cent_Cosine_res* performed slightly better than *Cent_Cosine_syntactic* (+4%). *Cent_LODS* performed also slightly better, but without a significant improvement. Overall, *LODS*-based approach is slightly better in centralized RS. On the other hand, in D3, *Cent_Cosine_res* is better than *Cent_Cosine_syntactic* (+9%), and with *Cent_LODS*, the improvement was more

[11] https://en.wikipedia.org/wiki/Student's_t-test.
[12] https://en.wikipedia.org/wiki/Statistical_significance.

significant (+30% comparing to *Cent_Cosine_syntactic*, and +23% comparing to *Cent_Cosine_res*). Overall, *LODS*-based approach is significantly better than the *cosine* based approaches in the centralized RS, and in contrast to D1 and D2, using LOD resources had a better impact.

Impact of Decentralization. In D1 and D2, the performances of *Dec-_Cosine_syntactic* is significantly lower than *Cent_Cosine_syntactic* ($p < 0.05$) (−19% in D1, and −35% in D2), and *Dec_Cosine_res* had not a significant improvement comparing to *Dec_Cosine_syntactic* (+4% in D1, and +7% in D2). However, *Dec_LODS* performed better than *Dec_Cosine_syntactic* and *Dec_Cosine_res* (+12% in D1, and +25% in D2), and when comparing to the centralized RSs, *Dec_LODS* had a competitive performances. Overall, in D1 and D2, *Cosine_Syntactic* was significantly impacted by the decentralization, and using LOD resources was not able to overcome the negative impact of decentralization. However, using *Dec_LODS* allowed to obtain significantly better results. In D3, *Dec_Cosine_syntactic* was also impacted by the decentralization (−9%). However, *Dec_Cosine_res* had a significant improvement comparing to *Dec_Cosine_syntactic* (+15%), and was able to maintain performances similar to those of *Cent_Cosine_res* (average difference less than 2%). Moreover, *Dec_LODS* performed significantly better than *Dec_Cosine_syntactic* and *Dec_Cosine_res* (+35% comparing to *Dec_Cosine_syntactic*, and +24% comparing to *Dec_Cosine_res*), and when comparing to the centralized RSs, *Dec_LODS* outperformed significantly both *Cent_Cosine_syntactic* and *Cent_Cosine_res* (+28% comparing to *Cent_Cosine_syntactic*, and +21% comparing to *Cent-_Cosine_res*). Overall, in D3, using LOD resources was able to overcome the negative impact of decentralization, and *Dec_LODS* was the most closer to *Cent_LODS*.

6 Conclusion and Discussion

We proposed a user-side mobile recommender system of POIs. Our solution can be seen as an intermediate between the existing siloed POIs registers and a highly decentralized solution. By doing so, issues related to data privacy and user preferences portability can be solved by design. Moreover, algorithm transparency can also be improved, since the user can switch to another RS without losing his preferences. Finally, several conclusions can be made from the conducted experiments.

Firstly, with the basic *cosine* approach, the negative impact of decentralization was important in all experiments. This can be explained by the difference in the amount of data that each RS (i.e. *centralized* and *decentralized*) was able to access in order to profile users and POIs. In fact, the centralized RS is able to use the whole database of profiles (users and POIs), and is, consequently, able to produce more accurate weights (TF-IDF) when learning profiles.

Secondly, using LOD resources had a slightly positive impact in all experiments. This can be justified by the increase of shared features between profiles once enriched with LOD resources. In fact, since categories can share some of LOD resources with other syntactically different categories, more common features can be discovered. For instance, the two categories *Art Museums* and *Art Galleries* are syntactically different, but they share the LOD resource *dbpedia:Art*. Thus, this approach was able to reduce the impact of decentralization. However, this reduction varied in the experiments. In fact, reduction was not significant in the local register, but was more important in cross-registers. This can be explained by the density of user profiles in cross-registers, where user preferences were learnt from different registers preferences, and consequently, profiles were richer.

Thirdly, *LODS*-based approach performed better than the *cosine*-based approaches. This observation is valid in all experiments, and can be supported by the fact the proposed *LODS*-based approach, in contrast to the *cosine*-based approaches, goes beyond the syntactical comparison by comparing also the non shared features, using *LODS* similarity measure, which operates at the semantic level. For instance, the category *Art Museums* has the LOD resources: {*dbpedia:Art, dbpedia:Museum*}, while *Art Galleries* has {*dbpedia:Art, dbpedia:Art_museum*}. Syntactically, they share only the resource *dbpedia:Art*. Therefore, the semantic similarity measure, *LODS*, is used to compare the unshared resources, *dbpedia:Museum* and *dbpedia:Art_museum*. Consequently, this approach was able reduce the impact of decentralization.

To conclude, we can say that relying on LOD and on an appropriate similarity measure can be a promising solution in our decentralized architecture. As future work, we want to: (1) conduct a user study experiments using a *Solid*-based server (such as *Inrupt*) to store user preferences. In fact, in this present work, we have only taken the main concepts from the paradigm of the decentralized Web, such as externalizing user data, but we didn't concretely implement the decentralized architecture; (2) compare results with other efficient centralized RSs; (3) generalize our solution, as we started to envision it in our earlier work [3].

References

1. Berners-Lee, T.: The web is under threat. Join us and fight for it. (2018). https://webfoundation.org/2018/03/web-birthday-29. Accessed 21 Dec 2019
2. Boja, U., Passant, A.: Weaving SIOC into the web of linked data. In: Proceedings of the Workshop on Linked Data on the Web (2008)
3. Boubenia, M., Belkhir, A., Bouyakoub, M.F.: A multi-level approach for mobile recommendation of services. In: Proceedings of the International Conference on Internet of Things and Cloud Computing, p. 40. ACM (2016). https://doi.org/10.1145/2896387.2896425
4. Cadwalladr, C., Graham-Harrison, E.: Revealed: 50 million Facebook profiles harvested for Cambridge Analytica in major data breach (Mar 2018). https://www.theguardian.com/news/2018/mar/17/cambridge-analytica-facebook-influence-us-election. Accessed 21 Dec 2019

5. Cheniki, N., Belkhir, A., Sam, Y., Messai, N.: LODS: a linked open data based similarity measure. In: 2016 IEEE 25th International Conference on Enabling Technologies: Infrastructure for Collaborative Enterprises, pp. 229–234. IEEE (2016). https://doi.org/10.1109/WETICE.2016.58
6. Cremonesi, P., Koren, Y., Turrin, R.: Performance of recommender algorithms on top-n recommendation tasks. In: Proceedings of the Fourth ACM Conference on Recommender Systems, pp. 39–46. ACM (2010). https://doi.org/10.1145/1864708.1864721
7. Heckmann, D., Schwartz, T., Brandherm, B., Schmitz, M., von Wilamowitz-Moellendorff, M.: Gumo–the general user model ontology. In: User Modeling 2005, pp. 428–432. Springer, Heidelberg (2005). https://doi.org/10.1007/11527886_58
8. Hochmair, H.H., Juhász, L., Cvetojevic, S.: Data quality of points of interest in selected mapping and social media platforms. In: LBS 2018: 14th International Conference on Location Based Services, pp. 293–313. Springer, Cham (2018). https://doi.org/10.1007/978-3-319-71470-7_15
9. Jaccard, P.: Étude comparative de la distribution florale dans une portion des Alpes et du Jura. Bulletin de la Société Vaudoise des Sciences Naturelles **37**, 547–579 (1901). https://doi.org/10.5169/seals-266450
10. Lops, P., de Gemmis, M., Semeraro, G.: Content-based recommender systems: state of the art and trends. In: Ricci, F., Rokach, L., Shapira, B., Kantor, P.B. (eds.) Recommender Systems Handbook, pp. 73–105. Springer, Boston (2011). https://doi.org/10.1007/978-0-387-85820-3_3
11. O'Donnell, A.: Why Sharing Your Location on Social Media Is a Bad Thing (2018). https://www.lifewire.com/why-sharing-your-location-on-social-media-is-a-bad-thing-2487165. Accessed 21 Dec 2019
12. Piao, G., Breslin, J.G.: Analyzing aggregated semantics-enabled user modeling on Google+ and Twitter for personalized link recommendations. In: Proceedings of the 2016 Conference on User Modeling Adaptation and Personalization, pp. 105–109. ACM (2016). https://doi.org/10.1145/2930238.2930278
13. Ricci, F., Rokach, L., Shapira, B., Kantor, P.B.: Recommender Systems Handbook. Springer, Boston (2011). https://doi.org/10.1007/978-0-387-85820-3
14. Riedl, J.: Personalization and privacy. IEEE Internet Comput. **5**(6), 29–31 (2001). https://doi.org/10.1109/4236.968828
15. Salton, G., Wong, A., Yang, C.S.: A vector space model for automatic indexing. Commun. ACM **18**(11), 613–620 (1975). https://doi.org/10.1145/361219.361220
16. Verborgh, R.: Re-decentralizing the web, for good this time. In: Seneviratne, O., Hendler, J. (eds.) Linking the World's Information: Tim Berners-Lee's Invention of the World Wide Web. ACM (2019). https://ruben.verborgh.org/articles/redecentralizing-the-web/
17. Zhang, J.: Anchoring effects of recommender systems. In: Proceedings of the Fifth ACM Conference on Recommender Systems, pp. 375–378. ACM (2011). https://doi.org/10.1145/2043932.2044010

Multicriteria Approach for Discovering IoT Services

Rabah Boucetti[1(✉)], Sofiane Mounine Hemam[1], and Ouassila Hioual[2]

[1] ICOSI Laboratory, Abbes Laghrour University-Khenchela, Khenchela, Algeria
boucetti.rabah30@gmail.com, sofiane.hemam@gmail.com
[2] LIRE Laboratory of Constantine, Abbes Laghrour University-Khenchela, Khenchela, Algeria
ouassila.hioual@gmail.com

Abstract. Nowadays, billions of Internet of Things (IoT) resources are deployed over the world in order to collect environmental information and react intelligently in real-time. In such environment, each IoT resource is characterized by its spatial location, energy, and functionality. To find and select the best IoT resource according to its characteristics and, that meet the user's QoS requirements is a real challenge. In this paper, we propose a new approach that takes into account both: IoT resource criteria and the user requirements in order to select the adequate IoT service that responds to the user request.

Keywords: Internet of Things · Service discovery · Services similarity · Services clustering · Quality of service

1 Introduction

The Internet of Things (IoT) is a new paradigm that involves connecting all types of devices to the internet like household appliances, smartphones, etc. and having a virtual view of the physical world capable of providing advanced services to end-users [1]. The introduction of the service-oriented architecture (SOA) into IoT has allowed these objects to expose their features as services (called IoT services). IoT services are the functionalities offered by interconnected heterogeneous objects abstract as services [2]. The geographical deployment of these objects will create intelligent spaces contain connected objects [3]. The advancement of this technology has generated a considerable number and diversity of IoT services. Therefore, a simple search to discover a service becomes a real challenge. How can a user find the IoT service that meets their requirements without any knowledge of their design, location, and even their existence? In particular, the considerable number of these services, the heterogeneity of objects, and their geographical distribution all contribute to the complexity of finding a relevant service for a given need [3].

Organizing IoT services to facilitate their discovery was the subject of several works. Among others, clustering services according to certain criteria are used to minimize the cost of the search. The process of finding a natural relationship between services and bundling similar services is called Clustering [4]. The Similarity between the IoT services

M. R. Senouci et al. (Eds.): CSA 2020, LNNS 199, pp. 208–219, 2021.
https://doi.org/10.1007/978-3-030-69418-0_19

is another problem that needs to be addressed before starting the process of clustering due to the diversity and heterogeneity of IoT services. IoT devices are constrained in many aspects such as energy, which can lead to the problem of the unavailability of the services on one hand, and are spread over a set of geographical locations which poses the spatial properties constraint on another hand [5], without forgetting the constraints in terms of quality of service (QoS) that will be imposed by the user.

Considering all criteria mentioned above, we will propose a multi-criteria approach for the discovery of IoT services. We will use the Google Similarity Distance method to determine the degree of similarity between the functionalities of IoT services. Then, we will group similar services in clusters, called Function Clusters (CF), by using the Spectral Clustering algorithm. We consider that the user must determine, in his request, the desired functionality, as well as the chosen location and the QoS preferences. In order to satisfy the user request, we compare at first the similarity of the desired functionality; in the user request with those of the main service of each cluster to define the cluster that contains the services that can satisfy the query. Then, we will eliminate IoT services whose energy is insufficient, as well as those that are outside the region of interest. Finally, the IoT services responding to the user's QoS preferences are retained, and the service closest to the chosen point by the user is elected as IoT service to respond to the user request.

The remainder of this paper is organized as follows: In Sect. 2, we review some related work. We give the definitions of the models used in Sect. 3, and we present our approach in Sect. 4. Section 5 concerns the case study in which we will simulate our approach, and we conclude our paper in Sect. 6.

2 Related Work

In this section, we relate some existing work in the context of discovery and similarity between IoT services in order to compare our work with the existing work. To reach this goal, we focus on our work presented in [6].

In [7], the authors claim that the number of connected objects will be huge and the search to find a particular service or sensor as in the traditional web will be the real challenge in the Web of Things (WoT). The authors propose a search method based on the similarity between the results offered by the sensors based on an algorithm that exploits the fuzzy sets to calculate the similarity between two sensors to obtain a list of corresponding sensors.

The authors in [8] present a research approach dedicated to the Web of Things based on three contributions: (1) The creation of semantic profiles for connected objects; (2) establishing similarities between these semantic profiles to cluster them and (3) calculating a score associating a "search context" with a query to select appropriate search algorithms. Their approach is based on the Semantic Web and uses ontologies for describing connected objects using machine-learning techniques.

In [9], the authors describe how a distributed management infrastructure could help users find smart devices. They propose making the location of an intelligent object its key dynamic property and using identifiers of logical places to hierarchically structure the nodes of the infrastructure. In addition, they describe a scalable search mechanism that treats requests as resources and has a caching system that greatly reduces the response time and the number of messages required when searching for devices. The proposed mechanism has advantages in terms of load balancing, since requests are automatically routed so as not to overload the overloaded parts of the infrastructure.

In [10], the authors introduce a primitive called "classification sensor" to efficiently search for sensors with a certain output state at the time of the query. The idea is to calculate for each sensor a probability of correspondence to the request and choose the sensor that has the greatest probability.

The literature [11] takes into account the energy constraint in wireless sensor networks. The authors propose a method to minimize the redundancy of the data captured by the sensors, minimizing the number of transmissions to the receiver in order to save energy. A filtering technique is used to identify the data entered periodically. The authors suggest a data aggregation model based on similarity functions and eliminate redundancy by preserving all data.

The authors in [12] propose an Energy-centered and QoS-aware service Selection Algorithm (EQSA), which consists of preselecting services that satisfy the user in terms of quality of service by using a lexicographic optimization strategy and relaxation of QoS constraints. Based on the relative dominance of Pareto services, the best-adapted services are selected. The dominance of a candidate service depends on its energy profile, its QoS attributes and the user's preferences.

In [13], the authors propose a service-oriented user-centric and event-aware framework (FASEM: Framework for Ambient Services and Events Monitoring). It is based on four aspects: ambient service discovery, invocation of ambient services, the selection of local ambient services and the selection of global ambient services. This control supports self-adaptation to unexpected changes through service replacement and rescheduling in the event of a failure.

In [14], authors use the Swarm network as a platform in a located discovery system for the Internet of Things. The discovery system takes advantage of the decentralized Swarm network and semantic mediation service to create a semantic-based location mechanism. To support location information, ontology has been created and a description of the service has been developed.

An approach to discovering IoT semantic services in a dynamic context and on a hierarchy of semantic gateways is proposed in [15]. An incremental clustering mechanism to create similar service groups has been defined in a gateway; this allows optimizing in terms of the number of clusters and the number of services per cluster.

The authors in [16] propose clustering-based and QoS-aware services composition algorithm (CQCA) for ambient intelligence environments. By using the *k-means* method, they partition the candidate services into clusters, where each cluster represents a level of QoS. They delete the services that show little promise in terms of QoS by using a utility function based on the characteristics of the resulting clusters. Then, to determine which services offer the level of QoS that meets the global QoS constraints, a lexicographic optimization method is used. Finally, a search tree is built to find near-to-optimal compositions.

The approach proposed in [17] is the closest to ours; the authors proposed a multi-index method to organize IoT services taking into account their functional specifications, spatial and temporal attributes, and develop a mechanism query based on the proposed multi-index method. They use Google Similarity Distance and Spectral Clustering to group services functionally into clusters, and SWC-tree (Skewness-aware Clustering Tree) to index services based on spatial attributes. Our approach differs in the treatment of spatial attributes where we will opt to calculate the Euclidean distance to choose at the end the most minimal distance from the target point in the user's request; and we also take into consideration the energy profile and user preferences (QoS) as important criteria in the search and selection of IoT services.

3 Models Definitions

Definition 1 (IoT Service, S-IoT)

We model an IoT service (S-IoT) as follows: *S-IoT = (ID, Fonct, QoS, Loc, EnP),* such as:

- *ID*: is the service identifier;
- *Funct*: defined the functional properties of the S-IoT, *Fonct* is a vector *[Desc, Inp, Out]*, where *Desc* is the description of the service's functionality, *Inp* is the set of input variables, *Out* is the set of output variables;
- *QoS*: is an n-tuple $[q_1, q_2, \ldots, q_n]$ where each q_i is a quality of service attribute, such as the cost, the response time and the reliability.
- *Loc*: spatial properties of the S-IoT, giving the geographical location of the service, $Loc = (x, y)$ where x, y are the geographical coordinates of the device.
- *EnP*: the energy profile of the S-IoT service.

Definition 2 (Functional Cluster, CF) [17]

The Functional Cluster (CF) is a cluster that contains similar IoT services according to their functional aspects, CF is a tuple: $CF = (ID, S\text{-}IoT_{main}, S\text{-}IoTs, S\text{-}IoT_{adj})$, where:

- *ID*: is the cluster identifier.
- $S\text{-}IoT_{main}$: it's the IoT service cluster principal (the center of the cluster), is the service that has the minimum value in the average of the degrees of similarity with the services of the same cluster. It identifies the functionality of the cluster.
- *S-IoTs*: set of IoT services grouped in the cluster.
- $S\text{-}IoT_{adj}$: it is a vector which contains the services of the cluster with the degree of similarity between them two by two.

Definition 3 (IoT service request, Q)

The user request for IoT services is defined as follow:

$$Q = (Q\text{-}ID,\ Q\text{-}Fonct,\ Q\text{-}QoS,\ Q\text{-}Loc), \text{ Such as:}$$

- *Q-ID*: request identifier;
- *Q-Funct*: defined the user functional need, *Q-Funct* is a 2-tuple *[Q-Desc, Q-Inp]*, which *Q-Desc* is the description of the user's need; *Q-Inp* is the set of input variables values of the desired IoT service.
- *Q-QoS*: set of constraints (preferences) imposed by the user on the non-functional properties (QoS) of the requested IoT service, n-tuple $[q_1, q_2, \ldots, q_n]$ where each q_i is a value of the QoS attribute of the S-IoT service.
- *Q-Loc*: is the location interested by the query, $Q\text{-}Loc = [(x_Q, y_Q), r]$, while x_Q and y_Q are the geographical coordinates of the location (point) targeted by the user and r a radius covering the area (region) of interest.

4 Proposed Approach

Our approach consists in grouping IoT services in clusters according to their functional similarities with the definition of the center's clusters; by comparing the function similarity of the desired service in the user request with each cluster center. This comparison allows us to select the cluster that contains the needed IoT services. At this stage, we have to select the adequate IoT service, i.e. the IoT service available in terms of energy and fulfilling the criteria defined in the query (location and QoS). The energy profile allows us to check the IoT service availability, so the IoT services having an insufficient energy in the selected cluster will be eliminated. Then, we choose the IoT services that meet the constraints in terms of non-functional properties, i.e., the QoS imposed by the user, by checking each QoS attribute in the query with its counterpart in each available service. From the obtained sub-cluster, we calculate the Euclidean distance between each IoT service and the user position, the service that has the minimum distance will be chosen as the elected IoT service to respond to the user's request.

4.1 Services Similarity and Clustering

Clustering IoT services according to their functional specifications, allows us to group services with similar functionality in the same cluster. To calculate the functional similarity between services, we adopt the method presented in [17]. This method is based on the Google Similarity Distance [18], and it is according to formula (1):

$$SimFonc(S\text{-}IoT_1, S\text{-}IoT_2) = a * Sim(Desc_1, Desc_2) + b * Sim(Inp_1, Inp_2) + c * Sim(Out_1, Out_2) \quad (1)$$

Where $Sim(Desc_1, Desc_2)$ is the similarity between the descriptions of the concerned services, $Sim(Inp_1, Inp_2)$ is the similarity between their input parameters and $Sim(Out_1, Out_2)$ is the one between their output parameters. The factors a, b and c indicate the importance of each attribute (description, inputs, outputs) in the similarity calculation according to the application context such that $a + b + c = 1$. The Google Similarity Distance method [18] calculates the degree of similarity between two words or phrases based on the number of pages returned by the Google search engine.

After calculating the functional similarity between services, we use the spectral clustering algorithm [19] to group these services into clusters according to their functional specifications following these steps:

- Define the graph G, the vertices are the IoT services $S\text{-}IoT_i$ ($i = 1...n$, n is the number of IoT services), the functional similarities between the IoT services are the weights of the edges. The graph G is presented as an adjacency matrix $n \times n$, denoted W.
- Define another matrix $n \times n$, which we will call D whose diagonal values D_{ii} are the sum of the values appearing on each line of the matrix W, and the other positions are at 0. Then, we can obtain the Laplacian graph L where $L = D - W$.
- Calculate the first m eigenvectors V_j ($j = 1...m$) corresponding to the first m eigenvalues of the matrix L. Then, position these eigenvectors vertically (in column) thus forming the matrix F of dimension $n \times m$.
- Represent each S-IoT_i service with the corresponding m dimension vector (line) in matrix F, where each line represents the vector characterizing the corresponding IoT service (VC_i, $i = 1...n$).
- Apply the k-means algorithm by taking $k = m$ (m-means) on the data set to clustering it into m clusters according to the functional characteristics of the S-IoTs.

At the end of the clustering procedure, we obtain m clusters, the VC_i vector cluster is the same cluster for the corresponding $S\text{-}IoT_i$ ($i = 1...n$). As a result, services with similar functionally will be grouped into the same CF_j cluster functional ($j = 1...m$).

4.2 Mechanism of Request

4.2.1 IoT Service Availability

We consider that the IoT service is available once the necessary energy for the execution of the user request is sufficient. The energy profile of the S-IoT service allows us to judge whether it is available or not. The energy profile of a service (*S-IoT.EnP*) is described by two parameters: i) the energy cost which is the energy consumption during the execution of the service (*ECost*), ii) the autonomy which represents the level energy of the object offering the service (*EAut*). Finally, the energy profile of a service is defined as the ratio between its energy consumption and its autonomy [12]:

$$S\text{-}IoT.EnP = ECost/EAut \tag{2}$$

From formula (2) it can be deduced that if: $S\text{-}IoT.EnP \leq 1$, then the S-IoT service is available, otherwise the service is not available.

4.2.2 IoT Service Location

The S-IoT service is linked to the device providing this service; physically this device is in a specific location, a point or covering a region. So the spatial properties for an IoT service are essential for the functioning of the IoT system. As mentioned before, each IoT service is assigned to a point $S\text{-}IoT.Loc = (x, y)$. In the query, the user specifies a point and a radius $Q.Q\text{-}Loc = [(xQ, yQ), r]$, thus making it possible to define a region if $r > 0$, or a point if $r = 0$. We use the calculation of Euclidean distance in the check if the requested service is at the desired place.

$$Dist_i = \left[\left(x_Q - S - IoT_i.x\right)^2 + \left(y_Q - S - IoT_i.y\right)^2 \right]^{1/2} \tag{3}$$

From the formula (3) we deduce that: if $Dist_i \leq r$, the service S-IoT$_i$ is in the region concerned by the request, otherwise S-IoT$_i$ is outside the zone targeted by the user.

4.2.3 QoS Processing

The QoS attribute vector *(S-IoT.QoS)* of a service defines its non-functional properties. $[q_1, q_2, \ldots, q_n]$ where n is the number of QoS attributes. We consider that the QoS constraints imposed by the user in his request are the same defined in the IoT service, i.e. the attributes of *S-IoT.QoS* and *Q.Q-QoS* are equivalent (each attribute q_i in *S-IoT.QoS* is the same attribute q_i in *Q.Q-QoS*). In general, QoS attributes are classified into two categories, positive attributes (availability and reliability for example) and negative attributes (such as cost and response time) [16]. Positive attributes have a positive impact on IoT service quality, i.e. as the value of the attribute increases, the overall IoT quality of service increases. In contrast, the negative attributes more the value of the attribute increases, the overall quality of IoT services decreases. Therefore, the user tends to maximize the positive attributes and will impose a minimum threshold (minimum barrier) accepted for these attributes, while a maximum threshold (maximum barrier) for the negative attributes is desired.

- The jth QoS attribute is a positive attribute: *S-IoT.QoS (j)* $\geq$ *Q.Q-QoS (j);*
- The jth QoS attribute is a negative attribute: *S-IoT.QoS (j)* $\leq$ *Q.Q-QoS (j).*

4.2.4 Processing of Request

The processing of the query is the final phase for our discovery process. *Algorithm 1* allows us to elect the IoT service S-IoT in order to satisfy the user request by considering the criteria mentioned above. Firstly, the algorithm makes a comparison between the query functionality and the functionalities of the main IoT services of the functional clusters in order to select the cluster corresponding to the desired functionality in the request. In a second step, the algorithm verifies in this cluster the services that are available according to their energy profile and their locations relative to the source of the query. Third, in the subgroup result of the second step, the algorithm selects the services that satisfy the user's preferences compared to the QoS of each S-IoT. At the end, we will have a subset of IoT services having the same functionality, available in terms of energy, in a specific location, and satisfied the QoS constraints imposed by the user. Once the candidate services satisfying these criteria are selected, the algorithm elect the IoT service having the minimum distance from the point chosen by the user, i.e. it selects the closest object (device) to this point.

Algorithm 1 *DescovryS-IoT*

```
Require: S-IoT Service Directory Clustered Functionally & IoT service request (Q)
Ensure: The candidate IoT service
RS, SC: sets of S-IoT
k ← number of CF
j ← GetSimFonc(Q.Q-Fonct, CF_1.S-IoT_main)
q ← 0; m ← 0
for (i=2, i ≤ k, i++)
   q ← GetSimFonc(Q.Q-Fonct, CF_i.S-IoT_main)
   if q < j then
     j ← q
     m ← i
   end if
end for
RS ← CF_m.S-IoTs
n_rs ← number of S-IoT in RS
for (i=1, i ≤ n_rs, i++)
   if (RS.S-IoT_i.EnP ≤ 1) & (Dist_i ≤ r) then SC ← SC U RS.S-IoT_i end if
end for
RS ← Ø
n_sc ← number of S-IoT in SC
n ← length of the vector Q.Q-QoS
for (i=1, i ≤ n_sc, i++)
  for (j=1, j ≤ n, j++)
   if (Q.Q-QoS(j) is a positive attribute) & (S-IoT_i.QoS (j) ≥ Q.Q-QoS(j)) then RS ← RS U SC.S-IoT_i end if
   if (Q.Q-QoS(j) is a negative attribute) & (S-IoT_i.QoS (j) ≤ Q.Q-QoS(j)) then RS ← RS U SC.S-IoT_i end if
  end for
end for
p ← number of S-IoT in RS
if p ≥ 1 then Return S-IoT where RS.S-IoT.Loc Minimum Dist
Else Return " Request Failed "
End if
```

5 Case Study and Simulation

We give a scenario in the context of the smart city as an example to illustrate our approach. The functional and non-functional description (QoS) of these S-IoTs used in such a scenario, as well as their spatial distributions and their energy profiles are summarized in Table 1.

Since it is almost impossible to take into account all the non-functional attributes, some predominant ones are selected to characterize the quality of service. For reasons of simplicity we choose the three attributes most studied in the literature: the execution time (T), the execution cost (C) and the reliability (R) as the components of the quality of service QoS vector, QoS = [T, C, R].

Table 1. *S-IoTs* description

S-IoT	*ID*	*Fonct*			*QoS*			*Loc*		*EnP*
		Desc	*Inp*	*Out*	*T*	*C*	*R*	*x*	*y*	
$S\text{-}IoT_1$	01	Measuring air quality	Time_Area	Air_Quality	120	115	0.89	1	7	0.55
$S\text{-}IoT_2$	02	Wind Measure	Time	Direction, Level	195	70	0.77	4	3	0.20
$S\text{-}IoT_3$	03	Monitor	Authorise_Code	Street_Video	200	70	0.75	4	6	1.93
$S\text{-}IoT_4$	04	Measuring wind	Time	Direction, Level	85	200	0.96	6	2	1.21
$S\text{-}IoT_5$	05	Probing air quality	Time_Area	Air_Quality	125	118	0.92	8	7	0.66
$S\text{-}IoT_6$	06	Street Monitor	Authorise_Code	Video	130	52	0.98	3	8	0.21
$S\text{-}IoT_7$	07	Air quality probe	Time_Area	Air_Quality	150	200	0.91	10	10	0.32
$S\text{-}IoT_8$	08	Monitor	Authorise_Code	Traffic_Vedio	95	83	0.91	6	7	0.52
$S\text{-}IoT_9$	09	Determining wind direction	Time	Wind_Direction	90	180	0.90	6	6	0.12
$S\text{-}IoT_{10}$	10	Measuring ambient temperature	Time	Temperature	80	110	0.97	11	8	0.01

1- The functional similarity between two services is calculated according to the formula (1). We consider in this scenario that the values of the weights *a*, *b* and *c* are 0.4, 0.4 and 0.2 respectively, which aims to show the importance of the description and the inputs compared to the outputs in the functional similarity calculation. More the degree of similarity is smaller; more the two IoT services are similar. The functional similarity results between the IoT services in our example are shown in Table 2; and the clustering stage generates four (04) functionality clusters; we schematize the result as shown in Fig. 1; The $\text{S-IoT}_{\text{main}}$ of each cluster is colored differently.

We consider the following request:

$$\boldsymbol{Q} = (\mathbf{01},\ [\boldsymbol{Monitor},\ \boldsymbol{Authorise\_Code}],\ [\boldsymbol{T} \leq \mathbf{200}, \boldsymbol{C} \leq \mathbf{100},\ \boldsymbol{R} \geq \mathbf{0.85}],\ [(\mathbf{4},\mathbf{7}),\mathbf{2}]);$$

With the formula (1), we calculate the similarity between *Q.Fonct* and $S\text{-}IoT_{main}.Fonct$ of each cluster: $Sim(Q,\ S\text{-}IoT_3) = 0.151$, $Sim\ (Q,\ S\text{-}IoT_4) = 0.295$, $Sim\ (Q,\ S\text{-}IoT_5) = 0.475$, $Sim\ (Q,\ S\text{-}IoT_9) = 0.513$. From these results, the cluster that contains the IoTs services that satisfy the user request is the cluster that contains the S-IoT_3 as the main one; in this case the CF_1 will be selected cluster.

Table 2. *S-IoTs* similarity matrix

	S-IoT₁									

	$S\text{-}IoT_1$	$S\text{-}IoT_2$	$S\text{-}IoT_3$	$S\text{-}IoT_4$	$S\text{-}IoT_5$	$S\text{-}IoT_6$	$S\text{-}IoT_7$	$S\text{-}IoT_8$	$S\text{-}IoT_9$	$S\text{-}IoT_{10}$
$S\text{-}IoT_1$	0.000	0.093	0.341	0.125	0.050	0.231	0.053	0.271	0.157	0.501
$S\text{-}IoT_2$	0.093	0.000	0.288	0.061	0.085	0.121	0.087	0.179	0.159	0.699
$S\text{-}IoT_3$	0.341	0.288	0.000	0.341	0.550	0.032	0.557	0.026	0.460	0.612
$S\text{-}IoT_4$	0.125	0.061	0.341	0.000	0.111	0.110	0.117	0.170	0.099	0.521
$S\text{-}IoT_5$	0.050	0.084	0.550	0.111	0.000	0.271	0.004	0.293	0.176	0.535
$S\text{-}IoT_6$	0.231	0.121	0.032	0.110	0.271	0.000	0.449	0.081	0.317	0.489
$S\text{-}IoT_7$	0.053	0.087	0.557	0.117	0.004	0.449	0.000	0.297	0.139	0.605
$S\text{-}IoT_8$	0.271	0.179	0.026	0.170	0.293	0.081	0.297	0.000	0.407	0.582
$S\text{-}IoT_9$	0.157	0.159	0.460	0.099	0.176	0.317	0.139	0.407	0.000	0.675
$S\text{-}IoT_{10}$	0.501	0.699	0.612	0.521	0.535	0.489	0.605	0.582	0.675	0.000

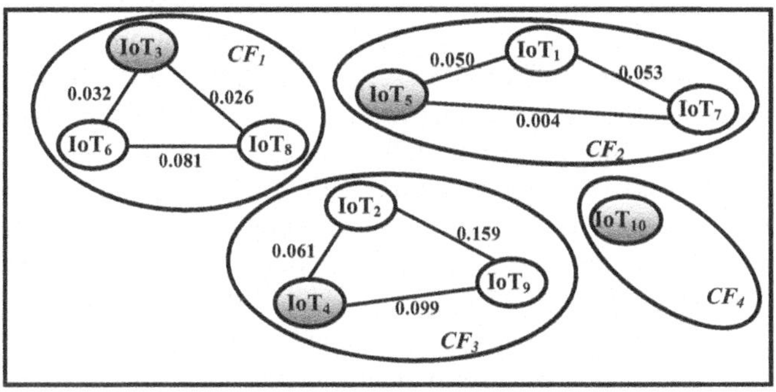

Fig. 1. Clustering result

2- In the second step, we will check the availability of CF_1 cluster services ($S\text{-}IoT_3$, $S\text{-}IoT_8$, $S\text{-}IoT_6$) in terms of energy based on their *EnP* attributes, and their location relative to the location source of the query (using the *Loc* attribute).

$S\text{-}IoT_3.EnP = 1.93 > 1$, then $S\text{-}IoT_3$ is unavailable (eliminated); $S\text{-}IoT_8.EnP = 0.52 \leq 1$, $S\text{-}IoT_8$ is available; $S\text{-}IoT_6.EnP = 0.21 \leq 1$, also $S\text{-}IoT_6$ is available.

$Dist_6 = [(4-3)^2 + (7-8)^2]^{1/2} = 1.414 \leq 2$, so $S\text{-}IoT_6$ is in the region of interest; $Dist_8 = 2 \leq 2$, of the same $S\text{-}IoT_8$ satisfies the constraint of the localization.

At the end of this step, we will have a sub-cluster of IoT services with features similar to those of the query, available in terms of energy and they are in the desired area. In this case, the {$S\text{-}IoT_6$, $S\text{-}IoT_8$} sub-cluster is retained.

3- The third step is the satisfaction check of the QoS preferences imposed in the request for each service in the sub-cluster selected in the previous step:

- *Execution time (T):* the execution time is a negative attribute, we tend to minimize it. In our scenario the user tolerable barrier is 200 ms. $S\text{-}IoT_6.QoS\ (T) = 130$ ms <

200, then S-IoT$_6$ satisfies the constraint; S-IoT$_8$.QoS (T) = 95 ms < 200, similarly, satisfied.
- *Execution cost (C):* the maximum barrier is 100 u. S-IoT$_6$.QoS (C) = 52 u < 100, satisfied. S-IoT$_8$.QoS (C) = 83 u < 100, satisfied also.
- *Reliability (R):* Reliability is a positive attribute that we want to maximize. So, impose a minimum value, 0.85 in our query. S-IoT$_6$.QoS (R) = 0.98 > 0.85, satisfied. S-IoT$_8$.QoS (R) = 0.91 > 0.85, satisfied.

We notice that both IoT services respond to non-functional constraints imposed by the user in his request, so they will be retained.

At the end, we choose the nearest IoT service to the location of the query source (*Dist* minimum), so the elected service for our query is **S-IoT$_6$**.

6 Conclusion

In this paper, we have proposed a multi-criteria approach allowing searching or discovering an IoT service that meet criteria jugged important and respond to the user request. These criteria are: (1) functionality as a user's need, (2) the availability of the service in terms of energy, (3) the spatial properties, (4) as well as the non-functional properties (QoS). To select the most adequate IoT service among the billions between them, we have proposed in this paper to combine between the Google Similarity Distance method and the spectral clustering algorithm to group the IoT services having similar functionally into clusters. Our proposed approach take into account the energy of the IoT service as an availability criterion, the geographical location and the vector of non-functional properties in terms of QoS, as important parameters. To show the reliability of our approach, we have made a case study on a scenario of the smart city, where we have simulated the discovery of an IoT service by taking into account the studied criteria. The obtained result was satisfactory. As a perspective, we aim to improve our approach in order to be used in the context of the IoT services composition to satisfy complex needs.

References

1. Al-Fuqaha, A., Guizani, M., Mohammadi, M., Aledhari, M., Ayyash, M.: Internet of Things: a survey on enabling technologies, protocols, and applications. IEEE Commun. Surv. Tutor. **17**(4), 2347–2376 (2015). https://doi.org/10.1109/COMST.2015.2444095
2. Papazoglou, M.P., Heuvel, W.J.: Service oriented architectures: approaches, technologies and research issues. Int. J. Very Large Data Bases **16**(3), 389–415 (2007). https://doi.org/10.1007/s00778-007-0044-3
3. Ben Fredj, S., Boussard, M., Kofman, D., Noirie, L.: A scalable IoT service search based on clustering and aggregation. In: IEEE International Conference on Green Computing and Communications and IEEE Internet of Things and IEEE Cyber, Physical and Social Computing, pp. 403–410 (2013). https://doi.org/10.1109/GreenCom-iThings-CPSCom.2013.86.
4. Can, G., Wang, S.: A novel node deployment and clustering scheme in wireless sensor networks. In: Second International Symposium on Electronic Commerce and Security, Nanchang, China (2009). https://doi.org/10.1109/ISECS.2009.69.

5. Xu, Y., Heidemann, J., Estrin, D.: Geography-informed energy conservation for ad hoc routing. In: Proceedings of the 7th Annual ACM/IEEE International Conference on Mobile Computing and Networking, Rome, Italy, New Orleans, LA (2001). https://doi.org/10.1145/381677.381685.
6. Boucetti, R., Hemam, S. M., Hioual, O.: QoS-aware IoT services composition: a survey. To appear in Distributed Sensing and Intelligent Systems Proceedings of ICDSIS 2020 Conference, Agadir, Morocco (2020). ISBN 978-3-030-64258-7
7. Truong, C., Römer, K., Chen, K.: Fuzzy-based sensor search in the Web of Things. In: 3rd IEEE International Conference on the Internet of Things, Wuxi, pp. 127–134 (2012). https://doi.org/10.1109/iot.2012.6402314
8. Christophe, B., Verdot, V., Toubiana, V.: Searching the 'Web of Things'. In: IEEE Fifth International Conference on Semantic Computing (2011). https://doi.org/10.1109/icsc.2011.69
9. Mayer, S., Guinard, D., Trifa, V.: Searching in a web-based infrastructure for smart things. In: 3rd IEEE International Conference on the Internet of Things, Wuxi, pp. 119–126 (2012). https://doi.org/10.1109/iot.2012.6402313
10. Elahi, B.M., Römer, K., Ostermaier, B., Fahrmair, M., Kellerer, W.: Sensor ranking: a primitive for efficient content-based sensor search. In: International Conference on Information Processing in Sensor Networks, pp. 217–228 (2009). https://doi.org/10.1145/1602165.1602186
11. Bahi, J. M., Makhoul, A., Medlej, M.: Data aggregation for periodic sensor networks using sets similarity functions. In: 7th International Wireless Communications and Mobile Computing Conference, Istanbul, pp. 559–564 (2011). https://doi.org/10.1109/iwcmc.2011.5982594
12. Khanouche, M.E., Amirat, Y., Chibani, A., Kerkar, M., Yachir, A.: Energy-centered and QoS-aware services selection for Internet of Things. IEEE Trans. Autom. Sci. Eng. 1256–1269 (2016). https://doi.org/10.1109/TASE.2016.2539240
13. Yachir, A., Amirat, Y., Chibani, A., Badache, N.: Event-aware framework for dynamic services discovery and selection in the context of ambient intelligence and Internet of Things. IEEE Trans. Autom. Sci. Eng. **13**(1), 85–102 (2016). https://doi.org/10.1109/TASE.2015.2499792
14. Calcina-Ccori, P.C., De Biase L.C.C., De Oliveira C.E.L., Fedrecheski, G., da Silva, F.C., Zuffo, M.K.: Location-aware discovery of services in the IoT: a Swarm approach. In: Global IoT Summit (GIoTS), Aarhus, Denmark, pp. 1–6 (2019). https://doi.org/10.1109/giots.2019.8766389.
15. Ben Fredj, S., Boussard, M., Kofman, D., Noirie, L.: Efficient semantic-based IoT service discovery mechanism for dynamic environments. In: IEEE 25th Annual International Symposium on Personal, Indoor, and Mobile Radio Communication (PIMRC), Washington, DC, pp. 2088–2092 (2014). https://doi.org/10.1109/pimrc.2014.7136516
16. Khanouche, M.E., Attal, F., Amirat, Y., Chibani, A., Kerkar, M.: Clustering-based and QoS-aware services composition algorithm for ambient intelligence. Inf. Sci. **482**, 419–439 (2019). https://doi.org/10.1016/j.ins.2019.01.015
17. Du, C., Zhou, Z., Ying, S., Niu, J., Wang, Q.: An efficient indexing and query mechanism for ubiquitous IoT services. Int. J. Ad Hoc Ubiquitous Comput. **18**(4), 245–255 (2015). https://doi.org/10.1504/ijahuc.2015.069060
18. Cilibrasi, R., Vitányi, P.: The Google similarity distance. IEEE Trans. Knowl. Data Eng. **19**(3), 370–383 (2007). https://doi.org/10.1109/TKDE.2007.48
19. von Luxburg, U.: A tutorial on spectral clustering. Stat. Comput. **17**(4), 395–416 (2007)

Selecting IoT Services Under Uncertain QoS

Etchiali Abdelhak[1,2], Hadjila Fethallah[1,2(✉)], and Merzoug Mohammed[1,2]

[1] Computer Science Department, Tlemcen University, Tlemcen, Algeria
{abdelhak.etchiali,fethallah.hadjila,mohammed.merzoug}@univ-tlemcen.dz
[2] LRIT Laboratory, Tlemcen, Algeria

Abstract. With the expansion of internet of things (IoT) technology, the number of deployed devices and their related services will rise rapidly. In this context, the selection of physical services (i.e., IoT services) that best meet the user's requirements is becoming a major issue. In this paper, we handle the selection of IoT services by assuming that their quality of service (QoS) is uncertain and constantly fluctuates. To do so, we leverage a stochastic dominance relationship that allows for ranking the existing services. The obtained results are very promising and confirm the adequation of our proposal to the IoT area.

Keywords: IoT · Service oriented architecture · Uncertainty · Quality of service · Stochastic dominance · Weighted vote

1 Introduction

The internet of thing (IoT) is an extension of the actual internet in which physical objects that surround as will connect to the internet and exchange data with other objects and humans. According to [4], IoT is "the pervasive presence around us of a variety of things Or objects – such as Radio-Frequency IDentification (RFID) tags, sensors, actuators, mobile phones, etc. – which, through unique addressing schemes, are able to interact with each other and cooperate with their neighbors to reach com-mon goals." The RFID group specifies the internet of thing as: "the worldwide net-work of interconnected objects uniquely addressable based on standard communication protocols." According to the European research cluster on the internet of thing (IERC),[1] IoT is a "dynamic global network infrastructure with self-configuring capabilities based on standards and interoperable communication protocols; physical and virtual things in an IoT have identities and attributes and are capable of using intelligent interfaces and being integrated as an information network." [10]. The US National Intelligence Council (NIC)[2] expects that by 2025 Internet nodes may be present in everyday

[1] http://www.internet-of-things-research.eu/.
[2] https://www.dni.gov/nic/NIC_me.html.

M. R. Senouci et al. (Eds.): CSA 2020, LNNS 199, pp. 220–230, 2021.
https://doi.org/10.1007/978-3-030-69418-0_20

things such as paper documents, food packages, Home Furnishings. Our life and economy can be ameliorated by the introduction of IoT technology. For instance, in the context of healthcare [14,17,18,20], IoT services can monitor the medical state of patients in real time and send the sensed information to specialized medical centers. In smart environments such as homes and offices [16], IoT services can monitor the heat system and lighting conditions according to the preferences of humans; these preferences can be learned automatically by the IoT framework. Cisco's Internet Business Solutions Group[3] predicts that 50 billion things will be connected to the Internet in 2020. Consequently, the number of physical services (also termed IoT services) will increase since each physical object is manipulated with one or more services. Hence, the selection of IoT services that best meet the user's requirements (such as QoS) is a primordial issue in the era of ubiquitous computing. In contrast to traditional web services that run on sophisticated servers [5], the IoT services run on resource-limited environments (which have less CPU computation power, less RAM, less Flash memory, and less energy). Moreover, the mobility, the limited battery of IoT devices, and their restricted sensing coverage entail an uncertainty about the network connectivity and consequently the criteria such as availability/reliability will be touched. Therefore, the QoS of IoT services is intrinsically uncertain and depends on environmental conditions. To select the most pertinent IoT services for a given request, we propose a solution based on a set of heuristics; the key ideas of our contribution can be summarized as follows:

1. We adopt the Physical Service Model (PSM) proposed in [11,12] to describe physical services (and which is shown in Sect. 3).
2. Based on PSM, we specify five qualities of service (QoS) criteria (available time, service area, time of processing or latency, reliability, and reputation) for handling the selection problem. These attributes constitute the most dominant characteristics of physical services.
3. We use a stochastic dominance relationship to compare the uncertain QoS of IoT services. We then perform a weighted vote to derive the Top-K IoT services that best meet the user's needs.

The reminder of the paper is organized as follows: Section 2 presents the existing works on IoT service selection. Section 3 introduces the proposed framework and the selection mechanism. Section 4 shows the empirical results and finally Sect. 5 concludes the paper and gives future directions.

2 State of the Art

Several models are proposed in the literature to describe the IoT services. Firstly, we note that the OWLS (ontology web language for services) model is less convenient to physical services since it does not handles the primordial features such as mobility, spatio-temporal attributes, constrained power/computation,

[3] http://share.cisco.com/internet-of-things.html.

and QoS uncertainty. To face these concerns, the research community has proposed additional models to handle the inherent attributes of the IoT technology. In Haller et al. [8], the authors proposed an IoT taxonomy that formalizes the fundamental concepts of the area. The proposed hierarchy provides five concepts: Augmented Entity, User, Device, Resource and Service. De et al. [2] enriched the IoT components with a semantic description that covers the following concepts: Entity, Resource and Service. It is interesting to note that this framework does not take into account the spatiotemporal attributes. The Incubator Group on Semantic Sensor Networks of W3C proposed a semantic sensor network (SSN) taxonomy [1] that enhances the core concepts of sensors (e.g., capabilities, measurement processes, observations, and deployments). Concerning the service selection using QoS, we distinguish two types of approaches: those that address traditional web services such as [6,7,9,13,15,19] and those that address IoT services such as [3,11,12,21]. [19] tackle the web service selection issue by handling uncertain QoS. The proposed framework leveraged the probabilistic dominance and constrained optimization to search the compositions that best meet the users' requirements. [13] address Web service selection with presence of noise. To this end, they used a set of heuristics to divide the global constraints into local constraints. Similarly, [9] handle the uncertain QoS by considering Gaussian probability distributions. The authors used thresholds to decompose the global constraints into local constraints. The latter thresholds are computed by dividing the global constraint constant in proportion to the averages of QoS properties of each task. After that, the approach performs a local search to derive a near optimal solution. The framework presented in [11,12] handles the IoT service selection by sorting the candidate services using four QoS attributes (availability, service area, latency, and reputation). More precisely, these authors leveraged the absolute dominance to filter out the services that are less effective. The work of [3] used five QoS attributes (reliability, security, business, mobility, and heterogeneity) to tackle the IoT service selection in the context of manufacturing. More specifically, the authors leveraged a multi-criteria decision approach termed analytic hierarchy process (AHP) to fuse the above-mentioned criteria.

3 Proposed Framework

3.1 IoT Model

To describe IoT services, we adopt the physical service model (PSM) proposed in [11]. PSM contains three main concepts which are defined below:

- **Device**: it represents a hardware (embedded in a real world environment) that ensures connectivity to the Internet. Each device exposes a set of services and has a temporal attribute that shows its availability time. Additionally, it possesses a deployment location (e.g., longitude and latitude). Figure 1 shows the most important features of devices.

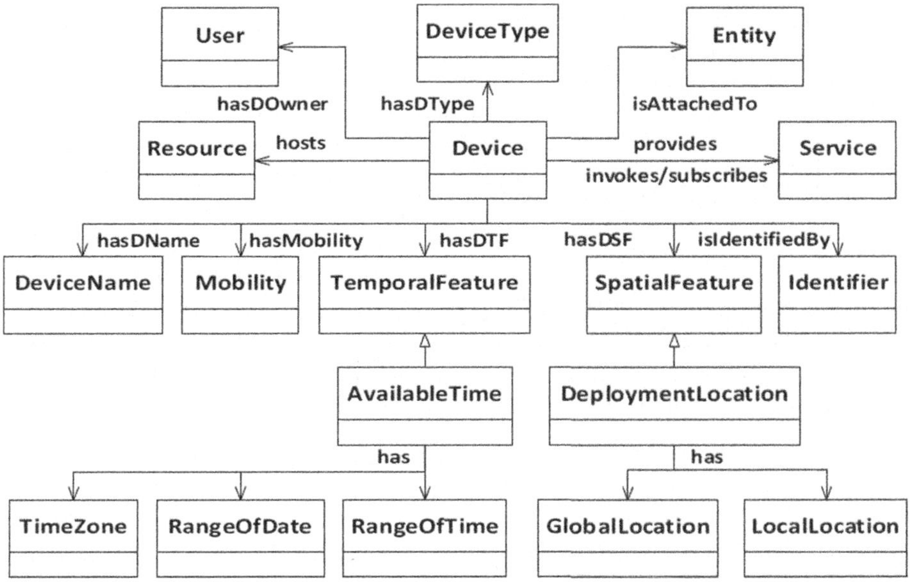

Fig. 1. Device description [11]

- **Resource**: it is a computational unit of a device that provides information about, or actuation capabilities on physical objects. Furthermore, Resources have a spatial attribute that provides the operational range of the hosted devices.
- **Service**: it provides an interface that controls or accesses information related to re-sources. The services have functional attributes such as inputs, outputs, pre-conditions, effects, invocation model (i.e., REST or SOAP) and nonfunctional attributes such as QoS. A subset of QoS attributes can be inferred from the spatiotemporal characteristics of the corresponding resources and devices. We also notice that the services can be accessed and monitored from the Internet.

3.2 Proposal Description

To select the IoT services based on the aforementioned QoS criteria (i.e., available time, service area, latency, reliability, and reputation), we propose two ranking mechanisms: the first one handles the QoS which is sensitive to the user's request (such as available time or service area) and the second one handles the remaining QoS (e.g., latency, reliability, and reputation).

Processing of Available Time/Service Area. Since the IoT devices are running with limited batteries, their availability is not always ensured. Additionally IoT devices can be unavailable due to privacy or security concerns. To measure the matching degree between the user's requested time interval (denoted as

Time(i)) and the provided time interval of the IoT service (denoted as Time(j)), we use the following formula:

$$AT_{i,j} = \begin{cases} 1, & \text{if the request interval time is included in the service time.} \\ 0, & \text{if the intersection time between i and j is empty.} \\ \dfrac{Time(i) \cap Time(j)}{range(Time(i))} & \text{, elsewhere.} \end{cases} \tag{1}$$

For instance, if the user's time interval is [9H:30 Min 9H:45 Min] and the IoT service time intervals are [9H:40 Min 10H:45 Min] and [11H:30 Min 12H:30 Min], then matching degree between i and j is 5/15 = 0.3. Similarly, we compute the matching degree between the requested location (denoted as loc(i)) and the service area of the IoT service (denoted as (loc(j)) as follows:

$$SA_{i,j} = \begin{cases} e^{-SquareDistance(loc(i),loc(j))}, \text{ if } Distance(loc(i), loc(j)) \leq range_j. \\ 0, \text{ if } Distance(loc(i), loc(j)) > range_j. \end{cases} \tag{2}$$

The symbol loc(i) (respectively loc(j)) denotes the latitude and the longitude coordinates of i (respectively j). If the distance between the device j (e.g., a sensing node) and the requested location i is beyond the sensing range of j ($range_j$), then the matching degree is 0, otherwise we use the exponential-based similarity to evaluate the proximity. Once the degrees $AT_{i,j}$ and $SA_{i,j}$ are computed, we rank the candidate IoT services according to the following Equation:

$$RankingScore(j) = AT_{i,j} * SA_{i,j} \tag{3}$$

The more the computed score is high, the better the rank. We use a multiplicative aggregation to penalize the compensation effect between the two criteria (i.e., if one of them is 0, the aggregated score is 0).

Processing of Latency/Reliability/Reputation. These criteria are considered as uncertain QoS; this means that they constantly fluctuate and therefore they can be treated as statistical samples which follow an unknown probability distribution. We assume that these data are collected from users' feedbacks (which are available over social networks) or third-party systems. Our purpose consists of choosing the services having the highest (uncertain) QoS performance. To do so, we leverage a stochastic dominance relationship that allows for comparing two QoS samples. Before the presentation of the stochastic dominance, we normalize the above-mentioned criteria using a set of equations.

The normalized reputation of the service j (denoted as Rep_j) is computed as follows:

$$Rep_j = \frac{rep_j - rep_{min}}{rep_{max} - rep_{min}} \tag{4}$$

where rep_j is the unnormalized rating (or reputation) of the IoT service j, rep_{max} (rep_{min}) represents the maximum (minimum) value of the candidates services.

The normalized reliability of the service j (denoted as Rel_j) is computed as follows:

$$Rel_j = \frac{rel_j - rel_{min}}{rel_{max} - rel_{min}} \tag{5}$$

Where rel_j is the unnormalized reliability of the IoT service j, rel_{max} (rel_{min}) represents the maximum (minimum) value of the candidates services.

The normalized latency of the service j (denoted as Lat_j) is computed as follows:

$$Lat_j = \frac{lat_{max} - lat_j}{lat_{max} - lat_{min}} \tag{6}$$

Where lat_j is the unnormalized latency of the IoT service j, Lat_{max} (lat_{min}) represents the maximum (minimum) value of the candidates services. To compare the QoS samples Q_r and Q'_r (i.e., two l-dimensional vectors that represent the rth QoS attribute) we apply the following equation:

$$StocDom(Q_r, Q'_r) = \frac{1}{l} \sum_{i=1}^{l} Step(Q_r(i), Q'_r(i)) \tag{7}$$

where l represents the sample size and the function Step is defined as (we assume that the QoS criteria are positive):

$$Step(Q_r(i), Q'_r(i)) = \begin{cases} 1, \text{ if } Q_r(i) \geq Q'_r(i). \\ 0, \text{ otherwise }. \end{cases} \tag{8}$$

Roughly speaking, a vector Qr stochastically dominates another vector Q'_r if $StocDom(Q_r, Q'_r) > StocDom(Q'_r, Q_r)$. For instance, if $Q_r = <0.5, 0.6, 0.8>$ and $Q'_r = <0.5, 0.5, 0.9>$, then $StocDom(Q_r, Q'_r) = 2/3 = 0.66$.

In what follows, we define a weighted vote function (denoted as WV) to determine the Top-K services that have the highest performance in terms of the rth QoS attribute.

$$WV(S_j.Q_r) = \frac{1}{size(dataset) - 1} \sum_{j' \neq j} StocDom(S_j.Q_r, S_{j'}.Q_r) \tag{9}$$

Equation 9 sums the dominance score between the QoS (i.e., Qr) of the current service (i.e., Sj) and the QoS of the remaining services (i.e., $S_{j'}.Q_r$). We normalize the score by dividing on the dataset size (that is, the number of candidate services). Moreover, we define the aggregated vote (denoted as AV) of a IoT service j as the product of the different QoS votes:

$$AV(S_j) = \prod_{r \in \{1,..,d\}} WV(S_j.Q_r) \tag{10}$$

The higher the aggregated vote of Sj, the better its rank. Now, each IoT service j has two scores: RankingScore(j) and AV(j); to establish the final ranking, we use the first score (i.e., RankingScore(j)) since it is more sensitive to the users requirements, then if we have a tie (i.e., there are two services S_j and S_m such that RankingScore(j) = RankingScore(m)), we use the second score (that is AV(j)), in this case, the preferred service is the candidate that maximizes the aggregated vote. In the following, we show a running example that elucidates the principles of our proposal, our scenario contains:

- A dataset having 03 IoT services: dataset = $\{S_1, S_2, S_3\}$.
- We have five QoS attributes $(Q_1, Q_2, Q_3, Q_4, Q_5)$, where Q_1 represents the availability time (see Eq. 1), Q_2 represents the service area (see Eq. 2), Q_3 represents the reputation (see Eq. 4), Q_4 represents the reliability (see Eq. 5), Q_5 represents the latency (see Eq. 6).
- We have three QoS realizations for Q_3, Q_4, Q_5 (i.e., l = 3).
- $S_1.Q_1 = 0.9$, $S_1.Q_2 = 0.4$, $S_1.Q_3 = <0.6, 0.4, 0.9>$, $S_1.Q_4 = <0.8, 0.8, 0.9>$, $S_1.Q_5 = <0.7, 0.6, 0.9>$.
- $S_2.Q_1 = 0.6$, $S_2.Q_2 = 0.6$, $S_2.Q_3 = <0.7, 0.4, 1>$, $S_2.Q_4 = <0.9, 0.8, 0.9>$, $S_2.Q_5 = <0.7, 0.7, 0.8>$.
- $S_3.Q_1 = 0.5$, $S_3.Q_2 = 0.6$, $S_3.Q_3 = <1, 0.9, 1>$, $S_3.Q_4 = <1, 1, 0.9>$, $S_3.Q_5 = <0.8, 1, 1>$.
- We aim to calculate the Top-2 services that best meet the user's need.

We first compute the rank based on Q_1 and Q_2: RankingScore(1) = 0.9 * 0.4 = 0.36, RankingScore(2) = 0.6 * 0.6 = 0.36, RankingScore(3) = 0.5 * 0.6 = 0.30.

Since we have a tie for S_1 and S_2, we compute the aggregated vote for both S_1 and S_2.

$WV(S1.Q3) = (StocDom(S_1.Q_3, S_2.Q_3) + StocDom(S_1.Q_3, S_3.Q_3))/2 = (1/3 + 0)/2 = 0.16$.

$WV(S1.Q4) = (StocDom(S1.Q4, S2.Q4) + StocDom(S1.Q4, S3.Q4))/2 = (2/3 + 1/3)/2 = 0.5$.

$WV(S1.Q5) = (StocDom(S1.Q5, S2.Q5) + StocDom(S1.Q5, S3.Q5))/2 = (2/3 + 0)/2 = 0.33$.

$WV(S2.Q3) = (StocDom(S2.Q3, S1.Q3) + StocDom(S2.Q3, S3.Q3))/2 = (1 + 1/3)/2 = 0.66$.

$WV(S2.Q4) = (StocDom(S2.Q4, S1.Q4) + StocDom(S2.Q4, S3.Q4))/2 = (1 + 1/3)/2 = 0.66$.

$WV(S2.Q5) = (StocDom(S2.Q3, S1.Q5) + StocDom(S2.Q5, S3.Q5))/2 = (2/3 + 0)/2 = 0.33$.

Thus, AV(1) = 0.16 * 0.5 * 0.33 = 0.02, AV(2) = 0.66 * 0.66 * 0.33 = 0.14, and $Top - 2(services) = <S2, S1>$.

4 Experimental Study

In this work, we use a random process to generate the IoT data set (a uniform law is used to produce the QoS realizations). The range of each parameter is shown in Table 1. The experiments are conducted on a desktop computer with Intel Core i3-4005U CPU@1.70 GHz, running on Ubuntu 16.04 (32-bit). The algorithms are developed with Netbeans IDE 8.1.

Table 1. Description of parameters.

Parameter	Description	Domain
N	The number of IoT services	{100,...,500}
D	The number of QoS attributes	{5}
L	The number of QoS realizations	{2,...,21}
K	The size of the returned list	{5}

In what follows, we study the run time with respect to n and l; additionally, we evaluate the quality of Top-5 results for a default configuration of n, d and l. Figure 2 shows the behavior of our ranking algorithm (Eqs. 3 and 10) with respect to the number of IoT services. Since our approach has a polynomial complexity, the increase of run time is almost linear for $n \leq 500$. The experiment also confirms the scalability of our approach for large datasets.

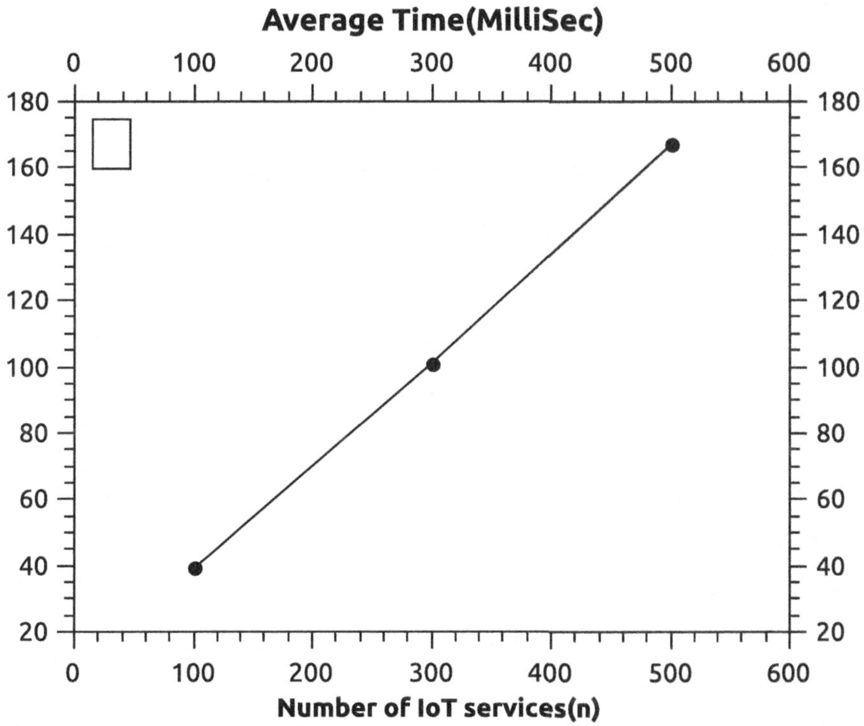

Fig. 2. Run time vs. n (for l = 7).

In Fig. 3, we depict the behavior of the ranking algorithm with respect to the number of QoS realizations l. We observe that the growth of time is almost linear with respect to l (especially for $l \leq 50$).

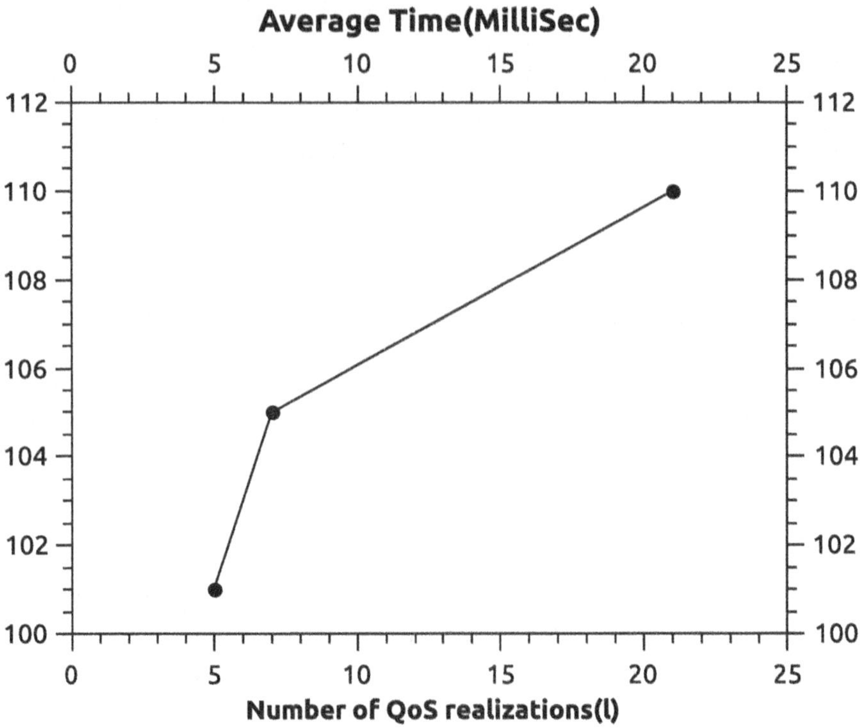

Fig. 3. Run time vs. l (for n = 300).

To assess the quality of the Top-k results, we demonstrate the ranking score and the aggregated vote in Table 2.

Table 2. Top-K IoT services (for l = 21, n = 300, k = 5).

Position	Ranking score (based on Q_1 and Q_2)	Aggregated vote
1	0.65	0.0289
2	0.65	0.0232
3	0.65	0.0193
4	0.64	0.0189
5	0.64	0.0167

We observe that the Top3 services are equivalent in terms of Q1 and Q2; therefore, we break the tie through the use of stochastic dominance. We notice that the increase of n will entail a lot of tied results. In addition, we observe that the stochastic dominance is more discriminative and ensures less ambiguous ranking (even if n is large). Finally, we note that the complexity of the stochastic dominance and the weighted vote is only polynomial and this makes them more suitable for large collections of IoT services.

5 Conclusion

In this paper, we have proposed an approach for selecting IoT services. The key idea consists of matching the QoS criteria of each IoT service with the user's need. To this end, we used a stochastic dominance function as well as aggregating schemes to rank the existing services. In future works, we plan to address additional QoS aspects such as privacy and security concerns. Furthermore, we plan to evaluate the proposed framework on real world datasets.

References

1. Compton, M., Barnaghi, P., Bermudez, L., García-Castro, R., Corcho, O., Cox, S., Taylor, K.: The SSN ontology of the W3C semantic sensor network incubator group. Web Semant. Sci. Serv. Agents World Wide Web **17**, 25–32 (2012)
2. De, S., Barnaghi, P., Bauer, M., and Meissner, S.: Service modeling for the Internet of Things. In: Proceedings of the Federated Conference on Computer Science and Information Systems (FedCSIS), pp. 949-955 (2011)
3. Durão, L.F.C., Carvalho, M.M., Takey, S., Cauchick-Miguel, P.A., Zancul, E.: Internet of Things process selection: AHP selection method. Int. J. Adv. Manuf. Technol. **99**(9–12), 2623–2634 (2018)
4. Giusto, D., Iera, A., Morabito, G., Atzori, L. (eds.) The Internet of Things, Springer (2010). ISBN 978-1-4419-1673-0
5. Hadjila, F.: Composition et interopération des services web sémantiques. These de doctorat (2014)
6. Halfaoui, A., Hadjila, F., Didi, F.: Qos-aware web services selection based on fuzzy dominance. In: IFIP International Conference on Computer Science and Its Applications, pp. 291–300. Springer, Cham (2015)
7. Halfaoui, A., Hadjila, F., Didi, F.: QoS-aware web service selection based on self-organising migrating algorithm and fuzzy dominance. Int. J. Comput. Sci. Eng. **17**(4), 377–389 (2018)
8. Haller, S., Serbanati, A., Bauer, M., Carrez, F.: A domain model for the Internet of Things. In: Proceedings of the Green Computing and Communications (GreenCom), IEEE and Internet of Things (iThings/CPSCom), IEEE International Conference on and IEEE Cyber, Physical and Social Computing, pp. 411–417 (2013)
9. Hwang, S.Y., Hsu, C.C., Lee, C.H.: Service selection for web services with probabilistic QoS. IEEE Trans. Serv. Comput. **8**(3), 467–480 (2015)
10. IERC: Coordinating and building a broadly based consensus on the ways to realize the internet of things in Europe (2013). http://www.internet-of-things-research.eu/pdf/Poster_IERC_A0_V01.pdf
11. Jin, X., Chun, S., Jung, J., Lee, K.H.: IoT service selection based on physical service model and absolute dominance relationship. In: 2014 IEEE 7th International Conference on Service-Oriented Computing and Applications (SOCA), pp. 65–72. IEEE (2014)

12. Jin, X., Chun, S., Jung, J., Lee, K.H.: A fast and scalable approach for IoT service selection based on a physical service model. Inf. Syst. Front. **19**(6), 1357–1372 (2017)
13. Kim, M., Oh, B., Jung, J., Lee, K.H.: Outlier-robust web service selection based on a probabilistic QoS model. Int. J. Web Grid Serv. **12**(2), 162–181 (2016)
14. Lane, N.D., Georgiev, P., Qendro, L.: DeepEar: robust smartphone audio sensing in unconstrained acoustic environments using deep learning. In: Proceedings of the ACM International Joint Conference on Pervasive and Ubiquitous Computing (UbiComp), pp. 283–294 (2015)
15. Mohammed, M., Chikh, M.A., Fethallah, H.: QoS-aware web service selection based on harmony search. In: IEEE ISKO-Maghreb: Concepts and Tools for knowledge Management (ISKO-Maghreb), 2014 4th International Symposium, pp. 1-6 (2014)
16. Nakamura, S., Shigaki, S., Hiromori, A., Yamaguchi, H.,Higashino, T.: A model-based approach to support smart and social home living. In: Proceedings of the ACM Internationall Joint Conference on Pervasive and Ubiquitous Computing (UbiComp), pp. 1101–1105 (2015)
17. Rabbi, M., Aung, M.H., Zhang, M., Choudhury, T.: MyBehavior: automatic personalized health feedback from user behaviors and preferences using smartphones. In: Proceedings of the ACM Int'l Joint Conference on Pervasive and Ubiquitous Computing (UbiComp), pp. 707–718 (2015)
18. Rahman, T., Adams, A.T., Ravichandran, R.V., Zhang, M., Patel, S.N., Kientz, J.A., Choudhury, T.: Dopplesleep: a contactless unobtrusive sleep sensing system using short-range Doppler radar. In: Proceedings of the ACM International Joint Conference on Pervasive and Ubiquitous Computing (UbiComp), pp. 39–50 (2015)
19. Remaci, Z.Y., Hadjila, F., Didi, F.: Selecting web service compositions under uncertain QoS. In: 2018 International Conference on Computer Science and Its Applications (CIIA), Oran, Algeria (2018)
20. Sun, X., Lu, Z., Hu, W., Cao, G.: SymDetector: detecting sound-related respiratory symptoms using smartphones. In: Proceedings of the ACM International Joint Conference on Pervasive and Ubiquitous Computing (UbiComp), pp. 97–108 (2015)
21. Yang, C., Shen, W., Lin, T., Wang, X.: IoT-enabled dynamic service selection across multiple manufacturing clouds. Manuf. Lett. **7**, 22–25 (2016)

HST-101: A New Secure SCADA Transport Protocol

Tarek Cherifi[1(✉)], Kenza Challal[1], and Walid Cherifi[2]

[1] Electrical Engineering and Industrial Computing Department, The National High School of Technology (ENST), Cite Diplomatique Dergana, Algiers, Algeria
{tarek.cherifi,kchallal}@enst.dz

[2] Ecole Militaire Polytechnique, Algiers, Algeria
walid.cherifi.alg@gmail.com

Abstract. Securing SCADA systems against cyber-attackers is one of the most important elements of critical infrastructure protection. It requires specific techniques different from those used by IT systems. This paper proposes a new approach to secure the IEC 60870-5-101 SCADA transmission protocols against external cyber-attacks targeting cooperate network segments, especially passive attacks and modification and fabrication attacks. Our approach is based on using a new real-time perfect secrecy crypt-system implementation to secure T-101 frames.

Keywords: SCADA security · IEC 60870-5-101 protocol · HST-101 protocol · Entropy accumulator · Seed reinjection

1 Introduction

Supervisory Control And Data Acquisition systems (SCADA systems) are real-time industrial control systems (ICS) mainly used to manage large geographically wide-spread assets and technical locations of other physical systems. Electrical networks, Water and waste-water canalizations, transport and traffic control, oil and gas transportation are samples of critical infrastructures that explore SCADA systems to get information about their installations states, to control the industrial devices, and to troubleshoot their vital processes. As centralized cyber-physical systems, SCADA architecture is based on three principal segments: (1) the main SCADA network segment, (2) the field devices segment, and (3) the cooperate network segment. The modern main SCADA network segment, known as Master Terminal Units (MTU), is a local area network that interconnects servers, workstations and monitors allowing operators to supervise and to control the physical system. The field devices segment represents the set of devices installed on remote technical assets including the remote terminal units (RTU), programmable logical controllers (PLC), intelligent electronic devices (IED), sensors and actuators. Finally, the cooperate network segment is the telecommunication networks ensuring the communication between the two other segments [17].

M. R. Senouci et al. (Eds.): CSA 2020, LNNS 199, pp. 231–240, 2021.
https://doi.org/10.1007/978-3-030-69418-0_21

In general, a SCADA network segment is based on one MTU that communicates with many RTUs. Each RTU ensures local communication with a set of other field equipment like IEDs, PLCs, sensors and actuators. In the monitor direction, the RTU collects, organizes then sends data received from other field equipment to the MTU; The MTU translates, presents and saves all necessary data to operators and supervisors. However, in the control direction, the MTU sends request and control instructions to the RTU that analyzes them then responds or retransmits them to other field equipments to be executed.

In addition to ensuring a real-time supervision and control to almost all modern critical infrastructures, they also can offer an instant analyze of the past and current events, archive important data, generate rapports, simulate situations with real parameters, share important information with other systems. These functionality and others make the management of the critical infrastructures strongly dependent on the SCADA systems. Any incident or malfunction of these systems can provide an important human, strategic, governmental, social, economic, or environmental accidents. Today, more than 70% of SCADA incidents are caused by external cyber-attackers [8,11,12].

This paper proposes a new approach to secure the IEC 60870-5-101 SCADA transmission protocols against external cyber-attacks targeting cooperate network segments, especially passive attacks and modification and fabrication attacks. Our approach is based on using a new real-time perfect secrecy cryptsystem to secure T-101 frames.

2 SCADA Security Overview

Securing SCADA systems against cyber-attackers is one of the most important elements of critical infrastructure protection. The nature of attacked SCADA segments defines the favorite kind of SCADA cyber-attack: In general, current SCADA network segments and field segments are local networks situated in control centers or technical locations with limited physical access. Attacking them needs direct communication, via local physical connections or via other connected corporate systems. Almost all attacks targeting them are internal. However, cooperate segments use commune WAN technologies and open protocols. They are more vulnerable to external attacks. The attack's goals analysis shows that the cooperate segment's attacks are composed of three main sets of basic attacks: (1) the passive attacks, (2) denial of service attacks, and (3) modification and fabrication attacks. Passive attacks, as traffic analysis attack, eavesdropping attack, brute-force search attack and rainbow table attack, are attacks on the system's confidentiality. They intercept exchanged data to collect important information or to prepare other active attack. Denial of service attacks (DOS attacks) are active attacks on the system's availability. They perturb and interrupt the SCADA transmission protocol. Modification and fabrication attacks message modification attack, information disclosure, masquerade attack, man-in-the-middle attack, and replay attack, are active attacks on the integrity and authenticity of the system. Attackers are able to introduce messages in order to get more data, to transmit false information or to execute control instructions [4,9,10,12,19].

In order to secure SCADA systems, IT security solutions are often used. However, Standard firewalls, user authentication tools, encryption algorithms as other highly developed IT security solutions cannot provide a sufficient level of SCADA security. Securing SCADA systems against cyber-attacks requires specific techniques different from those used by IT systems. Despite the fact that SCADA systems evolution has exploited the IT system's development, they do not have the same security requirements: Their security objectives, security architectures, hardware and software technologies and quality of service requirements are different: The main IT system's security objective is protecting the integrity, the confidentiality and the availability of data. However, SCADA systems have, as a priority, to secure services ensuring the human safety, the system normal operation and the protection of the installation [4,8,11,13].

Security architectures of IT systems and that of SCADA systems are, also, different. In IT systems, protecting data is the most important. Therefore, network centers, that contain important data servers, need more protection than edge nodes. However, the important SCADA services are, in general, controlled by the SCADA network segments (network centers), communicated by cooperate network segments, and ensured by field segments (edge nodes). So, every segment of the SCADA system must be highly protected [4].

Operating systems, human-machine interfaces, database management systems, communication protocols are simples software technologies making a difference between IT systems and SCADA systems. Usually, IT systems use powerful servers that use software technologies more adapted to saving, managing, exploiting, and presenting their high data volume. On the other side, SCADA uses real-time systems with limited resources. They should exploit optimized efficient software technologies adapted to their hardware architecture to ensure their real-time tasks and to manage their energy consumption [4].

From a quality of service point of view, SCADA systems require strict respect of temporal constraints. They refuse any delay or failure in control applications. Troubleshooting by rebooting are, also, unaccepted. However, IT systems can support tolerances for delay of data exchange, and occasional failures [4].

In order to secure SCADA systems, more than 40 regional, national or international standards, containing recommendations, guidelines and good practices have been defined. Our study of the 15 standards, namely, IEC62443, IEC62351, CIP-002 to CIP-011, SP800-82, NISTIR-7176, NISTIR-7628, DHS LANGUAGE, DHS STRATEGIC, DOE 21STEPS, GAO Protection, CPNI PRACTICE, GB T30976.1-2-2014, GB T26333-2010, GB T18272.1-8-2000, and MIIT GUIDE, qualified to be updated, specialized in ICS, used in large geographical areas, and defined by organizations international or nations heavily involved in modern cyber warfare has shown us that the security aspects of industrial control networks and IT networks have been confused in the majority of cases. As a result, no standard has taken into account the specificities of the cooperation network segments used by industrial control networks [3–5].

3 IEC 60870-5-101 Frames

The IEC 60870-5-101, known by T-101 protocol, is the first international non-routable SCADA open transport protocol. Its first version was published in 1995 and its last update was published in 2016. It is considered as the origin of other protocols such as the IEC 60870-5-104 and the pure 2.4. It is one of the most used SCADA transport protocol in Europe, North Africa, the Middle East, China and several regions of the world [1,5,15,16,18].

Based on the three layers Enhanced Performance Architecture. The T-101 physical layer accepts the use of the V.24/V.28 communication interface, the X.24/X.27 interface, or any other serial communication interface. It is adapted to the point-to-point topology, the point-to-multipoint topology, the star topology, the bus topology and the ring topology [5,18].

The T-101 link layer proposes the use of the balanced communication mode with the point-to-point topology and the unbalanced mode with the other topologies. It allows the use of one of the three FT 1.2 frame formats defined by the IEC 60870-5-1 document and presented in Fig. 1. It uses the signal control character (1 byte) to sign the presence of the sender device, the fixed length frames (04,05 or 06 bytes) to communicate without sending data, and the variable length frames (from 13 to 261 bytes) to send data [5,18].

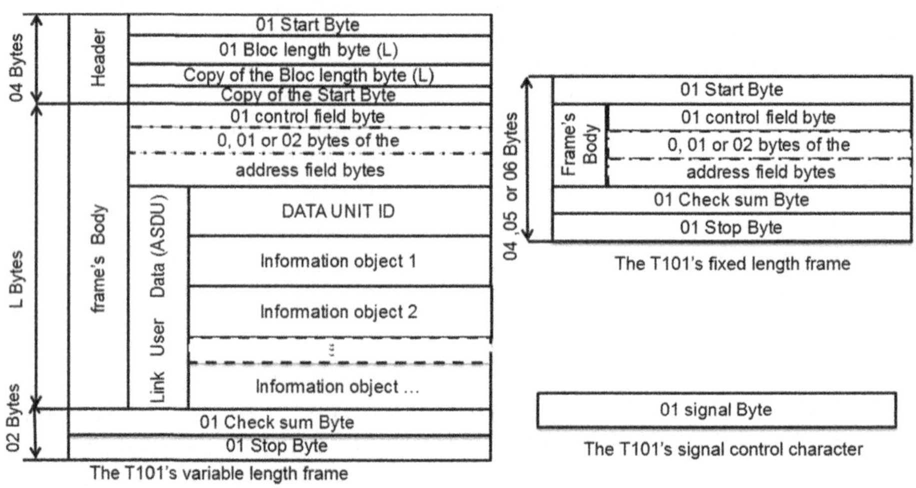

Fig. 1. The three T-101 frame formats.

The T-101 link layer addresses the main SCADA field equipments with 0,1 or two bytes depending on the density of the SCADA network. When the communication is point-to-point topology, the protocol doesn't need the use of addressee bytes. It uses one byte when the number of RTUs is less than 256 RTU. Using two bytes allows addressing 65536 units.

The application layer provides a simple representation of application service data units (ASDU) embedded on the frame link used data. The ASDU is composed of two parts: the data unit identification that defines the information object type supported by the frame and the information objects that associate to each information object address its information elements [1,5,15,16,18].

4 The Highly Secure T-101 Protocol (HST-101)

In order to highly secure T-101 protocol communication, we developed a new crypt-system to crypt the variable length T-101 frames (holding data) and fixed length T-101 frames. Our crypt-system is designed from two parts: the first one transforms the standard T-101 frame into a Modified T-101 frame. It completes the size of the communication T-101 frame to achieve the size of the longest T-101 frame (261 bytes) by adding a random noise with a studied distribution. The distribution of the added noise is calculated to tend the frame source distribution towards a uniform distribution [4]. Ciphering some fixed length T-101 frames helps to obtain a uniform distribution of the frame source and hide variable length frames holding data [4].

The second part tries to implement the perfect secrecy principle. First, it transforms the modified T-101 frame to a modified T-101 frame with a time tag. It replaces the fourth byte, having initially the same value of the first byte; by the value of the second byte then it replaces the second and the third bytes, having the same value of the fourth bytes, of the standard T-101 protocol by two key synchronizer bytes that time tag. Figure 2 presents the structures of the modified T-101 frame, the modified T-101 with time tag, and the highly secure T-101 frame.

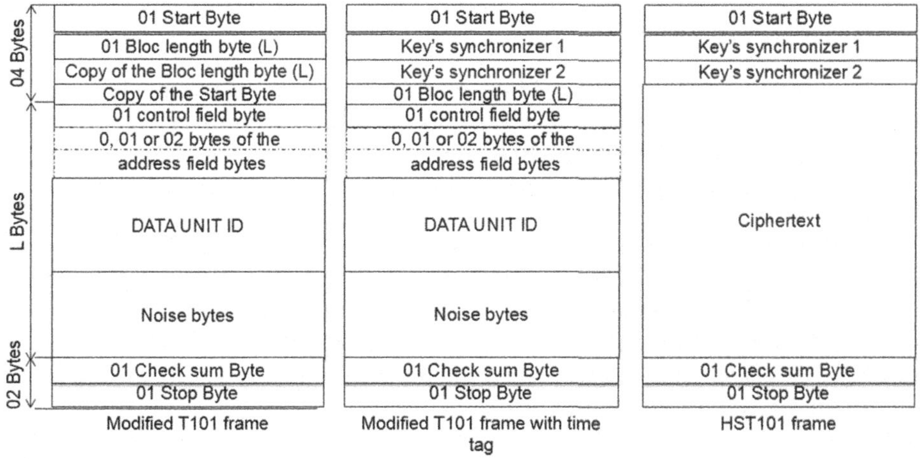

Fig. 2. Modified T-101 frame, Modified T-101 frame with time tag, and the HST-101 frame structures.

The two key synchronizer bytes are used to select the one time pad keys used to encrypt the body of the modified T-101 frame with time tag frame (byte 4 to byte 259) and to fight against the replay attacks. Our One Time Pad crypt-system (OTP system) explores a slow CS-PRNG and three independent simple encryption functions. The CS-PRNG generates periodically three key tables of 256 keys. The size of each key equals the frame body size (256 bytes).

The first and the second encryption function XOR, respectively, the T-101 frame body with the selected keys from the first key tables according to the values of the T-101 frame checksum and the first key synchronized byte. The third encryption function is a permutation according to the second key synchronizer byte. Figure 3 represent the structure of our one time pad crypt-system.

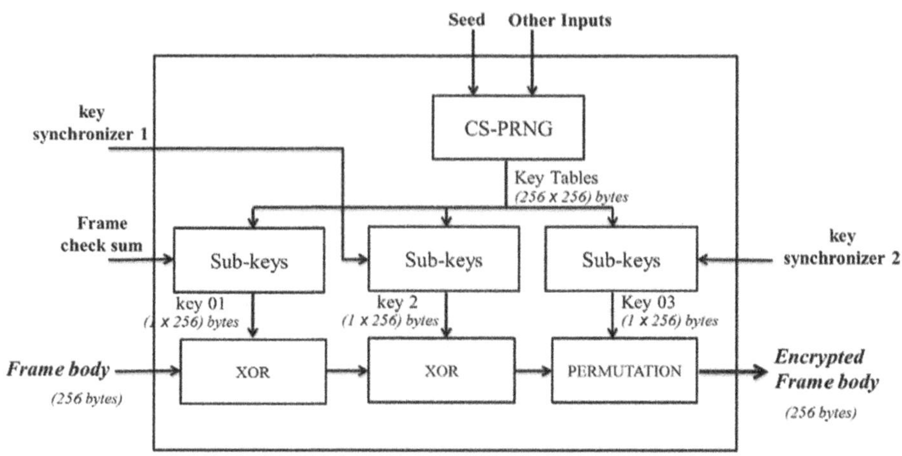

Fig. 3. The One Time Pad encryption system.

The design of our One Time Pad encrypt-system was mainly focused on the design of a new cryptographically secure pseudo-random number generator (CS-PRNG) from another CS-PRNG. The purpose of this perspective would be to increase the complexity of the number generator to better resist brute force attacks.

Based on previous researches on other cryptographically secure pseudorandom number generators and the advantages they provide as well. Yarrow for the entropy accumulator which selects the most independent random numbers possible [7]. For-tuna for its concept of re-injection of some random numbers calculated as seeds and to set up the numbers found in table form so that they can be reused in data encryption [6].

Our new CS-PRNG, presented in Fig. 4, is mainly composed of two other cryptographically secure pseudo-random number generators, connected to an entropy accumulator that is in charge of transposing the calculated random numbers and sending these values to the seed generator. The latter re-injects certain values as seeds from the initial sub-generators (automatic re-injection of seeds). As a result, the operation is automatically repeated in a loop. The entropy accumulator groups the pseudo-random numbers into a data table stored as a file of random and independent sub-keys. These sub-key tables can be reused in crypto-systems for message encryption operations.

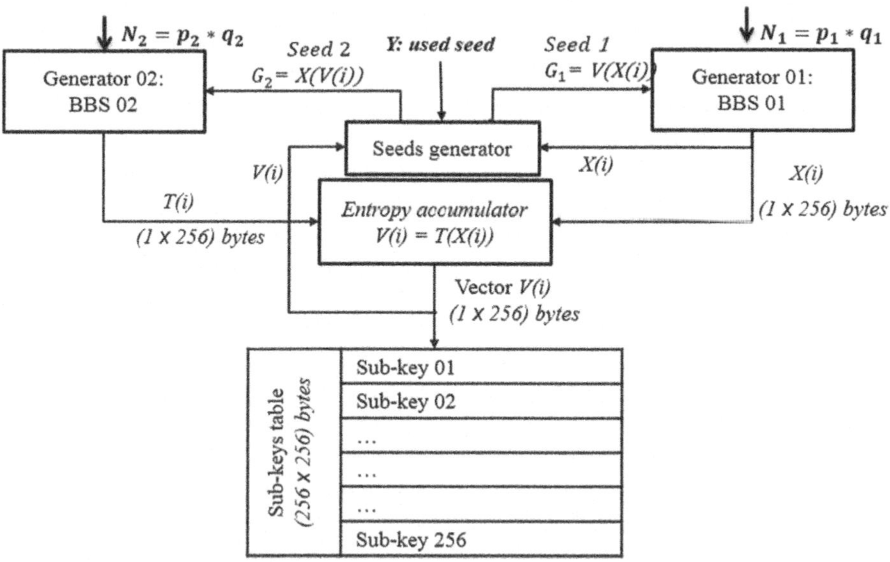

Fig. 4. The general structure of the used CS-PRNG.

In our case, we used two identical Blum-Blum-Shub (BBS) as the main CS-PRNG. We chose the BBS for its advantages: its simple algorithm to implement, the complexity of inverting its function, and the advantage provided by its considerable time processing, which is the reason for the resistivity of its algorithm to cryptanalystic attacks [2,14].

The inputs of our system are the inputs of the BBS generators: G_1 and N_1 inputs of the first generator. G_2 and N_2 inputs of the second generator. Generators 01 and 02 generate random integer vectors of size (1×256) bytes, which are respectively X_i and T_i. The vector V_i is the result of transposing the elements of the vector T_i as a function of the positions of the elements of the vector X_i according to the relation: $V(i) = T(X(i))$.

After transposition, the use of the variable Y is shown in its contribution to the choice of seeds for the first and second generator. The latter two will be changed according to V_i, X_i and Y according to the relations: $G_1 = V(X(Y))$ and $G_2 = X(V(Y))$.

A new V_i vector is generated after each seed variation (after each iteration of the loop).

A table of size (256 × 256) bytes is filled when $256V_i$ vectors are generated. The result of transposing the rows and columns of the previous table according to X_i and T_i is the final table which is our table of random keys.

5 Analysis of the Used PRNG Test Results

In addition to increasing generator complexity by using the principles of seed reinsertion, the use of the entropy accumulator and the safety characteristics of BBS [2,14], we will study the three statistical characteristics of the numbers generated by the previous generator, namely the mean, variance and periodicity, to validate the independence of the values. We were also interested in calculating the response time of our generator and comparing it with that of the BBS to check its resistance to brute force and rainbow attacks.

The results were obtained from the analysis of the 65536-bytes key tables by directly programming the generator with MATLAB 2015 on a DELL Intel(R) Celeron(R) CPU N3060 computer, 1.60 GHz and 1.60 GHz CPU, 4 GB RAM, 64-bit operating system, card, on Windows 10 Pro (10.0.17134) with a high priority of the MATLAB software execution process.

To do this, we manipulated the input parameters of the generator, define the intervals to which they belong, the conditions to which they are subjected, and the modification procedures to carry out the study.

The results of the study on different pseudo-random number tables generated by the new cryptographically secure pseudo-random number generator in

Table 1. Samples of the PRNG test results.

Constants	$G_1 = 237$ $G_2 = 106$ $Y = 161$		
Variables	$N_1 = 683 \times 4943$ $N_2 = 4507 \times 36187$	$N_1 = 54419 \times 77383$ $N_2 = 22123 \times 104399$	$N_1 = 2311 \times 56467$ $N_2 = 109987 \times 118399$
Means	120.45	128.19	127.77
Standard deviation	75.57	74.49	74.22.23
Response time (seconds)	133.52	146.41	174.44
Constants	$G_1 = 198$ $N_1 = 23 \times 223$ $Y = 161$		
Variables	$G_2 = 75$ $N_1 = 37027 \times 21123$	$G_2 = 191$ $N_1 = 12934 \times 59611$	$G_2 = 30$ N1=32203 × 1307
Means	120.91	126.33	127.77
Standard deviation	83.40	73.33	74.52
Response time (seconds)	142.6	152.97	128.91
Constants	$N_1 = 54419 \times 77383$ $N_2 = 22123 \times 104399$ $Y = 161$		
Variables	$G_1 = 11$ $G_2 = 67$	$G_1 = 29$ $G_2 = 102$	$G_1 = 52$ $G_2 = 226$
Means	127.56	117.45	126.23
Standard deviation	75.41	69.37	78.17
Response time (seconds)	150.83	151.91	146.63
Constants	$G_2 = 26$ $N_2 = 47779 \times 27983$ $Y = 161$		
Variables	$G_1 = 75$ $N_1 = 36107 \times 71443$	$G_1 = 79$ $N_1 = 109987 \times 97687$	$G_1 = 37$ $N_1 = 42023 \times 2687$
Means	127.89	127.61	126.22
Standard deviation	74.84	73.63	76.12
Response time (seconds)	144.78	148.27	201.83

previous four cases, shows a good dispersion of the calculated random numbers from 0 to 256, with a considerable standard deviation (in the order of 75) significant in front of the mean. We also notice a considerable time processing ranging from a minimum of 2 min and 8s (128 s) to a maximum of 3 min and 21 s (201 s). Finally, the graphs of the distribution of random numbers of the different tables generated show a absence of period for any example of the four cases considered out of 65536 values each (Table 1).

Therefore, we have a cryptographically secure pseudo-random number generator that is complex, slow, and not periodic. The complexity gives a good dispersion of the calculated values, therefore random and independent values. It is also a slow generator compared to the sub-generator that composes it, i.e. the BBS, which has an already considerable computation time of about one second (1s) for a vector of 256 random numbers of one byte each.

Thus, at the end of the study and analysis of the results obtained concerning the pseudo-random number tables produced by the new CS-PRNG, we conclude that we have designed a cryptographically secure pseudo-random number generator that is complex, robust, and slow serving as the key table generation source for a fast crypt-system.

6 Conclusion

We present, over this paper, our approach developed to secure the IEC 60870-5-101 open protocol. Our approach consists of crypting the transmitted frames with a one time pad crypt-system. The crypt-system is based on two independent simple functions (XOR, permutation) to respond to the real time constraints.

To ensure better keys independence, the used keys are selected by the use of information related to the moment of sending the frames (key synchronizers bytes) and dependent of the frame checksum.

As future work, we propose to study other CS-PRNG used to generate the pseudo independent keys. In addition, we propose compact versions of HST-101 for equipment with limited computing resources; and introduce, adapt and test our approach with other SCADA protocols.

References

1. Aghaee, S.: Random Stream Cipher. Ph.D. thesis, Växjö University, School of Mathematics and Systems Engineering (2007)
2. Blum, L., Blum, M., Shub, M.: A simple unpredictable pseudo-random number generator. SIAM J. Comput. **15**(2), 364–383 (1986). https://doi.org/10.1137/0215025
3. Cherifi, T.: Developpement d un protocole scada temps reel securise a base de l iec 60870-5-101. Ph.D thesis.Ecole Nationale Polytechnique d Alger (2019)
4. Cherifi, T., Hamami, L.: A practical implementation of unconditional security for THE IEC 60780–5-101 SCADA protocol. Int. J. Crit. Infrastruct. Prot. **20C**, 68–84 (2018). https://doi.org/10.1016/j.ijcip.2017.12.001
5. Clarke, G.R., Reynders, D., Wright, E.: Practical modern SCADA protocols: DNP3, 60870.5 and related systems. Newnes (2004)

6. Ferguson, N., Schneier, B.: Practical Cryptography, vol. 23. Wiley, New York (2003)
7. Kelsey, J., Schneier, B., Ferguson, N.: Yarrow-160: notes on the design and analysis of the yarrow cryptographic pseudorandom number generator. In: International Workshop on Selected Areas in Cryptography, pp. 13–33. Springer (1999)
8. Knowles, W., Prince, D., Hutchison, D., Disso, J.F.P., Jones, K.: A survey of cyber security management in industrial control systems. Int. J. Crit. Infrastruct. Prot. **9**, 52–80 (2015). https://doi.org/10.1016/j.ijcip.2015.02.002
9. Kraft, M.B., Marks, E.: US Government Counterterrorism: A Guide to Who Does What. CRC Press, Boca Raton (2016)
10. Markoff, J.: Cyberwar: Old trick threatens the newest weapons, 26 October 2009
11. MIIT, P.: Guide to industrial control systems information security protection (2016)
12. Nazir, S., Patel, S., Patel, D.: Assessing and augmenting SCADA cyber security: a survey of techniques. Comput. Secur. **70**, 436–454 (2017)
13. Robinson, M., Jones, K., Janicke, H.: Cyber warfare: issues and challenges. Comput. Secur. **49**, 70–94 (2015). https://doi.org/10.1016/j.cose.2014.11.007
14. Sidorenko, A., Schoenmakers, B.: Concrete security of the Blum-Blum-Shub pseudorandom generator. In: IMA International Conference on Cryptography and Coding, pp. 355–375. Springer (2005)
15. Smaiah, S., Khellaf, A., Cherifi, T.: The implementation of SCADA open protocol IEC60870-5-101 on ARDUINO UNO board. In: 2015 4th International Conference on Electrical Engineering (ICEE), pp. 1–6. IEEE (2015). https://doi.org/10.1109/INTEE.2015.7416844
16. Smaiah, S., Khellaf, A., Cherifi, T.: The implementation of SCADA open protocol PUR 2.4. In: 2015 First International Conference on New Technologies of Information and Communication (NTIC), pp. 1–6 (2015). https://doi.org/10.1109/NTIC.2015.7368746
17. Stouffer, K., Falco, J., Scarfone, K.: Guide to industrial control systems (ICS) security. NIST Special Publication 800(82), 16–16 (2011)
18. TC57, I.: Transmission protocol - section 101: Companion standard for basic telecontrol tasks (1995)
19. Zhou, X., Xu, Z., Wang, L., Chen, K.: What should we do? A structured review of SCADA system cyber security standards. In: 2017 4th International Conference on Control, Decision and Information Technologies (CoDIT), pp. 0605–0614. IEEE (2017). https://doi.org/10.1109/CoDIT.2017.8102661

A Multi-view Clustering Approach for the Recommendation of Items in Social Networking Context

Lamia Berkani[1,2](✉), Lylia Betit[1,2], and Louiza Belarif[2]

[1] Laboratory for Research in Artificial Intelligence (LRIA), Algiers, Algeria
[2] Department of Computer Science, USTHB University, Bab Ezzouar, 16111 Algiers, Algeria
{lberkani,lbetit,lbelarif}@usthb.dz

Abstract. The review of the literature shows that the clustering-based approaches suffer from relatively low accuracy and coverage. We propose in this article a multi-views clustering approach for the recommendation of items in social networks. First, users are iteratively clustered from the views of both rating patterns and social information which includes different features, namely: friendship, trust and influence. Different clustering algorithms have been used (Kmedoids and CLARANS algorithms). Then, based on the multi-view clustering, recommendations are generated according to different hybridization. In order to evaluate our approach, experiments have been conducted using the well-known FilmTrust dataset. The results we have obtained show that our approach outperforms the existing related work approaches and baselines.

Keywords: Multi-view clustering · Social recommendation · Collaborative filtering · User's profile · Friendship · Trust · Influence

1 Introduction

Clustering-based approaches are being demonstrated to be efficient and scalable to large-scale data sets. As a dimension-reduction method, they are capable of alleviating the sparsity of rating data [1]. However, although clustering techniques have been demonstrated to be efficient and scalable to large-scale data sets, the clustering-based approaches gain less attention in recommender systems. Even if some works reported that by applying more advanced clustering method, the accuracy can be further improved and even outperform the other collaborative filtering (CF) approaches [2, 3]. To address this problem, Guo et al. [4] developed a multi-view clustering method through which users are iteratively clustered from the views of both rating patterns (user similarity) and social trust relationships (social similarity).

We focus in this paper by this same issue for the social recommendation of items. The proposed items' recommendation algorithm considered a multi-views clustering approach (MV), where users are clustered from the views of both user similarity (using users' assessments on items) and social relationships. In this paper we have formalized

M. R. Senouci et al. (Eds.): CSA 2020, LNNS 199, pp. 241–251, 2021.
https://doi.org/10.1007/978-3-030-69418-0_22

the social information that includes the list of friends, trust and influence. Different clustering algorithms have been used, including: Kmedoids and CLARANS algorithms. Furthermore, in order to determine the influencing medoids of the community, the Degree of Centrality heuristic have been used.

The remainder of this paper is organized as follows: Sect. 2 presents some related work on clustering-based recommender systems. Section 3 proposes a novel social recommender system using a multi-views clustering approach. Then in Sect. 4, experiments are conducted and analyzed. Finally, Sect. 5 highlights the most important results and draws some future perspectives.

2 Literature Review

Few works have tried to integrate social relationships into clustering-based methods with the aim of better performance of CF. Sun et al. [5] proposed a social regularization approach that incorporates social network information, namely the users' friendships and rating records (tags) for the prediction of the recommendations. They used a bi-clustering algorithm to identify the most suitable group of friends for generating different final recommendations. DuBois et al. [6] combined a correlation clustering algorithm and trust models together to derive trust from the connection distance in a trust network. However, only limited improvement is observed. On the other hand, Selvi and Sivasankar [7] considered a model-based CF based on a fuzzy c-means clustering approach.

According to Guo et al. [4] previous clustering-based approaches suffer from relatively low accuracy and, especially, coverage. To alleviate these issues, they developed a multi-view clustering method through which users are iteratively clustered on the basis of rating patterns in one view and social trust relationships in the other [4]. Sheugh and Alizadeh [8] proposed a multi-view clustering based on Euclidean distance, merging similarity and trust relationships including explicit and implicit trusts. He et al. [9] proposed a web items recommendation system based on a multi-content clustering CF model. According to the authors, a multi-view clustering can be applied to the mining of the similarity and relevance of items to improve the classic CF. Different views such as user ratings and user comments have been considered and users' preferences were analyzed by their historical interaction features and additional behavior features for an appropriate recommendation.

3 MV-Based Recommendation Approach

We present in this section our recommendation approach which proposes a novel multi-views clustering technique (MV). Figure 1 illustrates the general structure of our recommendation system, including the multi-views clustering-based recommendation as well as other hybrid algorithms. First, we start with a representation of the user's profile and then we expose the different algorithms of our recommendation system: (1) the CF (i.e. standard and clustering-based CF algorithms); (2) the Social filtering (SocF) (i.e. the standard and clustering-based SocF algorithms); and (3) the hybrid algorithm (HybF), including the weighted HybF (i.e. standard and clustering-based HybF) and the multi-view clustering-based HybF algorithm.

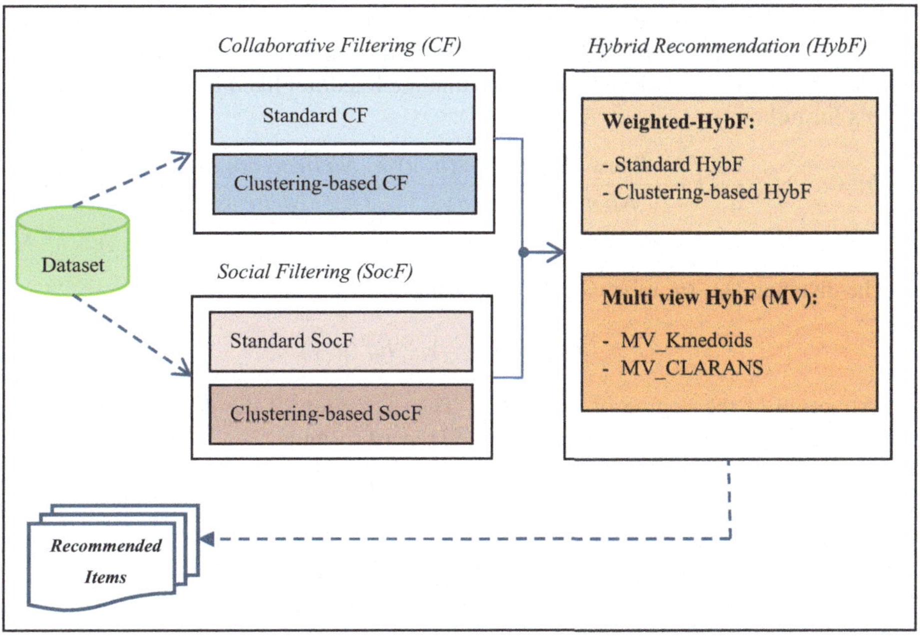

Fig. 1. General structure of our recommendation system

3.1 User's Profile Modeling

We consider that each user is characterized by a set of information including the following: assessments made on social network resources (reviews), the list of friends, trust and influence. Our recommendation algorithm is based on a hybridization of the two algorithms: collaborative filtering (CF) and social filtering (SocF). The similarity between two users is based on collaborative and social distance calculation.

Collaborative Distance. The similarity calculation between two users, u and v, is based on the evaluation history, and uses the Pearson correlation function [10]. The distance between two users, u and v, denoted $D_{Sim}(u, v)$ is calculated as follows:

$$D_{Sim}(u, v) = 1 - Sim_{Pearson}(u, v) \tag{1}$$

Social Distance. In order to determine the social relationship between users, we use three features, namely: trust, influence, and friendship:

1. Extraction of the degree of Trust (u, v): There are several algorithms for the calculation of trust. We have chosen the six-level method presented by Guo [4], which calculates how much two users, u and v trust each other, considering a distance equal to six. The distance $D_{Trust}(u, v)$ is calculated as follows:

$$D_{Trust}(u, v) = 1 - Trust(u, v) \tag{2}$$

2. Extraction of the degree of influence $Inf(u, v)$: The degree of influence between two users u and v is calculated taking into account the number of items that are liked/not liked by u and v according to the items that are liked/not liked by u, as presented by this formula:

$$Inf(u, v) = \frac{1}{2} * \left(\frac{Nb\,Items\,liked\,Comm(u, v)}{Nb\,items\,liked\,(u)} + \frac{Nb\,Items\,not\,liked\,Comm\,(u, v)}{Nb\,items\,not\,liked\,(u)} \right) \tag{3}$$

The distance $D_{Inf}(u, v)$ is calculated as follows:

$$D_{Inf}(u, v) = 1 - Inf(u, v) \tag{4}$$

3. Extraction of the degree of friendship: The extraction of the degree of friendship between two users is calculated with the Jaccard formula:

$$Friendship(u, v) = \frac{|F_u \cap F_v|}{|F_u \cup F_v|} \tag{5}$$

where: F_u is the set of friends of u and F_v is the set of friends of v.
The distance $D_{Friendship}(u, v)$ is calculated as follows:

$$D_{Friendship}(u, v) = 1 - Friendship(u, v) \tag{6}$$

3.2 Collaborative Filtering

We considered two variants of the CF: (1) the standard CF; and (2) The CF based on clustering technique. These algorithms will be presented as follows:

Standard CF. The standard CF allows detecting the community of the active user u based on the distances that separate him from the other users (D_{Sim}). The prediction will be based on the neighbors who are most similar to the active user. We chose the user-based filtering. This type of filtering starts by fully loading the usage matrix into memory (including user ratings on items), and then two important steps are executed:

- Selection of the K closest neighbors according to a given similarity threshold (if D_{Sim} $> =$ threshold), where the value of the threshold is fixed according to the density of the database and varies in the interval $[-1, 1]$.
- Prediction based on the ratings of the K closest neighbors.

Clustering-Based CF. This algorithm detects the community of the active user u using one of the clustering techniques. The prediction will be based on users' evaluations that belong to the same cluster as the active user. The CF algorithm based on the Kmedoids classification, for example, has been implemented according to the following two variants:

- The first variant of the algorithm consists in applying Kmedoids clustering.

- The second variant of the algorithm consists in applying the Kmedoids clustering, however when choosing the initial medoids we used a "Degree of centrality" heuristic to determine the influencing medoids of the community.

The CF algorithm based on Kmedoids is as follows:

```
Algorithm 1: CF-based Kmedoids

Input: Similarity Distance D_Sim, Active user U_a, Item j,
Number of medoids K, Heuristic: Boolean.

Output: Prediction for the active user on the item j.

BEGIN

  // Select Init_Medoids set according to the Heuristic

  1: If Heuristic = false then Init_Medoids = Random se-
     lection of K initial medoids.

     Else Init_Medoids = Selection of initial K medoids ac-
     cording to the "Degree of centrality" heuristic.

     End if.

  2: Apply the Kmedoids algorithm using Init_Medoids set
     and D_Sim for the computation of the distances;

  3: Identify the cluster of U_a.

  4: Predict (U_a, j).

END.
```

3.3 Social Filtering

As for the CF, the SocF algorithm consists of two variants: (1) the standard SocF; and (2) the Clustering-based SocF. The SocF considers the three features: trust, influence and friendship to calculate the social distance. A weighted formula was used to calculate the social distance, as follows:

$$D_{Soc} = \beta_1 D_{Trust} + \beta_2 D_{Inf} + \beta_3 D_{Friendship} \quad (7)$$

where: $\beta_1, \beta_2, \beta_3$: represent the importance weights related, respectively, to trust, influence and friendship, with: $\beta_1 + \beta_2 + \beta_3 = 1$.

Standard SocF. The pseudo algorithm below illustrates the standard SocF steps:

```
Algorithm 2: SocF ()

Input: Active user U_a, the distances: D_Trust, D_Inf, D_Friendship,
the item j, the number of k nearest neighbors to consider
for prediction.

Output: Prediction (U_a, j).

BEGIN

  1: Construct D_Soc table using the weighted D_Soc formula
     (Equation 7);

  2: Generate the clustering configuration based on D_Soc ;

  3: Select the K closest neighbors in terms of D_Soc;

  4: Apply the prediction (U_a, j) based on the assessments
     of the K nearest neighbors;

END.
```

Clustering-Based SocF. This algorithm is similar to the clustering-based CF. The only difference consists to replace the collaborative distance D_{Sim} by the social distance D_{Soc}.

3.4 CF and SocF Hybridization

The Hybrid filtering algorithm (HybF) combines the two filtering algorithms (CF and SocF) to consider both views (similarity view and social view) at the same time. We implemented the hybrid algorithm using two different methods, namely: (1) the weighted hybrid method (W-HybF); and (2) the multi-view clustering hybrid method (MV-HybF).

The Weighted HybF (W-HybF). This algorithm combines the CF and SocF algorithms according to the following formula:

$$D_{Hyb} = \alpha * D_{Sim} + (1 - \alpha) * D_{Soc} \tag{8}$$

where: α is the importance degree of the CF; and intuitively, $(1 - \alpha)$ is the importance degree of the SocF.

The detail of this formula is represented as follows:

$$D_{Hyb} = \alpha * D_{Sim} + (1 - \alpha) * \left(\beta_1 D_{Trust} + \beta_2 D_{Inf} + \beta_3 D_{Friendship}\right) \tag{9}$$

Therefore, we will have different hybrid algorithms: (1) the standard weighted hybrid algorithm; and (2) the weighted clustering-based hybrid algorithm.

1. *The standard weighted hybrid algorithm:* This algorithm starts by calculating the distances between the active user U_a and all the other users by applying the previous formula (Formula 9), in order to identify the k nearest neighbors to U_a. Then the prediction function will be applied.

2. *The weighted clustering-based hybrid algorithm:* The weighted hybrid algorithm with clustering proceeds according to the following steps:

- Calculate the distances between U_a and the other users using D_{Hyb};
- Generate a final configuration of clustering;
- Identify the cluster associated with the active user;
- Calculate the predictions using the harmonic average weighted prediction formula considering the users of the cluster where U_a belongs.

The MV-HybF. We adapted the multi-view algorithm presented by Guo et al. [4], with improvements and extensions. Our proposal consists to combine the social and similarity information. This hybridization method is, however, different from that mentioned in the weighted hybridization algorithm (W-HybF) because with this method we will not have to compute once the distance by considering a weighted combination between the two pieces of information. However we will calculate the distance D_{Sim} to make a classification according to the similarity and separately we calculate the distance D_{Soc} and we will make a classification according to a weighting of the social information. These two classification processes run in an alternative way and end with an integration phase to generate a final multi-views classification. The clustering-based MV is executed according to the following steps:

- *Step 1:* This step consists in enriching the social information of the MV algorithm considered in Guo et. al [4] (where only trust information has been taken into account), adding influence and friendship.
- *Step 2:* This step allows the selection of medoids randomly or using the "Degree of centrality" heuristic. The algorithms generated by this method are named as follows: Adapted_MV_Kmedoids and Adapted_MV_KMedoids_H; (with or without heuristic). In this step, the classification will be done once according to the social view, and another time according to the similarity view while passing the resulting medoids from each classification step to the next classification that will follow. Thus two clustering configurations of different views C_{Soc} and C_{Sim} will be built.
- *Step 3:* Apply an integration algorithm.
- *Step 4*: With this hybridization technique, a user can belong to at most two clusters, due to the integration step (step 3). Once the final configuration is obtained, we will only have to identify the community of the user U_a to be able to calculate its prediction on the item j to recommend. The latter can belong either to a single cluster "C" and thus the generation of prediction $P^{C}_{U_a,j}$ will be done with the formula of the harmonic mean, or it will belong to the intersection of two clusters, in this case the prediction generation will proceed in two ways:

1. *The Harmonic average method:* let us start with the most trivial method that will be noted *AVG*, which consists in calculating the average of the two predictions: $P^{C_1}_{U_a,j}$ and $P^{C_2}_{U_a,j}$ applying the formula of the harmonic mean and considering the two clusters C_1 and C_2 separately where U_a appears, as follows:

$$AVG = \frac{1}{2}\left(P^{C_1}_{U_a,j} + P^{C_2}_{U_a,j}\right) \tag{10}$$

2. *The SVR regression-based method:* in order to improve our results, we applied a second method for the generation of the prediction of the active user, which belongs to the intersection of two clusters. This method involves using supervised classification for the prediction, using the SVR regression technique.

4 Experiments

In order to evaluate our approach, experiments have been conducted using the well-known FilmTrust[1] dataset.

4.1 Dataset Description

FilmTrust is a database collected from the website that allows users to share movie ratings. It explicitly specifies other users as trusted neighbors. We used the dataset provided by Guo et al. [4] where the ratings are assigned in a range of 0.5 to 4 with a step of 0.5. This database contains 2853 trusted links linking 1508 users, making 35,497 reviews on 2071 listings. The density of this database is equal to 1.14.

4.2 Evaluation Metrics

We used the following two metrics: the MAE metric (Mean Absolute Error) and the RMSE metric (Root Mean Square Error):

$$MAE = \frac{\sum_{u,i\in\omega}\left|r_{u,i} - p_{u,i}\right|}{|\omega|} \tag{11}$$

$$RMSE = \sqrt{\frac{\sum_{u,i\in\omega}\left(r_{u,i} - p_{u,i}\right)^2}{|\omega|}} \tag{12}$$

where:

- ω: is the set of test evaluations and | ω | is the size of the test assessments;
- $r_{u,i}$: represents the evaluations of the user u on the item i;
- $p_{u,i}$: represents the prediction of the evaluations of the user u on the item i.

4.3 Results

Evaluation of the CF. We have evaluated the CF with a variation of the similarity threshold (CF-Threshold) from 0.1 to 1. The best value obtained is for the CF-threshold equal to 0.6. This value will be considered in the rest of our evaluations.

Contribution of Social Information. To better explore the contribution of social information to the recommendation and to set our parameters, we evaluated all the social information, namely: trust, influence and friendship with different weights and varying the number of clusters K. The results are presented in Table 1, where we have highlighted in bold the best obtained values with the number of clusters $K = 95$. These values will be considered in the rest of our evaluations.

[1] www.FilmTrust.com.

Table 1. Evaluation of the social information

Weights	K	MAE	RMSE
$\beta_1 = 0.33$, $\beta_2 = 0.33$, $\beta_3 = 0.33$	95	0.685208	0.8812
$\beta_1 = 0.25$, $\beta_2 = 0.5$, $\beta_3 = 0.25$	**95**	**0.68460**	**0.8825**
$\beta_1 = 0.5$, $\beta_2 = 0.25$, $\beta_3 = 0.25$	95	0.684779	0.8810
$\beta_1 = 0.25$, $\beta_2 = 0.25$, $\beta_3 = 0.5$	95	0.684782	0.8811

Evaluation of the Standard HybF. We tested different weights that allowed us to determine the best values between the CF and the SocF. Table 2 shows these results:

Table 2. Evaluation of the standard HybF

Weights	K	MAE	RMSE
$\alpha = 0.5$	100	0.64174	0.8648
$\alpha = 0.3$	100	0.64173	0.8647
$\alpha = 0.7$	**100**	**0.63491**	**0.8575**

Contribution of Clustering on the Recommendation. We have evaluated the clustering-based CF and the clustering-based SocF. Table 3 summarizes the results we have obtained using the clustering Kmedoids algorithm with the different recommendation algorithms (CF, SocF and HybF):

Table 3. Comparison between CF, SocF and HybF algorithms

Algorithm	Without heuristic	With heuristic
CF using D_{Sim}	MAE: 0.7004 RMSE: 0.8946	MAE: 0.6913 RMSE:0.8712
SocF using D_{Soc}	MAE: 0.7378 RMSE:0.9371	MAE: 0.701 RMSE:0.88
HybF ($\alpha * D_{Sim} + (1 - \alpha) * D_{Soc}$)	**MAE: 0.6887** **RMSE: 0.8767**	**MAE: 0.691** **RMSE: 0.8752**

We can see that hybridization has significantly improved the CF algorithm in terms of MAE and RMSE (for both cases, using the heuristic or without using it).

Clustering-based recommendation algorithms (see Table 4):

Table 4. Evaluation of the MV-based recommendation algorithm

Multi-Views approaches	K	MAE	RMSE
MV_Kmedoids_H	50	0.6287	0.7091
MV_CLARANS_H	30	**0.5675**	0.7413
MV_CLARANS_Trust	10	0.6115	0.8051
MV_Kmedoids-Trust [4]	10	0.67271	0.8604

- The adapted MV with Kmedoids using the heuristic "Degree of centrality" and considering all the social information, i.e. trust, influence and friendship. This algorithm is noted by: MV_Kmedoids_H.
- The MV with Kmedoids and considering only Trust information, which is equivalent to the approach proposed by Guo et al. [4]. This algorithm will allow us to perform a comparison of our algorithms with the existing related work algorithms.
- The adapted MV with CLARANS clustering algorithm using the heuristic "Degree of centrality" and considering all the social information. Furthermore we have also considered the MV with CLARANS where the social information based on the trust information.

We can see the contribution of the MV algorithm compared to the existing related work. The use of friendship and influence in addition to trust to model social information has contributed to this improvement. Furthermore, the use of the degree of centrality heuristic has also enhanced the results, especially with the CLARANS algorithm which has given the best results compared to the Kmedoids algorithm.

5 Conclusion

This paper proposes a recommendation approach which applies a multi-views clustering method for the recommendation of items. Different clustering algorithms have been used. The multi-views approach (combining the similarity view and the social view) proved to be the best in terms of MAE and RMSE, compared to the hybrid and the clustering-based hybrid algorithms. The results we have obtained are promising especially when using CLARANS algorithm and the degree of centrality heuristic. In our future work, we plan to test our approach with other databases and compare it with other existing approaches [11]. Moreover, we envisage to further improve the recommendation accuracy: (1) using optimization techniques to have the best partitioning for a given clustering such as in [12]; (2) applying deep learning techniques [13]; and (3) integrating other features like users' interaction behavior and items popularity such as in [14].

References

1. Pham, M., Cao, Y., Klamma, R., Jarke, M.: A clustering approach for collaborative filtering recommendation using social network analysis. J.UCS J. **17**, 583–604 (2011)
2. Bellogín, A., Parapar, J.: Using graph partitioning techniques for neighbour selection in user-based collaborative filtering. In: Proceedings of the 6th ACM Conference on Recommender Systems (RecSys), pp. 213–216 (2012)
3. Salah, A., Rogovschi, N., Nadif, M.: A dynamic collaborative filtering system via a weighted clustering approach. Neuro Comput. **175**, 206–215 (2016)
4. Guo, G., Zhang, J., Yorke-Smith, N.: Leveraging multi-views of trust and similarity to enhance clustering-based recommender systems. KBS J. **74**, 14–27 (2015)
5. Sun, Z., Han, L., Huang, W.G., Wang, X., Zeng, X., Wang, M., Yan, H.: Recommender systems based on social networks. J. Syst. Softw. **99**, 109–119 (2015)
6. DuBois, T., Golbeck, J., Kleint, J., Srinivasan, A.: Improving recommendation accuracy by clustering social networks with trust. Recommender Syst. Social Web **532**, 1–8 (2009)
7. Selvi, C., Sivasankar, E.: A novel optimization algorithm for recommender system using modified fuzzy c-means clustering approach. Soft Comput. **23**, 1–16 (2017)
8. Sheugh, L., Alizadeh, S.H.: Merging similarity and trust based social networks to enhance the accuracy of trust-aware recommender systems. J. Comput. Rob. **8**(2), 43–51 (2015)
9. He, X., Kan, M.Y., Xie, P., Chen, X.: Comment-based multi-view clustering of web 2.0 items. In: International World Wide Web Conference, Seoul, Korea, pp. 771–781. ACM (2014)
10. Adomavicius, G., Tuzhilin, A.: Toward the next generation of recommender systems: a survey of the state-of-the-art and possible extensions. TKDE **17**, 734–749 (2005)
11. Li, W., Ye, Z., Xin, M., Jin, Q.: Social recommendation based on trust and influence in SNS environments. Multimedia Tools Appl. **76**(9), 11585–11602 (2017)
12. Berkani, L.: Social-based collaborative recommendation: bees swarm optimization based clustering approach. In: The 9th International Conference on Model & Data Engineering (2019)
13. Deng, S., Huang, L., Xu, G., Wu, X., Wu, Z.: On deep learning for trust-aware recommendations in social networks. IEEE Trans. Neural Netw. Learn. Syst. **28**(5), 1164–1177 (2016)
14. Lai, C.H., Lee, S.J., Huang, H.L.: A social recommendation method based on the integration of social relationship and product popularity. Int. J. Hum Comput Stud. **121**, 42–57 (2019)

Image and Speech Processing

Analysis of the Influence of the Arabic Fricatives Vocalic Context on Their Spectral Parameters

Aicha Zitouni[1](✉), Leila Falek[1], Mourad Abbas[2], Mohamed Lichouri[2], and Khaled Lounnas[1]

[1] University of Science and Technology Houari Boumediene, FEI, LCPTS, Algiers, Algeria
{a.zitouni,l.falek,k.lounnas}@usthb.dz

[2] Computational Linguistics Department, CRSTDLA Research Center, Algiers, Algeria
{m.abbas,m.lichouri}@crstdla.dz

Abstract. An analysis of the right vowel context influence on their spectrum of fourteen Arabic fricative was presented in this study. The results were validated on the base of the recognition of these fricatives from a system based on Neural Network (Multilayer Perceptron MLP). The latter showed that the number of fricatives considered, the vowel context and the voiced/unvoiced type of fricatives used play a decisive role in the recognition rate. In addition, an improvement in this rate has been possible by associating the center of gravity (CoG) with the cepstral parameters (MFCC, ΔMFCC, ΔΔMFCC) of the features vector used. A reduction in execution time from 03 h to 18 min (reduction of 90%) was possible by reducing the size of the features vector from 39 parameters (13MFCC + 13ΔMFCC + 13ΔΔMFCC) to 14 parameters (13MFCC + 01CoG) without a great variation in the recognition rate.

Keyword: MFCC · CoG · Neural Network · Arabic fricatives · Speech recognition

1 Introduction

One of the important specificities of the Arabic language is its rich exclusively consonant alphabet. More complex than each other, Arabic consonants whose count is twenty-eight are generally classified on the basis of their articulation place which is in direct relation with the consonant spectral representation (fundamental frequency and formants). Vowels that distinguish them in a text are represented by diacritical symbols. In this study, we looked at the influence of the right vocalic context of the consonant on its spectral parameters, namely the center of gravity (CoG) and the cepstral parameters (MFCC). The study validation was made on the basis of the recognition of the various fricatives considered which are 14. These latter, characterized by a large amount of spectral energy in high frequencies [1], justifies the CoG use as fricatives descriptor in a classic automatic consonant recognition system which uses the cepstral parameters as components of its input features vectors.

M. R. Senouci et al. (Eds.): CSA 2020, LNNS 199, pp. 255–264, 2021.
https://doi.org/10.1007/978-3-030-69418-0_23

The interest of this work is firstly, on the language used, which is the Arabic language, whose number of studies is relatively limited compared for example to Latin languages, and secondly, in the number of fricatives considered in our recognition system. Indeed, no study until now has considered such a high number of fricatives (14 fricatives). The majority of them are limited to four [2] or at most five [3] fricatives. The validation tool used is a consonant automatic recognition system based on neural network (MLP) whose descriptor vector is composed of MFCC and CoG parameters (13MFCC, 13ΔMFCC, 13ΔΔMFCC, 01CoG).

This study will allow us: to show the effect of the vocalic context on the spectral structure of these sounds and to better separate the fricatives between them by identifying those that are the most difficult to recognize perceptually as: س [s], ص [sˤ]. In addition, to reduce the significant execution time resulting from the length of the descriptor vector: 39 parameters (13MFCC, 13ΔMFCC, 13ΔΔMFCC), a study of the descriptors type influence on the recognition rate limited their number to 14 (13MFCC, 01CoG) without much change on the result (see Table 2).

The rest of the paper will be organized as follows: Sect. 2 gives an overview on the state of the art, then the methodology of features extraction used for the purpose of consonant recognition was mentioned in Sect. 3. Section 4 describe the used corpus and methods used during the design of the system as well as the obtained results. We end the work by conclusion and perspectives in Sect. 5.

2 Related Work

Despite the high number of studies carried out in the field of speech signal processing, there is not much work done on fricatives and in particular on Arabic fricatives. Among the works that we have found in the literature on this field, we cite a work done in 2000 by Henry Fu et al. [3] where they proposed an algorithm for the classification of unvoiced fricatives based on using moments computed from the spectra of speech signals: the five English unvoiced fricatives used /f/, /th/, /s/, /sh/, and /h/, were be partially separated.

In 2012, Chelali and Djeradi [4] have developed an Arabic fricatives speech/speaker recognition system using MFCC analysis and the VQ-LBG algorithm in order to analyze the effect of compression in an acoustic parameterization phase.

In 2013, Patgiri and al. [2] used a Recurrent Neural Network (RNN) to describe the recognition of four fricatives from Assamese speech, where a feature vector is generated from center of gravity (CoG), standard deviation (SD), skewness and kurtosis. Experimental results show around 96% recognition rate, which is much higher than the phoneme recognition strategy.

In 2014, Frid and Lavner [5] used kNN or SVM for classification of four unvoiced fricatives. The data are represented by two dimensionality reduction algorithms, namely, t-distributed Stochastic Neighbour Embedding (t-SNE) and Sequential For-ward Floating Selection (SFFS). This study gives a classification rate of almost 90% by a feature vector with as few as 3 dimensions.

In 2014, Frid and Lavner [1] used Support Vector Machine (SVM) for fricatives classification to examine the discrimination between voiced and unvoiced fricatives, classification of unvoiced fricatives, classification of voiced fricatives and the discrimination between sibilants and non-sibilants from combination of several acoustic time and frequency domain features.

3 Features Extraction

In this section, we present the different acoustic parameters used to create a reliable features vector which would allow obtaining a good recognition rate using the neural network (MLP).

3.1 Mel Frequency Cepstral Coefficients (MFCC)

MFCC is based on the human peripheral auditory system [2]. It is the most commonly used speech parameterization method in speech and speaker recognition because it solves the problem of frequency sensitivity of the hearing system while remaining in the human hearing range through the conversion on the Mel scale [6].

For each signal, we proceeded as follows [7]:

- Signal pre-emphasis using a high-pass filter;
- The speech signal is divided into a number of overlapping time windows (Hamming) of 20 ms with a frame period of 5 ms;
- Transformation in Mel scale using a filterbank of 24 Triangular filters;
- 13 MFCC are extracted from each speech frame.
- 13ΔMFCC + 13ΔΔMFCC (dynamic features) are calculated from each speech frame

3.2 Spectral Centroid

The spectral centroid is defined as the center of gravity of the magnitude spectrum of the Short Time Fourier Transform (STFT) [5]:

$$S_i = \frac{\sum_{k=0}^{k-1} M_i(k).k}{\sum_{k=0}^{k-1} M_i(k)} \quad (1)$$

Where $M_i(k)$ is the magnitude of the Fourier Transform.

In acoustic phonetics, the center of gravity is a good indicator of the frequency distribution of spectrum energy [8].

3.3 Neural Network

Artificial Neural Networks is a widely used method for recognition and classification of different speech types (type of consonants, voicing, language, etc.). In general, an ANN is a set of nodes that correspond to neurons and links that represent the connections and the flow of data between neurons. Connections are quantified by weights, which are dynamically adjusted during learning. The input vectors must be associated with a desired output, which is encoded as another vector, called the desired output vector [7].

To adjust the neural network parameters, we carried several experiments where we changed one parameter by fixing the others and choosing the optimal value obtained. The preserved values are:

- The number of iterations: 100;
- The value of the gradient: 10^{-15};
- The learning step: 0.01;
- The value of the global error (threshold): 0;
- The transfer function: the sigmoid (continuous and derivable function).

4 Experimental Setup and Results

We analyze in this section the feature extraction process describing the acoustic features MFCC and CoG, using neural network to obtain the recognition rate for each parameterization.

4.1 Linguistic Material

The corpus used in this work contains 14 fricatives Arabic.

ث ج ح خ ذ ز س ش ص ظ ع غ ف ه

The stimuli speech units are VCV (Vowel-Consonant-Vowel), where the left vocal is context /V/ is the vowel (/a/), /C/: consonant is fricative and the right vocalic context /V/ is one of the three Arabic vowels (/a/, /i/, /u/). Six Arabic native speakers (03 males and 03 females) repeat the same /VCV/ ten times. The data base includes 2520 speech units recorded at a sampling rate 44.1 kHz with Praat software. The total duration of the speech corpus is 45 min. The fricatives are extracted and their frequencies are filtered between 1000 Hz and 15000Hz (related to fricatives frequency interval).

4.2 Features Extraction (MFCC Coefficients and CoG)

The classification performance depends on the set of selected descriptors [8]. The MFCC feature extraction process was performed frame by frame. The frame length is set to 20 ms with a step of 5ms to extract 13MFCC coefficients per frame where the first represents Energy. For a same frame, ΔMFCC, ΔΔMFCC and CoG are calculated. The results of the center of gravity and the standard deviation for each fricative are shown in the Table 2 by changing the right vocalic context. Each input utterance is transformed into a matrix of 40 columns and N rows (N is number of frames). Each column is scaled to zero mean and unit variance. The N rows are divided into three portions (30%, 40%, 30%) to calculate the average of each portion and obtain a matrix of 40 columns and 03 rows [9]. The input matrix of each vowel has 2520 rows (14 fricatives*03 averages *10 repetitions*06 speakers), it is divided into two matrix, 80% for learning and 20% for test.

4.3 Center of Gravity and Standard Deviation

Table 1 shows the variation of the CoG with respect to three factors: the place of articulation, phonation and the right vocalic context. In Fig. 1 and Fig. 2, we also showed

Table 1. Center of gravity (Hz) and Standard deviation (Hz) by changing the right vowel context.

Place of articulation	Voiced fricative				Unvoiced fricative			
	Fricative	Right context	CoG	Standard deviation	Fricative	Right context	CoG	Standard deviation
Labiodental					ف [f]	/a/	6020	1069
						/i/	6568	1209
						/u/	6252	1217
Dental	ذ [ð]	/a/	431	176	ث [θ]	/a/	4771	2034
		/i/	557	387		/i/	4617	1995
		/u/	559	484		/u/	5542	1805
Alveolar	ز [z]	/a/	5724	1255	س [s]	/a/	7512	965
		/i/	6137	1028		/i/	7449	1015
		/u/	5769	649		/u/	6802	777
Postalveolar	ج [ʒ]	/a/	4372	838	ش [ʃ]	/a/	5338	453
		/i/	4510	798		/i/	5202	455
		/u/	4111	1455		/u/	5509	447
Uvular	غ [ʁ]	/a/	752	172	خ [χ]	/a/	4417	691
		/i/	843	242		/i/	4536	1087
		/u/	730	365		/u/	4339	940
Pharyngeals	ع [ʕ]	/a/	926	243	ح [ħ]	/a/	2663	563
		/i/	1057	292		/i/	2860	560
		/u/	871	243		/u/	2808	660
Laryngals	ه [h]	/a/	959	265				
		/i/	744	208				
		/u/	688	183				
Pharyngealized	ظ [ðˤ]	/a/	444	165	ص [sˤ]	/a/	7449	978
		/i/	637	528		/i/	7454	1197
		/u/	561	306		/u/	6998	777

that the CoG changes between a male and female speaker for the same fricative. For all these factors, we obtained for each fricative, an average value and standard deviation.

The CoG is not a reliable indicator to distinguish between fricatives. Results of ذ [ð] and its emphatic ظ [ðˤ] are in the same interval, as well as س [s] and its emphatic ص [sˤ].

Fricatives ذ [ð], ث [θ] and ظ [ðˤ],, which are dental sounds, have high standard deviations due to their large variation in pronunciation by the speakers.

Fricatives ح [ħ] and ش [ʃ] can be characterized by their CoG since their means are distant and their standard deviations are low.

The fricatives ز [z] and ج [ʒ] are voiced sounds but their CoG are high and approximate those of the unvoiced sound, which gives an overlap between ز [z] and ف [f] and between ج [ʒ] and خ [χ].

The CoG of the 03 fricatives ع [ʕ], غ [ʁ] and ه [h] are located in the low frequencies that are a little overlapping (see Fig. 3).

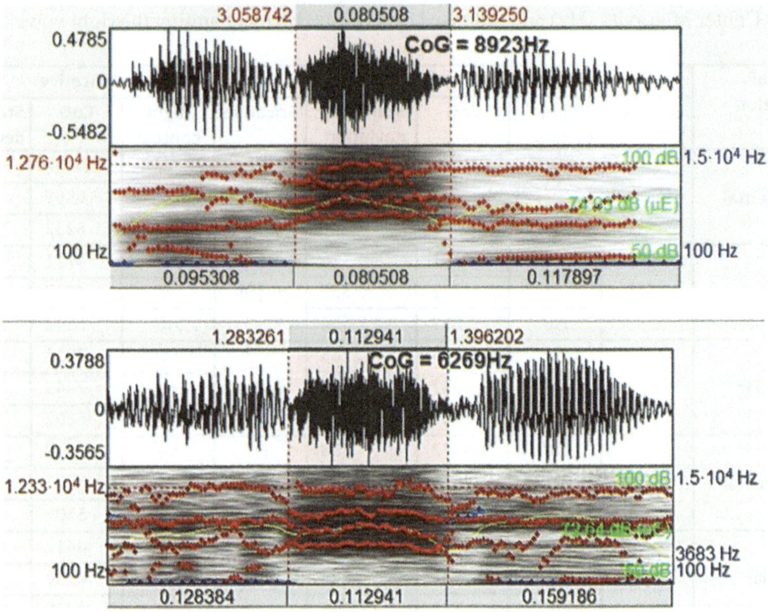

Fig. 1. Friction noise of fricative "س [s]" in vowel context (female: top, male: bottom)

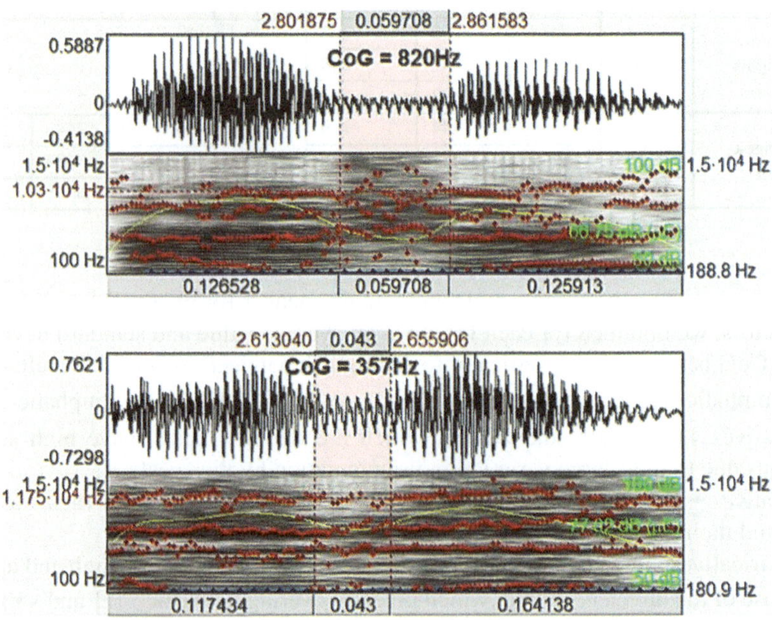

Fig. 2. Friction noise of fricative "ذ [ð]" in vowel context (female: top, male: bottom)

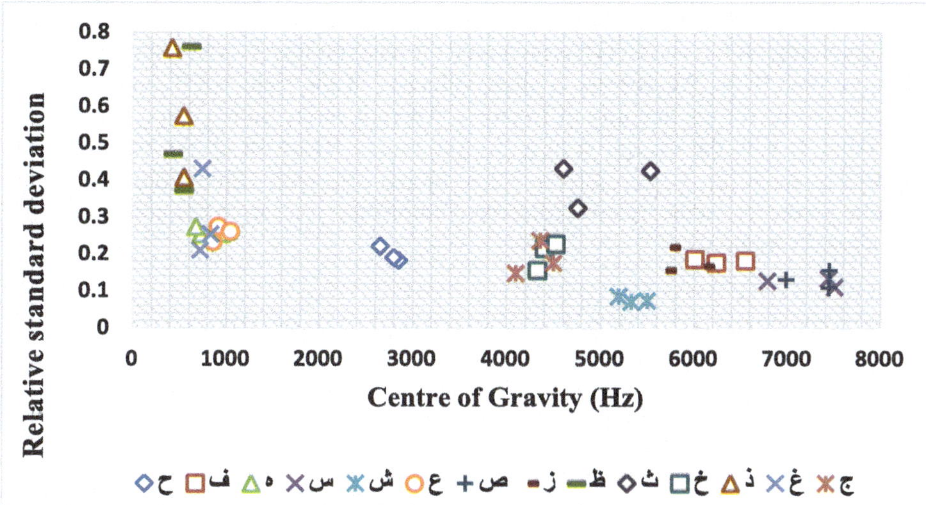

Fig. 3. Relative standard deviation as a function of CoG.

In the next part, we will combine MFCC parameters and CoG to obtain a best recognition rate by optimizing the features vector and the execution time.

4.4 Recognition Rate of Fricatives by Changing the Right Vowel Context and Features Vector

In this part, we aimed the impact of the right vowel context on the fricative with the change of the feature vectors. The recognition rate is calculated for each vowel using neural network. The results are summarized in Table 2. In general, each feature vector increases the recognition rate. We can see that the impact of the vowel /a/ is greater leading to a lower rate (92.26%), while for the other vowels we get a better rate: 93.06% for the vowel /i/ and 95.03% for the vowel /u/.

Table 2. Recognition rate of the various parameters

Features vector	Vowels		
	a	i	u
13 MFCC	86.11%	86.90%	89.48%
13 MFCC + 13ΔMFCC + 13 ΔΔMFCC	91.27%	90.48%	94.25%
13MFCC + 01CoG	91.67%	91.46%	94.64%
13MFCC + 13ΔMFCC + 13ΔΔMFCC + 01CoG	92.26%	93.06%	95.03%

We can also infer from this table, that by replacing **Δ**MFCC and **ΔΔ**MFCC by the CoG parameter, we can have the same performance while minimizing the execution time from 03 h (39 parameters) to 18 min (14 parameters).

For the rest of this article, the parameters chosen are MFCC + CoG (14 parameters).

Combining MFCC and CoG has increased the recognition rate where some fricatives are more than 97%. We observe a great influence of the right vocal context on the recognition rate for the majority of fricatives where it varies from a context to another (see Table 3) with the exception of ح [ħ], the rate is 100% for all the right contexts followed by the غ [ʁ], خ [χ] , ش [ʃ], ع [ʕ], ذ [ð] and ه [h] with a little difference, on the opposite for the rest of the fricatives, the recognition rate varies from a context to another with a great variation.

Table 3. Success rate of Arabic fricatives using 14 features (MFCC and COG)

Vowels / Fricatives	a	i	u	Mean
ث [θ]	78%	86%	97%	87%
ج [ʒ]	100%	83%	86%	90%
ح [ħ]	100%	100%	100%	100%
خ [χ]	97%	94%	100%	97%
ذ [ð]	100%	97%	100%	99%
ز [z]	83%	94%	100%	92%
س [s]	69%	81%	72%	74%
ش [ʃ]	100%	94%	97%	97%
ص [sˤ]	81%	89%	100%	90%
ظ [ðˤ]	100%	83%	92%	92%
ع [ʕ]	97%	97%	100%	98%
غ [ʁ]	92%	92%	94%	93%
ف [f]	89%	89%	94%	91%
ه [h]	97%	89%	92%	93%

4.5 Recognition Rate of Unvoiced and Voiced Fricatives

In this part, we have separated the voiced fricatives and the unvoiced fricatives.

Table 4. Recognition rate of unvoiced fricatives.

Fricatives / Vowels	ث [θ]	ح [ħ]	خ [χ]	س [s]	ش [ʃ]	ص [sˤ]	ف [f]
Context /a/	83%	100%	97%	92%	100%	89%	97%
Context /i/	97%	100%	94%	97%	100%	94%	94%
Context /u/	89%	100%	100%	92%	100%	100%	100%

Table 5. Recognition rate of voiced fricatives.

Fricatives / Vowels	ج [ʒ]	ذ [ð]	ز [z]	ظ [ðˤ]	ع [ʕ]	غ [ʁ]	ه [h]
Context /a/	100%	97%	94%	94%	97%	97%	100%
Context /i/	100%	94%	100%	83%	100%	97%	100%
Context /u/	100%	97%	100%	92%	100%	100%	100%

Tables 4 and 5 shows a very large increase in the recognition rate as the number of classes has been reduced (from 14 to 7 classes). Several fricatives achieved 100% in all right contexts (ج [ʒ], ش [ʃ], ح [ħ] and ه [h]).

A confusion was can observe between ث [θ] and ف [f], س [s] and its emphatic ص [sˤ] and between ذ [ð] and its emphatic ظ [ðˤ]. .

For the voiced fricatives, the system produced a recognition rate of 94%, 97% and 97% for the right contexts /a/, /i/ and /u/ respectively and a rate of 97%, 96% and 98% for the unvoiced fricatives and for the same right contexts.

4.6 Classification of Fricatives

In this part, all vowels are in the same matrix. Figure 4 shows that the recognition rate is lower for all fricatives when comparing the results obtained with the averages in Table 3. This confirms that the right vowel context has a great influence on the consonant's comportment and can change the distribution of their frequencies.

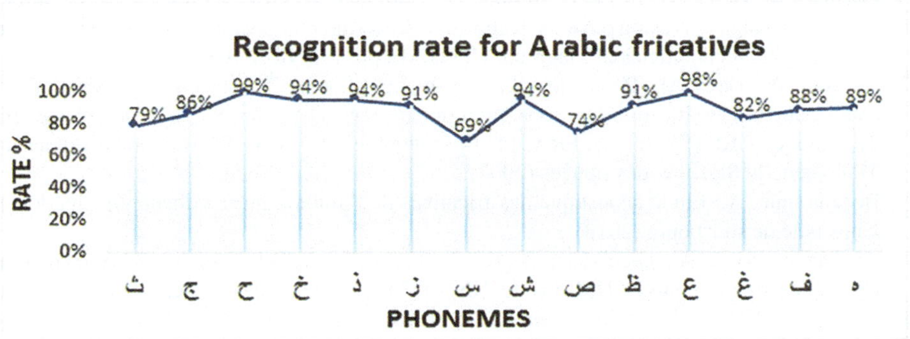

Fig. 4. Recognition rate obtained for Arabic fricatives

5 Conclusion and Perspectives

In this study, we have analyzed the impact of the right vocalic context of 14 Arabic fricatives placed in speech units: vowel-consonant-vowel /VCV/. The analysis tool used is a consonant recognition system based on the multilayer perceptron neural network (MLP). Results obtained made it possible to highlight the impact of the vowel context on the spectrum of each fricative studied. The great difficulty of this study lies in the insufficient bibliographical references in this domain. This has limited us in validating our results. In fact, the important number of Arabic fricatives studied has made our system more complex (14 classes considered in the MLP). In the literature, most studies do not exceed 5 fricatives.

Results obtained are satisfactory, however, improvement can be made by augmenting the number of speakers in order to use convolutional neural networks (CNNs) [10] which will bring more value for this study.

References

1. Frid, A., Lavner, Y.: Spectral and textural features for automatic classification of fricatives using SVM. In: IWSSIP 2014 Proceedings, pp. 99–102. IEEE, May 2014
2. Patgiri, C., Sarma, M., Sarma, K.: Recurrent neural network based approach to recognize assamese fricatives using experimentally derived acoustic-phonetic features. In: 2013 1st International Conference on Emerging Trends and Applications in Computer Science, pp. 33–37. IEEE, September 2013
3. Fu, H., Rodman, R., McAllister, D., Bitzer, D., Xu, B.: Classification of voiceless fricatives through spectral moments. Master's thesis, North Carolina State University (1999)
4. Chelali, F.Z., Djeradi, A.: MFCC and vector quantization for Arabic fricatives speech/speaker recognition. In: 2012 International Conference on Multimedia Computing and Systems, pp. 284–289. IEEE, May 2012
5. Frid, A., Lavner, Y.: Spectral and textural features for automatic classification of fricatives. In: XXII Annual Pacific Voice Conference (PVC), pp. 1–4. IEEE, April 2014
6. Lounnas, K., Demri, L., Falek, L., Teffahi, H.: Automatic language identification for Berber and Arabic languages using prosodic features. In: 2018 International Conference on Electrical Sciences and Technologies in Maghreb (CISTEM), pp. 1–4. IEEE, October 2018
7. Rabaoui, A., Davy, M., Rossignol, S., Lachiri, Z., Ellouze, N.: Sélection de descripteurs audio pour la classification des sons environnementaux avec des SVMs mono-classe. In: 21 Colloque GRETSI, Troyes, FRA, 11–14 septembre 2007. GRETSI, Groupe d'Etudes du Traitement du Signal et des Images (2007)
8. Benamranne, A.: Etude acoustique des fricatives de l'arabe standard (locuteurs algériens), Université de strasbourg (2013)
9. Ahcène, A., Aissa, A., Abdelkader, D., Khadidja, B., Ghania, D.H.: Automatic segmentation of Arabic speech signals by HMM and ANN. In: 2016 International Conference on Electrical Sciences and Technologies in Maghreb (CISTEM), pp. 1–4. IEEE, October 2016
10. Lounnas, K., Abbas, M., Lichouri, M.: Building a speech corpus based on Arabic podcasts for language and dialect identification. In: Proceedings of the 3rd International Conference on Natural Language and Speech Processing, pp. 54–58 (2019)

Topological Analysis for Osteoporosis Assessment in X-Ray Images Using Random Forest Model

Khaled Harrar(✉)

LIST Laboratory, Faculty of Technology, University M'Hamed Bougara of Boumerdes, 35000 Boumerdes, Algeria
khaled.harrar@univ-boumerdes.dz

Abstract. This paper deals with the characterization of the texture of the bone. The objective is to combine clinical parameters and topological attributes extracted from X-Ray images for osteoporosis assessment. A total of 120 women are included in this study, divided into two populations, composed of 60 healthy subjects, and 60 osteoporotic patients. Nine features are involved and a random forest model is used to differentiate the patients. Different configurations are tested, trained, and validated using *k*-fold cross-validation technique with different values of *k*. The aim is to seek the combination giving the best accuracy for discriminating between the populations. Several classifiers are also tested and compared. The obtained results affirm that the tenfold cross-validation technique with the random forest model combining the topological and the clinical parameters outperform the other classifiers, and provide high accuracy (ACC = 87,5%) demonstrating the efficiency of this model.

Keyword: Topological analysis · Texture · Osteoporosis · Random forest

1 Introduction

Osteoporosis is a sickness of the bone tissue, resulting in a loss of the trabecular network connectivity. In the clinical routine, doctors use the bone mineral density (BMD) measurement with dual-energy X-ray absorptiometry (DXA) for osteoporosis assessment [1], and it is recognized as the gold standard to diagnose the disease. Therefore, BMD provides only 60% of this disease [2]. The microarchitecture plays a crucial role in bone characterization and has proven as a complementary parameter to the BMD [3]. To assess the microarchitecture, doctors use a bone biopsy, which is an invasive method. Texture analysis allows characterizing microarchitecture and provides an assessment of the osseous fragility [4, 5].

Various studies were conducted on the bone texture analysis for osteoporosis assessment. Kawashima et al. [6] worked on head CT images to find osteoporosis cases from healthy bone density. They used histogram and Law's features. Their finding reveals that the texture parameters demonstrated significant differences ($p < 0.01$) between normal bone density and osteoporosis. Harrar et al. [7] proposed a method for the porosity

M. R. Senouci et al. (Eds.): CSA 2020, LNNS 199, pp. 265–275, 2021.
https://doi.org/10.1007/978-3-030-69418-0_24

quantification in radiographic bone images. Two methods were used, star volume and fractal lacunarity. The study involved 80 healthy cases and 80 pathological patients. The results showed good discrimination between the populations. Areeckal et al. [8] presented a low-cost pre-screening technique for screening osteoporosis, using wrist and hand images. The co-occurrence matrix and the morphological gradient method were used to separate 134 Indian subjects and 65 swiss samples population. The obtained results suggest that the weighted KNN classifier showed the best test accuracy of 78%.

Most of the presented works deal with texture analysis consider only the gray level intensity of the radiographs to characterize the bone. Therefore, just a few studies were conducted on the microstructure of bone architecture.

The aim of this paper is the assessment of osteoporosis using a topological technique. In one hand, different attributes of the bone X-Ray images are extracted including the number of termini and nodes, the number of branches whose extremities are nodes, the number of branches whose extremities are nodes and termini, and the number of trees. On the other hand, clinical parameters of the patients including age, bone mass index, bone mineral content, and bone mineral density, are combined with the topological attributes. Several configurations are trained and validated using *k*-fold cross-validation method, and the random forest model.

This paper is organized as follows: In Sect. 2, the dataset is described. Section 3 provides the adopted methodology, and Sect. 4 presents results and discussion. Finally, a conclusion is given in Sect. 5.

2 Dataset Description

A total of 120 women were enrolled in this study, divided into two groups 60 control subjects (healthy) with age 68.51 ± 10.21 SD, and 60 osteoporotic cases with fractures (pathological) aged 70.98 ± 10.78 SD. All the images were provided from patients of Orleans' hospital. The site was chosen on the calcaneus bone (heel) using a digital X-ray apparatus (Medical D3A Systems, BMA™, Orleans, France) [3]. The same technique of acquisition was used for all the persons, with the distance of the focal set to 1.15 m, the exposure time set to 20 mAs, and the voltage to 55 kV. Two markers were used for each patient to give the same ROI [9, 10].

It was located by the operator on the image (A and B in Fig. 1a). The ROI gave images of 1.6 × 1.6 cm^2 (Fig. 1a) which were coded into 16 bits and sized to 400 × 400 pixels. Figure 1 illustrates a sample of an ROI of a control (healthy) subject and a pathological patient.

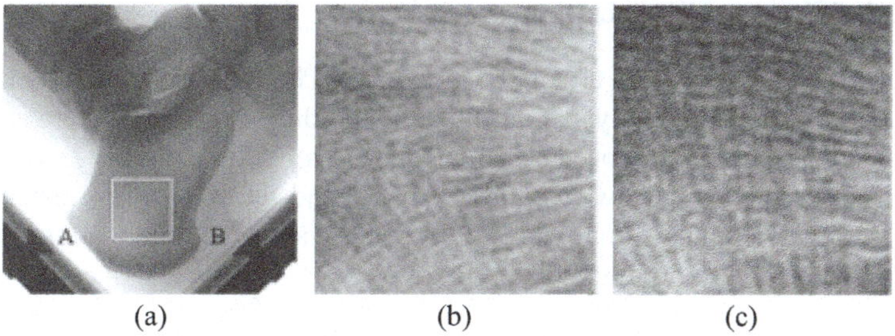

Fig. 1. X-ray images of the calcaneus with the markers A and B (a), ROI of a normal case (b), and ROI of a pathological patient (c)

3 Adopted Methodology

To distinguish the two populations using X-ray radiographs, different processes should be performed including filtering images, trabecular bone segmentation, binarization, skeletonization, extraction, and classification models. Figure 2 shows the system classification used in this paper.

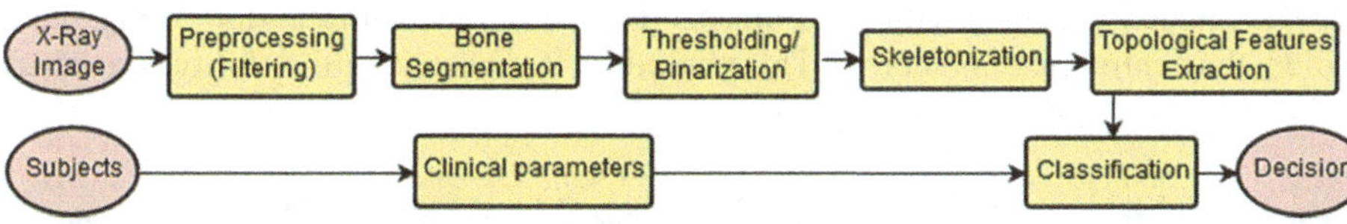

Fig. 2. Flowchart representing the system of the classification

3.1 Preprocessing

This step aims to provide a high quality of X-ray images since they are affected by impulse noise. In this study, a median filter was used. It consists of replacing each gray level pixel with the median value of neighboring gray levels pixels. Median filtering is usually used in digital image processing, and it was chosen in this study because it preserves the structure of the bone while removing noise.

3.2 Trabecular Bone Segmentation

To perform topological analysis, images should be segmented. This step allows separating the bone from marrow space. For this, we have developed a new variant of the DoG filter called a semi-DoG filter. It consists of applying a Gaussian filter to the original image, and then a subtraction of the blurred image from the original one is done. In this study, the used Gaussian kernel is (sigma = 10, core size = 5). This new filter allows detecting the whole trabecular bone.

3.3 Thresholding and Binarization

The images of the previous step are on the gray level. The images should be binarized to get an efficient topological analysis. For this, thresholding was used, where a gray level greater than a threshold is supposed as a bone and set to 255. For a gray level less than a threshold is supposed as marrow (medullary space) and set to 0. In this paper, the Otsu approach [11] was retained. Finally, a pruning step was performed to remove residual artifacts; it consists on removing isolated pixels.

3.4 Skeletonization

An important step for a better analysis is skeletonization. The skeletonization of an image is realized by a sequential thinning, where the image is no longer modified. The skeletonization of X by element B is expressed as:

$$X \circ \left\{B^i\right\} = [(X \circ B^1) \circ B^2] \circ B^3 \ldots . \tag{1}$$

The skeletonization consists of "reducing" the bone to a shape representing in a simplified way the bone [12]. The skeletonization passes by two steps:

Peeling: It is supposed that the marrow space is 0 and the trabecular network pixel is 1. The image is scanned with 3 × 3 masks (Fig. 3). If any configuration of Fig. 3.a is verified, the central pixel is set to 0. This procedure is repeated until stability is reached.

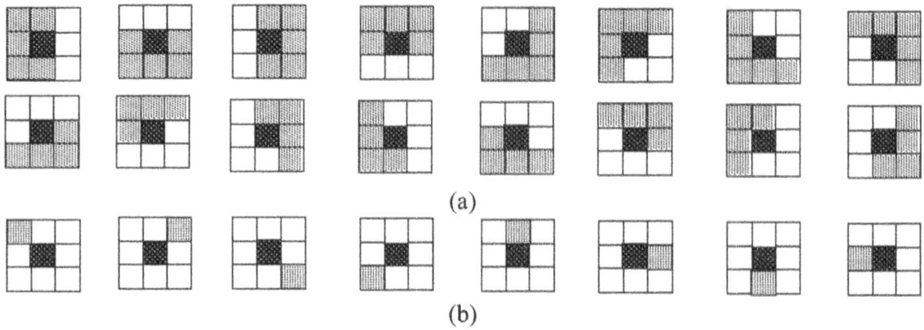

Fig. 3. Masks used for skeletonization. (a) peeling with 16 masks, (b) cleaning with 8 masks

Cleaning: If the boundaries of the object are noisy, post-processing is necessary. The trabecular network skeleton provides topological attributes. The result of the skeletonization is shown in Fig. 4.

3.5 Topological Attributes Extraction

The topological analysis allows for obtaining a connectivity assessment of the trabecular network. The topological analysis consists of counting the clusters of the trabeculae. The so-called "connectivity" attributes are calculated from the skeleton of the bone image.

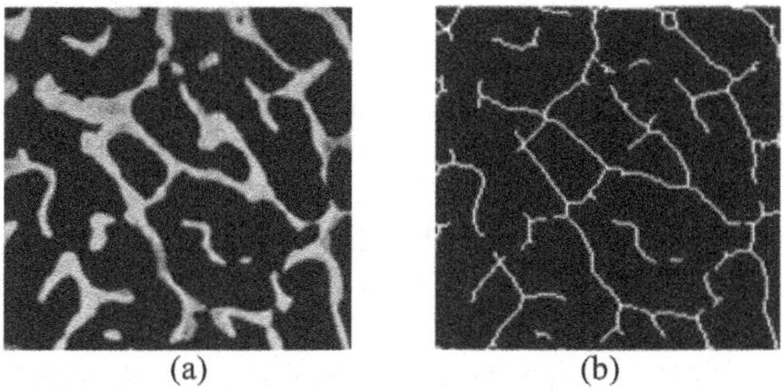

Fig. 4. Skeletonization. (a) Magnified ROI of a patient's bone, (b) result of the skeletonization

These parameters give information on trabecular profiles. In addition, they estimate the connections on 2D sections by an analysis called "strut analysis". This provides Nodes (*Nd*), free or termini (*Tm*), and branches:

- *Number of nodes (Nd)*: A pixel in the skeletonized image is supposed as a node if it has more than two neighbors (Fig. 5). A node can be assumed as a case of a junction.
- *Number of Termini (Tm)*: A pixel of the skeleton is supposed as termini if it has only one neighbor (Fig. 5).
- *Number of branches node-to-node (NNd)*: This parameter estimates the number of branches whose extremities are only nodes.
- *Number of branches node-to-termini (NNdTm)*: This attribute estimates the number of branches whose extremities are nodes and termini.
- *Number of trees (T)*: This counts the trees, which are composed of many branches.

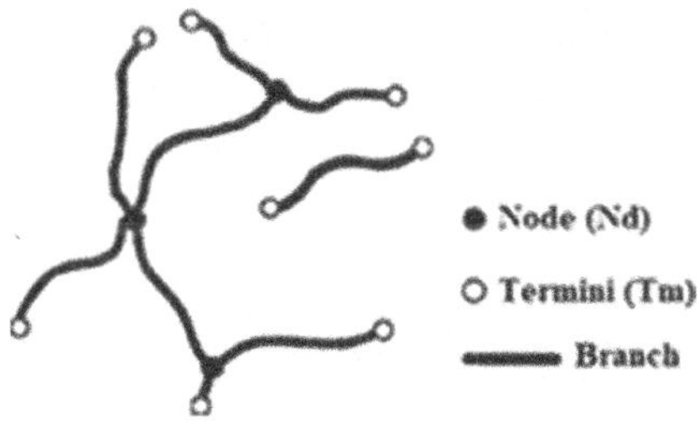

Fig. 5. Trabecular network with the connectivity parameters

The complexity of some X-ray images makes finding the connectivity parameters difficult. We have found some cases with several neighbors, and are not considered as nodes, this is mainly due to the discretization and the trabecular network complexity. To seek the real nodes, we propose a method, which allows carrying out several operations on the nodes' image. First, the image is thresholded, and the binary image is then skeletonized. The next step is to remove the small chains and count pixels (as node) which has two neighbors. Finally perform a thinning for the image to reduce the size of the clusters of neighboring nodes to a pixel. This will eliminate many false alarms (false nodes).

3.6 Clinical Parameters

In this study, different parameters of patients are involved. They are combined with the topological attributes for screening the disease:

- *Age*: this parameter informs on the probability of the bone disconnectivity over time. The group of the healthy subjects are aged 68.51 ± 10.21 SD, and the group of the pathological subjects are aged 70.98 ± 10.78 SD.
- *Bone Mass Index (BMI)*: this attribute is related to the height and the weight of the patient. To calculate the *BMI*, the mass is divided over the squared height (in kg/m^2).
- *Bone Mineral Content (BMC)*: this metric quantifies in grams, the bone mineral estimated in a specific area.
- *Bone Mineral Density (BMD)*: The *BMD* was estimated at the hip region, using the absorptiometry technique DXA (Hologic Delphi). A high protocol of standardization was set for all the patients.

3.7 Random Forest Classifier

Random forest classifier [13] is a meta estimator belonging to ensemble algorithms. It uses decision trees and creates a set from a randomly selected training subset. It aggregates the votes from different decision trees and predicts the output based on the majority of votes generated from each decision tree. It improves the predictive accuracy and controls over-fitting. The choice of this classifier was motivated by its performance in medical applications.

The algorithm to train a random forest uses a technique named ‘bagging’ to tree learner. Let a training set $X = X_1, \ldots, X_n$, and responses $Y = Y_1, Y_2, \ldots, Y_n$. The bagging method is based on selecting (M times) a ranm sample, replace the training set, and fits trees to these samples. If $m = 1, \ldots, M$, the first step is to sample m training examples from X, Y called X_m, Y_m. The second step is to train a classification tree f_m on X_m, Y_m. Prediction of a new sample x' is made by selecting the majority vote of classification trees, with:

$$\hat{f} = \frac{1}{M} \sum\nolimits_{m=1}^{M} f_m(x') \qquad (2)$$

3.8 Training Testing and Validating the Model

To train and validate the model, the k-fold cross-validation technique was used. The dataset is sampled into k independent folds, with $k-1$ folds are utilized to train the model and the rest is devoted to the test. This is repeated until all folds are utilized once as a test set. The output of the system of classification is calculated by averaging over the obtained results from each test set. In this paper, we have tested several values of k. A confusion matrix was used for the discrimination between the groups including: TN = Healthy subjects correctly classified, FP = Pathological patients incorrectly classified, TP = Pathological patients correctly classified, FN = Healthy subjects incorrectly classified, SE = Standard Error, Precision = Positive Predictive Value, Recall = True positive rate, F1 = Measure of incorrectly classified, AUC = Area Under Roc Curve, ACC = Classification Accuracy, with Acc = (TP + TN)/(TN + FP + TP + FN). The evaluation of classification was studied with the non-parametric statistical Wilcoxon test (*p*-value) [14]. A test was significant if $p < 0.05$.

4 Results and Discussion

Figure 6 illustrates the repartition of the most discriminative parameters. An important overlapping between clinical variables is noticed, leading to difficult classification (Fig. 6a). On the other hand, there was less overlapping for topological attributes, where the discrimination between the populations was easier (Fig. 6b). This demonstrates the capability of the topological attributes to predict bone fragility.

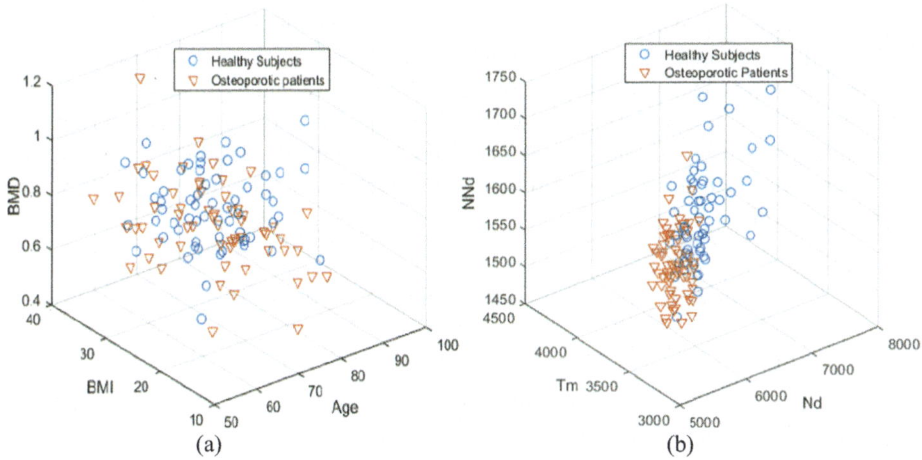

Fig. 6. Parameters representation, (a) clinical variables, (b) topological attributes

Table 1. A comparison between features to discriminate healthy and pathological groups

	Healthy subjects		Pathological patients		
Parameter	Mean	Standard deviation	Mean	Standard deviation	*p*-value
Age	58.61	10.21	70.98	10.78	0.183
BMI	24.80	3.47	26.81	4.53	0.014
BMC	29.90	5.42	27.02	6.30	0.007
BMD	0.83	0.12	0.73	0.14	1.04e−4

(continued)

Table 1. *(continued)*

Parameter	Healthy subjects Mean	Standard deviation	Pathological patients Mean	Standard deviation	*p*-value
Nd	6380.21	535.78	5661.13	269.25	2.28e−7
Tm	3479.91	123.29	3675.38	167.49	3.27e−6
NNd	1596.55	53.04	1574.56	44.21	0.024
NNdTm	1695.65	48.26	1754	64.33	2.71e−5
T	186.85	14.86	187.46	15.98	0.899

Table 1 shows the statistical significance metric (*p*-value) to discriminate between healthy and pathological groups for all parameters included in this study. The means and the standards deviations were also reported. The high performance to separate between the populations was achieved with *Nd*, providing the lowest value (*p*-value = 2.28e−7). Same findings for *Tm, NNdTm,* and *BMD* where they provided good performance in discriminating the two populations, which goes well with the results of other studies [15]. Therefore, *BMI* and *T* provided bad performance with a high *p*-value (0.183 and 0.899 respectively).

Table 2. Confusion matrix discriminating the two populations

Parameter	SE	TP	FP	FN	TN	Precision	Recall	F1	AUC	ACC
Age	0.052	29	20	31	40	0.591	0.483	0.532	0.570	0.575
BMI	0.050	37	25	23	35	0.596	0.616	0.606	0.629	0.600
BMC	0.050	33	17	27	43	0.660	0.550	0.600	0.642	0.633
BMD	0.047	39	16	21	44	0.709	0.650	0.678	0.705	0.691
Nd	0.023	44	1	16	59	0.977	0.733	0.838	0.934	0.858
Tm	0.037	40	7	20	53	0851	0.666	0.747	0.832	0.775
NNd	0.051	35	22	25	38	0.614	0.583	0.598	0.619	0.608
NNdTm	0.041	43	10	17	50	0.811	0.716	0.761	0.794	0.775
T	0.052	28	24	32	36	0.538	0.466	0.500	0.506	0.533

Table 2 illustrates the matrix of confusion for the parameters tested separately. The high performance was up to *Nd* with high accuracy (ACC = 85.8%) and (AUC = 93.4%). As for *Tm, NNdTm,* and *BMD,* they provided good results in discriminating the two populations where they achieved less error. These parameters demonstrated the ability to detect a high rate of correctly classified patients with less rate of false alarms. Same findings for the precision and the recall where we can notice a compromise between these metrics. The parameters revealed good performance except for *Age* and *T* where they provided bad results. This is due to much overlapping existing between the populations in estimating Age and T.

The study of the effect of the value k-fold in the cross-validation method with a random forest classifier was reported in Table 3. Different configurations and values of k were performed. Although all the values provided good performance, the high classification performance was accomplished with $k = 10$ for all the metrics (AUC = 94.3% and ACC = 87.5%). This is to state that the results depend on the number of folds. Comparing with the results given in Table 2, we can notice improvement of the separation accuracy; this is due to the combination of the clinical variable to the topological attributes.

Table 3. Effect of the cross-validation technique on discriminating the two populations with random forest classifier

k-fold	SE	TP	FP	FN	TN	Precision	Recall	F1	AUC	ACC
2	0.026	51	9	11	49	0.834	0.833	0.833	0.919	0.833
3	0.026	50	10	9	51	0.842	0.842	0.842	0.921	0.841
4	0.024	52	8	12	48	0.835	0.833	0.833	0.932	0.833
5	0.022	50	10	7	53	0.859	0.858	0.858	0.941	0.858
6	0.025	48	12	8	52	0.835	0.833	0.833	0.924	0.833
7	0.022	50	10	8	52	0.850	0.850	0.850	0.940	0.850
8	0.023	50	10	10	50	0.833	0.833	0.833	0.935	0.833
9	0.025	47	13	7	53	0.837	0.833	0.833	0.927	0.833
10	0.021	52	8	7	53	0.875	0.875	0.875	0.943	0.875

Table 4 shows a comparison between 5 classifiers most used. The value of k-fold was fixed to 10. Different configurations for the features were performed for all the classifiers, and the parameters providing better results were retained and reported in Table 4. As we can notice, random forest (RF) gave the best accuracy of classification between the populations for all metrics. This classifier is powerful due to the technique used in boosting the hierarchical searching trees. Same findings for Bayes network (BN) and Multilayer perceptron neural network (MLP), where they achieved good performance in classifying the patients with an accuracy of 84.1%. Support vector machine (SVM) and logistic regression (LR) classifiers provided good results with ACC = 81.6%, and 79.1% respectively. All the classifiers revealed good performance to separate healthy and pathological groups, with high accuracies and less false detections. Comparing to other studies, our work revealed promising results. In the study of [7] the accuracy was 75%, and in [8] ACC = 78%. Finally, in [16] the accuracy reached a rate of 79.3% using a basic configuration of deep learning.

Table 4. Comparative of the random forest to other models for $k = 10$ using all features

Classifier	SE	TP	FP	FN	TN	Precision	Recall	F1	AUC	ACC
RF	0.021	52	8	7	53	0.875	0.875	0.875	0.943	0.875
BN	0.024	53	7	12	48	0.844	0.842	0.841	0.930	0.841
MLP	0.028	50	10	9	51	0.842	0.842	0.842	0.906	0.841
SVM	0.039	47	13	9	51	0.818	0.817	0.816	0.817	0.816
LR	0.030	49	11	14	46	0.792	0.792	0.792	0.892	0.791

5 Conclusion

This study aimed to combine topological attributes of X-ray images to clinical parameters of patients suffering from osteoporosis. The goal was to demonstrate the capability of the topological analysis to discriminate between the populations and predict the bone weakness. The obtained results showed that the topological attributes outperformed the clinical variables used alone in hospitals and improved the classification rate. A random forest classifier was applied to mix the parameters and improve the efficiency of classification. It demonstrated the best accuracy compared to other classifiers widely used. In future work, we are planning to add structural parameters and fuse several classifier models.

References

1. Sweeney, A.T., Malabanan, A.O., Blake, M.A., Weinberg, J., Turner, A., Ray, P., Holick, M.F.: Bone mineral density assessment: comparison of dual-energy X-ray absorptiometry measurements at the calcaneus, spine, and hip. J. Clin. Densitometry **5**(1), 57–62 (2002)
2. Ammann, P., Rizzoli, R.: Bone strength and its determinants. Osteoporos. Int. **14**(Suppl 3), S13–S18 (2003)
3. Lespessailles, E., Gadois, C., Kousignian, I., Neveu, J.P., Fardellone, P., Kolta, S., Roux, C., Do-Huu, J.P., Benhamou, C.L.: Clinical interest of bone texture analysis in osteoporosis: a case control multicenter study. Osteoporos. Int. **19**(7), 1019–1028 (2008)
4. Piveteau, T., Winzenrieth, R., Hans, D.: Trabecular bone score (TBS) the new parameter of 2D texture analysis for the evaluation of 3D bone micro architecture status. Bone **48**, S176 (2011)
5. Harrar, K., Khider, M.: Texture analysis using multifractal spectrum. Int. J. Model. Optim. **4**(4), 336–341 (2014)
6. Kawashima, Y., Fujita, A., Buch, K., Li, B., Qureshi, M.M., Chapman, M.N., Sakai, O.: Using texture analysis of head CT images to differentiate osteoporosis from normal bone density. Eur. J. Radiol. **116**, 212–218 (2019)
7. Harrar, K., Jennane, R.: Quantification of trabecular bone porosity on X-ray images. J. Ind. Intell. Inf. **3**(4), 280–2085 (2015)
8. Areeckal, A.S., Kamath, J., Zawadynski, S., Kocher, M., Sumam, D.S.: Combined radiogrammetry and texture analysis for early diagnosis of osteoporosis using Indian and Swiss data. Comput. Med. Imaging Graph. **68**, 25–39 (2018)

9. Benhamou, C.L., Lespessailles, E., Jacquet, G., Harba, R., Jennane, R., Loussot, T., Tourliere, D., Ohley, W.: Fractal organization of trabecular bone images oncalcaneus radiographs. J. Bone Miner. Res. **9**(12), 1909–1918 (1994)
10. Harrar, K.: Trabecular texture analysis using morpho-clinical features and Bayes classifiers. In: Proceedings of the 6th International Conference on Image and Signal Processing and their Applications (ISPA), Mostaganem, Algeria, 24–25 November (2019)
11. Otsu, N.: A threshold selection method from gray-level histograms. IEEE Trans. Syst. Man Cybern. **9**(1), 62–66 (1979)
12. Thomsen, J.S., Laib, A., Koller, B., Prohaska, S., Mosekilde, L.I., Gowin, W.: Stereological measures of trabecular bone structure: comparison of 3D micro computed tomography with 2D histological sections in human proximal tibial bone biopsies. J. Microsc. **218**(5), 171–179 (2005)
13. Breiman, L.: Random forests. Mach. Learn. **45**(1), 5–32 (2001)
14. Wilcoxon, F.: Individual comparisons by ranking methods. Biometrics Bull. **1**(6), 80–83 (1945)
15. Harrar, K., Jennane, R.: Trabecular texture analysis using fractal metrics for bone fragility assessment. Int. J. Biomed. Biol. Eng. **9**(9), 683–688 (2015)
16. Grace, S., Kumar, D., Gautam, R., Sachin, M.: Osteoporosis detection using deep learning. Int. J. Mod. Trends Sci. Technol. **5**(3), 17–20 (2019)

A Comparative Study of Chest X-Ray Image Enhancement Techniques for Pneumonia Recognition

Sabrina Nefoussi[1(✉)], Abdenour Amamra[1], and Idir Amine Amarouche[2]

[1] Ecole Militaire Polytechnique, Bordj El-Bahri BP 17, Algiers, Algeria
nefoussicsabrina@gmail.com, amamra.abdenour@gmail.com
[2] Hopital Central de l'Armée, Ain Naadja, Algiers, Algeria
i.a.amarouche@gmail.com

Abstract. Unlike the identification of subjects with salient features in natural images, the visual similarity between the pathological features of Chest X-Ray (CXR) images complicates the distinction and interpretation of pneumonia signs. In this work, we aim to enhance Chest X-ray images by applying several image pre-processing techniques such as histogram equalization, contrast limited adaptive histogram equalization (CLAHE) and Unsharp mask. For instance, the fact of being the most performant natural image enhancement algorithms raises the question of whether they can achieve similar performance on CXR imagery. Hence, our objective is to investigate these enhancement techniques for the task of Pneumonia classification with deep neural architectures. To validate our findings, we provide a comparative study on the largest public pneumonia dataset of the Radiological Society of North America (RSNA) dubbed "Pneumonia Detection Challenge Dataset". The performance was measured in the two cases of balanced and imbalanced positive (pneumonia) and negative (no pneumonia) classes, where the preprocessing is intended to mitigate the bias toward the dominant class.

Keywords: Chest X-ray · Chest X-ray image enhancement · Pneumonia recognition · Deep convolutional networks

1 Introduction

Pneumonia is the leading infectious cause of death among children worldwide. Chest radiography (chest X-ray or CXR) is the most commonly used screening technique in the diagnosis of more than 14 chest diseases, like pneumonia. The great advantages offered by radiography include their low cost, wide availability, ease of use and especially its minimal radiation side-effect. Despite the ease of acquisition, the interpretation of a chest X-ray is the most challenging [1]. For instance, the correct interpretation of the CXR is difficult for physicians, given its wealth of medical information the overlap of the tissue in the chest X-ray.

M. R. Senouci et al. (Eds.): CSA 2020, LNNS 199, pp. 276–288, 2021.
https://doi.org/10.1007/978-3-030-69418-0_25

The large use of this imaging modality and the complexity of detecting pneumonia make computer-aided pneumonia detection systems a hot research topic since the system can help doctors in early diagnosis of pneumonia, thus saving the lives of many people around the world.

Pneumonia detection Convolutional Neural Networks (CNN) based is a promising research direction that has been attracting researchers. Nevertheless, several challenges have been related to the contrast of performance among machine learning approaches on various Chest X-ray datasets. For instance, machine learning algorithms typically encompass two main steps: (1) image preparation, and (2) pathology recognition or classification. The accumulation of large volumes of medical images has led to new opportunities for building accurate machine learning-based pathology recognition systems. Besides, these systems are negatively affected by data-related challenges such as weak contrast in images and class-imbalance. In this paper, we compare the impact, in terms of accuracy, precision-recall, F-score and ROC curve, of applying different image enhancement on the CXR image classification task. We consider pneumonia as a use case to validate our findings on real data.

The remainder of the paper is organized as follows: In the Sect. 2, we expose the most relevant related works. Afterward, in Sect. 3 we describe a theoretical primer about the different algorithms considered in our study. Then, the followed methodology is detailed in Sect. 4. In Sect. 5, we discuss the obtained results. We conclude the paper with Sect. 6 where we summarize our contribution and we address future directions.

2 Related Work

Our work is a part of a recent effort to harness advances in deep learning to improve computer-assisted diagnosis in medicine. During the current decade, the volume of clinical data in machine-readable format has grown significantly particularly in medical imaging. While previous generations of machine learning algorithms struggled to make effective use of high-dimensional data, modern neural networks have excelled on such tasks. For instance, several deep learning models have been successfully applied to recognize pathologies, locate abnormal regions or segment tumor/lesions in CXR images. In what follows, we will present the most relevant works in the field.

Several works have been oriented pneumonia recognition. The authors of [10] proposed an end-to-end classification model based on a CNN that takes a CXR image as input and outputs 14 probabilities of different pathologies including pneumonia. The proposed neural architecture was CheXNet, a 121-layer convolutional neural network trained on the ChestX-ray14 dataset. However, the proposed model returned relatively weak results of the classification of pneumonia. The authors in [12] have proposed a combination of weakly supervised learning algorithms and CNN to locate the infectious region in the CXR image. Their model had also a low pneumonia detection accuracy. Furthermore, authors in [6] proposed CXR image classification using a neural network to detect major lung diseases. For lung boundaries detection, they proposed simple image processing techniques like intensity and discontinuity-based method.

Preprocessing techniques have improved the classification performance: in [3] the authors proposed a model that combines three different architectures: (1) DenseNet-121: which checks the quality of the processed image, (2) Convolutional auto-encoder: performing the bone shadow exclusion and (3) CheXNet. Their model applies image preprocessing steps to extract the key features before feeding the data to the classifier. The authors in [11] utilized bilateral filtering for noise reduction and edge preservation. They perform optimal enhancement of the images using contrast-limited adaptive histogram equalization (CLAHE). These techniques helped in the detection of opacities related to pneumonia in the CXR images. The authors proposed a model leveraging the benefits of deep residual learning along with a separable convolution algorithm.

As RSNA pneumonia detection challenge[1] discloses interesting datasets, some researchers have applied advanced classification models. The authors in [4] proposed a model based on Mask-RCNN (based on ResNet50 and ResNet101) incorporating global and local features. Others [13] employed the robustness of large-scale databases where they proposed two convolutional neural networks, namely RetinaNet and Mask R-CNN. While different contributions have pros and cons, the authors in [15] used modified RNNs, then they exploit the conditional dependencies among abnormality labels to improve diagnosis results.

The above-mentioned studies have performed efficiently; however, we cannot objectively compare their results while the test conditions are not similar. In addition, all the previous works used imbalanced datasets, which perfectly reflects the nature of medical data. Nevertheless, the imbalance of classes impacts strongly the performance of the classification task. Also, works that used the trained neural networks as an auxiliary tool for the coarse pathology localization, the transfer learning in their cases may carry forward undesirable biases from tangentially related tasks. The use of image preprocessing methods certainly improves the quality of the data, and consequently the accuracy of classification, but it is first necessary to choose suitable preprocessing methods for each type of image to obtain the best results.

In the light of the discussed literature, we exclusively target pneumonia recognition in the largest labeled pneumonia dataset (1353 in Chest14-xray [14] vs 6012 in RSNA database (see Footnote 1)). Ideally, one may avoid the use of a pre-trained model when there is sufficient training data. In what follows we examine the impact of different image enhancement techniques for the task of pneumonia recognition.

3 Chest X-Ray Image Databases and Their Enhancement Techniques

In this section, we discuss the most used chest X-ray datasets in the literature for medical image analysis. Then, we focus on the important image preprocessing methods applicable to CXR.

[1] https://www.kaggle.com/c/rsna-pneumonia-detection-challenge.

3.1 Datasets

The performance of the proposed models based on deep learning is mainly related to the training step. Good training relies significantly on the quality and the size of the training dataset. On the other hand, the creation of a large annotated private image dataset is tedious and prone to error, the reason for which researchers rely on the publicly available datasets. Among the well-known datasets for pneumonia detection and classification we mention:

Indiana Dataset [2]: it consists of 7470 chest radiographs including the frontal and lateral images of disease annotations including pneumonia, the images of this dataset are available in the Digital Imaging and Communications in Medicine (DICOM) format.

Chest X-ray-14 Dataset [14]: it contains 112,120 frontal views of X-ray images of 30,805 patients and image labels of 14 different diseases (including atelectasis, consolidation, infiltration, pneumothorax, edema, emphysema, fibrosis, effusion, pneumonia, pleural thickening, cardiomegaly, nodule, mass, and hernia) where each image is labeled with a pathology. The images are in PNG format with a resolution of 1024×1024 pixels. Although this database is the largest in terms of images count, it contains only 1353 pneumonia cases against 60,412 pathology-free cases.

Mendeley [5]: Contains 5856 chest X-ray images, the specificity of this dataset is that it only contains 1 to 5 years old children X-ray images.

RSNA Pneumonia Detection Challenge Dataset (see Footnote 1): is publicly available, it contains 26,684 CXRs, including 6012 images with pneumonia finding available in DICOM format.

3.2 Image Preprocessing Techniques

Typically, image classification systems take input images for a series of preprocessing steps. Image preprocessing is intended to enhance the quality of the images and highlight the region of interest (ROI), in order to improve the classification accuracy. Typically, preprocessing techniques that can be applied on CXR include [9]: image enhancement, image segmentation, and bone suppression.

Enhancement: X-ray images usually have low contrast that results in degrading quality, thus affecting the accuracy of diagnosis. To surpass this problem, we enhance contrast using popular histogram equalization method, unfortunately, this technique could increase the noise, blur or remove key features in X-ray images. Local histogram equalization, as Adaptive Histogram Equalization (AHE) [8], can work well with X-ray images by adjusting the contrast locally. However, this method also suffers from noise. The Contrast Limited Adaptive Histogram Equalization (CLAHE) [16] defines a threshold when equalizing, thus adjusting the contrast without increasing noise in the image. This technique uses the clip limit parameter to limit the amplification by cropping the histogram at a predefined value. Other preprocessing techniques are used in order to highlight the structural features in the image like edge sharpening using an unsharp mask [7]. We can also improve the image quality by applying filters adding to the use of noise filtering.

Segmentation: it is used to separate the ROI, and clarify the different organ boundaries.

Bone Suppression: this technique is specially designed for X-ray imaging, it consists of the elimination of the bones such as ribs and clavicle, resulting in a clear image and improved classification performance.

4 Methodology

Our primary focus is the enhancement of the CXR images being passed into the convolutional neural network[2] in order to improve the classification performance. Generally, chest X-rays acquired from multiple sources vary in overall brightness and contrast. Histogram equalization is one of the commonly used normalization techniques. It is applied at a global level and contributes only to the standardization of brightness and contrast of the overall CXR image. In our work, we first test the impact of the application of this technique on classification. Image contrast normalization ignores the visibility of fine structures and small opacities. With the intention of overcoming such a shortcoming, we investigate local enhancement techniques to improve the visibility in CXR images with CLAHE. Besides, CLAHE normalization requires a set of hyper-parameters that strongly depend on the semantic content of the image. Hence, we test various values of clip limit, the values 1.5, 2.0, 4.0 and 150.0 have proven their efficacy in recent works on X-ray imaging.[3] We also tested the unsharp masking technique in order to sharpen the fine details of CXR images before feeding them into the classification network.

4.1 Data and Feature Description

In our work, we used the dataset proposed in RSNA Pneumonia Detection Challenge (see Footnote 1), since it is the largest in terms of pneumonia cases included (CXR images). Despite that, like most of the medical datasets, RSNA dataset suffers from imbalanced distribution of data between the two classes, with 6012 *pneumonia* images vs 20672 of *no-finding* images. This imbalance is considered a problem in the classification process as it leads to occur biased results. For this reason, we also decided to investigate the impact of imbalance on image classification. We performed two types of experiments on the same classification model (see Footnote 2). We trained this model on balanced and imbalanced datasets:

---

[2] https://www.kaggle.com/jonn-edtc/cnn-segmentation-connected-components/notebook.

[3] https://github.com/asalmada/x-ray-images-enhancement/blob/master/README.md.

1. **For the balanced dataset**, we used 7000 training samples (3500 with pneumonia finding, 3500 with no finding), 2000 test samples (1000 with pneumonia finding, 1000 with no finding) and 1000 samples for validation (500 with pneumonia finding, 500 with no finding).
2. **For the imbalanced dataset**, we randomly selected 7000 samples for training, including 1579 images with pneumonia findings. We selected 2000 samples, from the remaining data, containing 447 pneumonia images. Finally, we selected 1000 of validation with 217 pneumonia findings.

Data preparation: we down-sample the size of the original images by a factor of 0.25 (i.e. from 1024×1024 to 256×256) which should facilitate the training.

Data augmentation: for more training efficiency we used horizontal flip as augmentation, as it keeps lesion structure and texture features consistent with the original data.

Data preprocessing: In general, the classification performance of CXR images varies greatly under different conditions. Our work aims to help the classification model to distinguish pneumonia characteristic features from image noise. As mentioned above, we used three image enhancement techniques: *Histogram Equalization*, *CLAHE* with a variation of clip limit = 1.5, 2.0, 4.0, 150.0 and the *Unsharp Masking*.

4.2 Architecture

In order to assert whether preprocessing like histogram equalization, CLAHE and Unsharp masking have a positive impact on the quality of CXR images for the pneumonia classification system, we adopted the architecture (see Footnote 2) based on convolutional neural network. The latter was proposed for the RSNA Pneumonia Detection Challenge (see Footnote 1), since it is well-documented and criticized where the choices of all stages were formally validated. As presented in Table 1, the network consists of a number of residual blocks (Resblock in Table 1) with convolutions and down-sampling blocks (Downsample in Table 1) as well as max-pooling.

We carried out 14 different experiments, under special conditions. We tested the performance of the classification model in cases of:

1. Balanced dataset-Raw: in this case, we did not use any preprocessing, we used raw images of a balanced dataset.
2. Balanced dataset-Histogram Equalization: here we used the histogram equalization as preprocessing technique on CXR images of a balanced dataset.
3. Balanced dataset-CLAHE 1.5: we used balanced dataset images enhanced by CLAHE technique with a clip limit of 1.5.
4. Balanced dataset CLAHE 2.0: we used balanced dataset images enhanced by CLAHE technique with a clip limit of 2.0.
5. Balanced dataset CLAHE 4.0: we used balanced dataset images enhanced by CLAHE technique with a clip limit of 4.0.

Table 1. Classification model

Block/Layer (type)	Output shape	Connected to
input_1 (InputLayer)	(256, 256, 1)	
conv2d (Conv2D)	(256, 256, 32)	input_1[0][0]
Downsample	(128, 128, 64)	conv2d[0][0]
Resblock	(128, 128, 64)	max_pooling2d[0][0]
Resblock1	(128, 128, 64)	add[0][0]
Downsample1	(64, 64, 128)	add_1[0][0]
Resblock2	(64, 64, 128)	max_pooling2d_1[0][0]
Resblock3	(64, 64, 128)	add_2[0][0]
Downsample2	(32, 32, 256)	add_3[0][0]
Resblock4	(32, 32, 256)	max_pooling2d_2[0][0]
Resblock5	(32, 32, 256)	add_4[0][0]
Downsample3	(16, 16, 512)	add_5[0][0]
Resblock6	(16, 16, 512)	max_pooling2d_3[0][0]
Resblock7	(16, 16, 512)	add_6[0][0]
batch_normalization_20	(16, 16, 512)	add_7[0][0]
leaky_re_lu_20 (LeakyReLU)	(16, 16, 512)	batch_normalization_20[0][0]
conv2d_21 (Conv2D)	(16, 16, 1)	leaky_re_lu_20[0][0]
up_sampling2d (UpSampling2D)	(256, 256, 1)	conv2d_21[0][0]

6. Balanced dataset CLAHE 150.0: we used balanced dataset images enhanced by CLAHE technique with a clip limit of 150.0.
7. Balanced dataset Unsharp Mask: the last enhancement technique used in this case is the unsharp masking applied on CXR images of a balanced dataset.
8. Imbalanced dataset Raw: in this case, we used raw images of an imbalanced dataset.
9. Imbalanced dataset Histogram Equalization: here we used the histogram equalization as preprocessing technique on CXR images of imbalanced dataset.
10. Imbalanced dataset CLAHE 1.5: we used imbalanced dataset images enhanced by CLAHE technique with a clip limit of 1.5.
11. Imbalanced dataset CLAHE 2.0: we used imbalanced dataset images enhanced by CLAHE technique with a clip limit of 2.0.
12. Imbalanced dataset CLAHE 4.0: we used imbalanced dataset images enhanced by CLAHE technique with a clip limit of 4.0.
13. Imbalanced dataset CLAHE 150.0: we used imbalanced dataset images enhanced by CLAHE technique with a clip limit of 150.0.
14. Imbalanced dataset Unsharp Mask: the last enhancement technique used in this case is the unsharp masking applied on CXR images of an imbalanced dataset.

In each of the above-mentioned cases, we trained the model (see Footnote 2) for 40 epochs. However the model performance, in terms of loss and accuracy, stopped improving after about 25 epochs. As illustrated in Fig. 1, for example in cases (3) and (10), we noticed that the validation loss, after about 25 epochs, does not change while the training loss keeps decreasing. Also, the validation accuracy plateaued after about 30 epochs, however, the training accuracy keeps increasing. These results show signs of **Overfitting**, where the best classification model would be obtained at the first optimal validation accuracy.

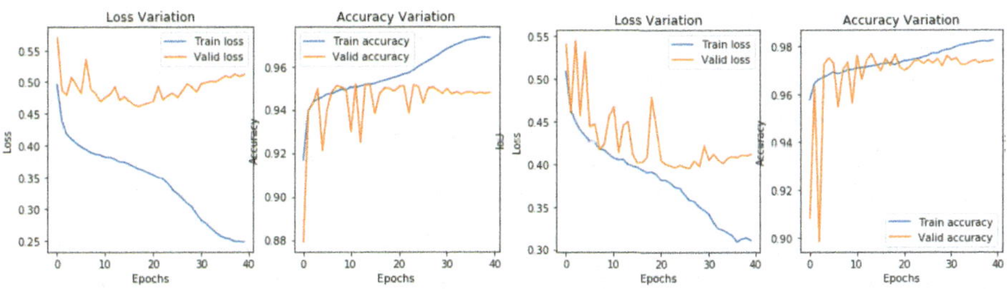

(a) Balanced CLAHE 1.5 -Case (3)- (b) Imbalanced CLAHE 1.5 -Case (10)-

Fig. 1. Training Results using CLAHE with clip limit = 1.5: (a) Accuracy and Loss variation during training for balanced data case, (b) Accuracy and Loss variation during training for imbalanced data case

5 Results and Discussion

In the evaluation of the trained model we used the standards metrics in order to assess the performance of the model. We used four metrics derived from the confusion matrix:

Accuracy: the ratio of correct predictions to the total number of predictions,

$$Acc = \frac{(TP + TN)}{(TP + TN + FP + FN)} \tag{1}$$

Precision or positive predictive value is the ratio of good positives from the total detected positives:

$$Pr = \frac{TP}{(TP + FP)} \tag{2}$$

Recall or sensitivity : corresponds to the probability of correctly classifying pneumonia sample in our case,

$$R = \frac{TP}{(TP + FN)} \tag{3}$$

F-score: is a measure that combines both precision and recall as determined in the formula:

$$F_{score} = \frac{2 \times Pr \times R}{(Pr + R)} \tag{4}$$

5.1 Results

Table 2 shows the model evaluation results for each experiment conducted through the fourteen different setups and according to the above-mentioned metrics. We evaluated the model classification performance on two different test sets. For the balanced dataset (first column in Table 2), we used test support of 1000 pneumonia samples and 1000 no-finding samples. Whereas For the imbalanced one (second column in Table 2), the test support contains 2000 CXR images including 447 pneumonia samples and 1553 no-finding samples.

Table 2. CXR images classification results for both balanced and imbalanced dataset.

Preprocessing technique	Balanced dataset				Imbalanced dataset			
Metrics	Acc	R	Pr	F-score	Acc	R	Pr	F-score
Raw	0.950	0.74	**0.80**	**0.77**	0.973	0.84	**0.91**	0.87
Histogram Equalization	0.951	**0.90**	0.65	0.75	0.972	0.89	0.88	0.88
CLAHE 1.5	0.950	0.77	0.77	**0.77**	0.973	0.89	0.89	0.89
CLAHE 2.0	0.950	0.77	0.76	**0.77**	0.973	0.88	0.88	0.88
CLAHE 4.0	0.950	0.84	0.69	0.76	0.974	**0.99**	0.78	0.87
CLAHE 150.0	0.948	0.76	0.75	0.76	0.973	0.90	0.88	0.89
Unsharp mask	**0.952**	0.87	0.70	**0.77**	**0.975**	0.92	0.88	**0.90**

The model performance illustrates a significant contrast between the balanced and imbalanced datasets. Imbalanced dataset accuracy revolves around 0.97, which was higher by about 0.02 than that of the balanced dataset. Also, the recall, precision, and F-score metrics are better by about 0.1 than those of the imbalanced dataset. For the preprocessing techniques (the 7 rows in Table 2) the unsharp mask has the best result in terms of accuracy, then comes CLAHE with clip limit = 4.0 in the second position. Regarding the recall metric, CLAHE 4.0 in the imbalanced case achieved the best result. For the balanced dataset, the recall achieves the best value in the histogram equalization case with a value of recall = 0.90. Contrariwise, the precision has been degraded by preprocessing in the two cases of the balanced and imbalanced dataset. For the balanced dataset case, it decreases from 0.80 to 0.65 using histogram equalization. Always considering precision in the case of the imbalanced dataset, it decreases from 0.91 to 0.78 when we used CLAHE 4.0 for example. Finally, for the F-score in imbalanced case, it was been mostly improved by the unsharp mask achieving the value of $F = 0.90$. For the balanced dataset: F-score remains at 0.77 by applying CLAHE 1.5, CLAHE 2.0 and unsharp mask enhancements; whereas it decreased in the remaining cases.

5.2 Discussion

The gap shown in Table 2 between the results of the balanced dataset and the imbalanced one proved the existence of performance bias for same metrics. These metrics do not faithfully reflect the performance in the imbalanced dataset. On the other hand, real data in the medical domain is mostly imbalanced, thus we have to use fitting metrics which takes into account the proportions between the different classes in the dataset.

The results show that the application of enhancement techniques such as histogram equalization, CLAHE, and unsharp mask, generally impacts the performance of classification with low precision and high recall, which indicates that the model tends to capture more pneumonia cases which are the more important. However, the classification model can easily be misled to predict more false-positive samples, which can be explained by the fact that enhancement methods could increase the noise that can be interpreted as clouds in CXR images.

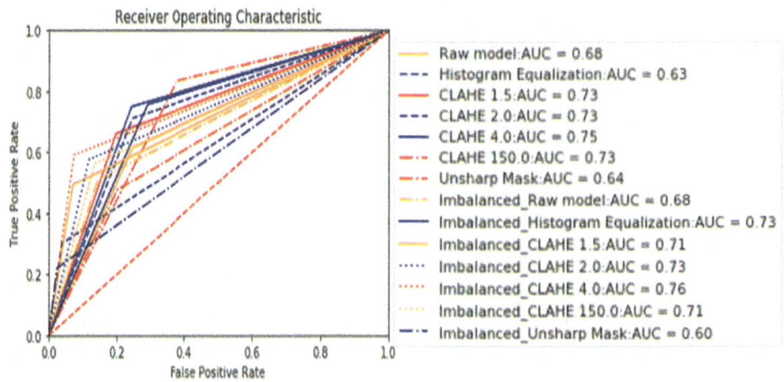

Fig. 2. Roc curves for various image preprocessing techniques, in balanced and imbalanced dataset

We notice that applying preprocessing techniques on CXR images before proceeding through the classification task can improve the model's performance. However, we have to choose suitable parameters to maximize the gain. In our case, we could not assert that one preprocessing technique is better than the other since each metric promotes the use of one technique instead of others. Nevertheless, we suggest combining them to obtain the best classification result. For example, using CLAHE 4.0 with the Unsharp mask will help to get advantages of the use of CLAHE 4.0 (improvement of accuracy and F-score values) plus those of the Unsharp mask (improvement of recall). Consequently, we have to find the best combination of the preprocessing techniques that improve the performance of the task in hands.

To be able to compare the performance of the model in the fourteen tested cases, we also evaluated each obtained model using the Receiver Operating Characteristic (ROC) curve. The y-axis of the ROC curve is the sensitivity (true positive rate or recall), which measures the capacity of the model to characterize the CXR images with pneumonia findings. The x-axis, however, represents the false positive rate FPR, which given by:

$$\begin{aligned} FPR = 1 - \text{Specificity} = \frac{1 - TN}{(TN + FP)} = \frac{FP}{(TN + FP)} \\ \text{where Specificity} = \frac{TN}{(TN + FP)} \end{aligned} \tag{5}$$

It measures the success of not flagging normal cases as abnormal. The area under the ROC curve (AUC) metric measures the performance of the model in terms of separability between CXR images with pneumonia finding and healthy cases. We calculated the AUC metric value of the model in the fourteen cases using ROC curves as shown in Fig. 2.

We noted the case (12) using CLAHE with a clip limit of 4.0 in the imbalanced dataset has the best measure of separability with AUC = 0.76. This preprocessing technique achieves also the best result in the balanced dataset with $AUC = 0.75$. The unsharp mask as a preprocessing technique has achieved the worst result of separability, with $AUC = 0.60$ on the imbalanced dataset and $AUC = 0.64$ on the balanced dataset. CLAHE technique performed well-using clip limit = 1.5, 2.0 and 150.0, where the obtained results were between 0.71 and 0.73. The histogram equalization enhances the capability of separability of the model in the imbalanced dataset case with $AUC = 0.73$, however, it degraded its capability in the balanced case to $AUC = 0.63$. When the model used the raw CXR images, in the two cases of the balanced and imbalanced dataset, the AUC metric achieves 0.68. As a result, we can conclude that using CLAHE as a preprocessing technique, before passing CXR images to the classification model, enhances remarkably the capacity of separability, mostly when we use clip limit parameter equals to 4.0.

6 Conclusion

Our main goal in this work is to improve the quality and the productivity of radiologists in the detection of pneumonia signs. This paper presented an ensemble of results of applying different image preprocessing techniques on chest X-rays before feeding them to the pneumonia detection neural network. We validated our results on the largest labeled public pneumonia dataset. We performed our tests using two types of dataset: balanced and imbalanced. This choice was motivated by the fact that imbalanced datasets represent faithfully the real distribution of the pneumonia disease among individuals, where the negative class is predominant.

After analyzing the performance of the network on pneumonia identification from preprocessed chest x-rays, we identified several directions for future extension of the presented work. First, even if medical imaging is the main information source for pneumonia identification, other clinical results such as specific clinical symptoms, blood test, pulse oximetry, sputum test, conditions the comprehensive diagnosis of this disease. Second, the contribution of preprocessing techniques appears better when combined. Third, we can augment a frontal chest x-ray with a lateral chest x-ray or/and computed tomography image, in order to obtain finer details of the lung field and ensure correct diagnosis of cases with uncertainties. Fourth, when we use medical imbalanced datasets in research, one should pay attention to the use of fitting metrics for evaluation purposes.

References

1. Delrue, L., Gosselin, R., Ilsen, B., Van Landeghem, A., de Mey, J., Duyck, P.: Difficulties in the interpretation of chest radiography. In: Comparative Interpretation of CT and Standard Radiography of the Chest, pp. 27–49. Springer (2011)
2. Demner-Fushman, D., Kohli, M.D., Rosenman, M.B., Shooshan, S.E., Rodriguez, L., Antani, S., Thoma, G.R., McDonald, C.J.: Preparing a collection of radiology examinations for distribution and retrieval. J. Am. Med. Inform. Assoc. **23**(2), 304–310 (2015)
3. Huynh, M., Nguyen, T., Tran, M.: Context learning for bone shadow exclusion in CheXNet accuracy improvement. In: 2018 10th International Conference on Knowledge and Systems Engineering (KSE), pp. 135–140, November 2018
4. Jaiswal, A.K., Tiwari, P., Kumar, S., Gupta, D., Khanna, A., Rodrigues, J.J.: Identifying pneumonia in chest x-rays: a deep learning approach. Measurement **145**, 511–518 (2019)
5. Kermany, D., Zhang, K.G.M.: Large dataset of labeled optical coherence tomography (OCT) and chest x-ray images,Mendeley data, v3 (2018). https://doi.org/10.17632/rscbjbr9sj.3
6. Khobragade, S., Tiwari, A., Patil, C.Y., Narke, V.: Automatic detection of major lung diseases using chest radiographs and classification by feed-forward artificial neural network. In: 2016 IEEE 1st International Conference on Power Electronics, Intelligent Control and Energy Systems (ICPEICES), pp. 1–5 (2016)
7. Malin, D.F.: Unsharp masking. AAS Photo Bull. **16**, 10–13 (1977)
8. Pizer, S.M., Amburn, E.P., Austin, J.D., Cromartie, R., Geselowitz, A., Greer, T., ter Haar Romeny, B., Zimmerman, J.B., Zuiderveld, K.: Adaptive histogram equalization and its variations. Comput. Vis. Graph. Image Process. **39**(3), 355–368 (1987)
9. Qin, C., Yao, D., Shi, Y., Song, Z.: Computer-aided detection in chest radiography based on artificial intelligence: a survey. BioMed. Eng. OnLine **17**, 1–23 (2018)
10. Rajpurkar, P., Irvin, J., Zhu, K., Yang, B., Mehta, H., Duan, T., Ding, D.Y., Bagul, A., Langlotz, C., Shpanskaya, K.S., Lungren, M.P., Ng, A.Y.: CheXNet: radiologist-level pneumonia detection on chest x-rays with deep learning. CoRR **abs/1711.05225** (2017)
11. Sarkar, R., Hazra, A., Sadhu, K., Ghosh, P.: A novel method for pneumonia diagnosis from chest x-ray images using deep residual learning with separable convolutional networks. In: Gupta, M., Konar, D., Bhattacharyya, S., Biswas, S. (eds.) Computer Vision and Machine Intelligence in Medical Image Analysis, pp. 1–12. Springer, Singapore (2020)
12. Sedai, S., Mahapatra, D., Ge, Z., Chakravorty, R., Garnavi, R.: Deep multiscale convolutional feature learning for weakly supervised localization of chest pathologies in x-ray images. In: Shi, Y., Suk, H.I., Liu, M. (eds.) Machine Learning in Medical Imaging, pp. 267–275. Springer, Cham (2018)

13. Sirazitdinov, I., Kholiavchenko, M., Mustafaev, T., Yixuan, Y., Kuleev, R., Ibragimov, B.: Deep neural network ensemble for pneumonia localization from a large-scale chest x-ray database. Comput. Electr. Eng. **78**, 388–399 (2019)
14. Wang, X., Peng, Y., Lu, L., Lu, Z., Bagheri, M., Summers, R.M.: ChestX-ray8: hospital-scale chest x-ray database and benchmarks on weakly-supervised classification and localization of common thorax diseases. In: 2017 IEEE Conference on Computer Vision and Pattern Recognition (CVPR), July 2017. https://doi.org/10.1109/cvpr.2017.369
15. Yao, L., Poblenz, E., Dagunts, D., Covington, B., Bernard, D., Lyman, K.: Learning to diagnose from scratch by exploiting dependencies among labels (2017)
16. Zuiderveld, K.: Contrast limited adaptive histogram equalization. In: Heckbert, P.S. (ed.) Graphics Gems IV, pp. 474–485. Morgan Kaufmann, Burlington (1994)

Automatic Brain Tumor Segmentation Using Multi-OTSU Thresholding and Morphological Reconstruction

Imane Mehidi(✉), Djamel Eddine Chouaib Belkhiat, and Dalel Jabri

DAC HR laboratory, Ferhat Abbas University, Setif, Algeria
{imane.mehidi,djamel.belkhiat,dalel.jabri}@univ-setif.dz

Abstract. Images segmentation aims to divide an image into several segments. They can be selected according to the composition of the region of interest, the types of tissues, and the functional zones [1]. In this paper, we present a new segmentation method, based on the multi-thresholding method and morphological reconstruction for brain tumor separation from Magnetic Resonance Imaging (MRI). Firstly, we use a pre-processing to enhance image contrast and quality by intensity adjustment. Secondly, the improved image is segmented using the multi-Otsu method and finally, a morphological reconstruction was performed with the appropriate structuring element parameter on the segmented image to determine the tumor. A comparison with some state-of-the-art algorithms demonstrated the efficiency of the proposed method with regard to accuracy.

Keywords: Multi-Otsu method · Morphological reconstruction · Tumor

1 Introduction

The segmentation of the medical image is one of the most important issues under consideration. It assists in the diagnosis and treatment of diseases. To achieve this, several methods have been suggested, such as thresholding methods. Thresholding is a technique commonly applied to image segmentation. Its main purpose is to classify the pixels of a particular image into two categories: those relating to an object and related to the background. Thresholding algorithms run automatically, without user intervention, and are unsupervised algorithms, but the presence of complex images has led to the need for a multilevel method [2].

In literature, several studies have been proposed for image segmentation using multi-threshold technique. For instance, Senthilkumaran and NandVaithegi [3] introduced a locally adapted threshold technique that removes the background using local mean and standard deviation. Furthermore, Dawngliana et al. [4] Suggested a hybrid method between the multilevel thresholding and level set

M. R. Senouci et al. (Eds.): CSA 2020, LNNS 199, pp. 289–300, 2021.
https://doi.org/10.1007/978-3-030-69418-0_26

for automatic brain tumor segmentation. In the same way, Sujan et al. [5] presented a threshold-based method, combined with some techniques of morphological image analysis, to detect a brain tumor in an MRI image. The first step in the proposed approach is to convert the MRI image to a grayscale image. Then, the noise was eliminated using different filtering techniques. Next, the resulting image was segmented using the Otsu threshold method. After that, morphological processes were applied to the segmented image for tumor detection. In addition, a segmentation method was developed by Ilhan et al. [6] to identify the tissues affected by cancer. The proposed approach has combined a range of techniques such as morphological process, pixel subtraction, segmentation using thresholding and image filtering. This method helps to make the segmented tumor area more visible to the health expert so that it is easy to identify during diagnosis. More recently, Mehidi et al. [7] proposed a method for brain tumors segmentation from MRI using an improved clustering method based on the K-Means (KM) algorithm. The main idea was to combine the Darwinian Particle Swarm Optimization (DPSO) technique, the KM algorithm and the morphological reconstruction operation (MR). The DPSO technique was applied to MRI medical images for the initialization of centroid clusters. Moreover, the MR filter was used to eliminate noise and generate more compact and well-separated clusters. In addition to the works mentioned above, further papers [8,9], dealing with image segmentation using multi-threshold technique, have been performed.

Hence, the aim of this work consists in the proposition of new image segmentation method that can identify and segment the tumor region with a high degree of accuracy. The proposed method can perform:

- Segment and identify a brain tumor of type multiple glioblastomas (GBM) in MRI medical imaging.
- Determine with precision the region of interest (ROI) tumor.

The remainder of this paper is organized as follows. In Sect. 2, we explain some basic concepts that help understand the proposed methodology. Section 3 presents a detailed explanation of the proposed method. Section 4 presents the experimental and evaluation parts. Finally, some conclusion is presented in Sect. 5.

2 Motivations

2.1 Pre-processing

The purpose of pre-processing is to improve some properties of an image for use in processing and analysis tasks. Thus, pre-processing is necessary before segmentation, which is often a difficult process. In this subsection, we show the pre-processing operations incorporated into the proposed method:

- *Remove of non-brain Tissue for MRI image*: Most approaches for brain tumor segmentation rely on a skull-stripping step before the actual segmentation is performed. Usually, in the medical image segmentation, the background, skull, scalp, eyes, and structures that do not matter are removed because this reduces the amount of memory used and increases the processing speed and, reduces the effect of these parts in the segmentation [10].

- *Adjust intensity*: It is an image enhancement technique that maps the intensity values of an A grayscale image to new values in B. This operation increases the contrast of the output image B. Figure 1 illustrates the concept of this operation, where the contrast of the image at the top is low, the intensity values being limited to a part of the range as we see in the histogram, while the image below shows a higher contrast, with values of intensity that fill the entire intensity range [0, 255]. In the high-contrast image, darker shadows and highlights are brighter.

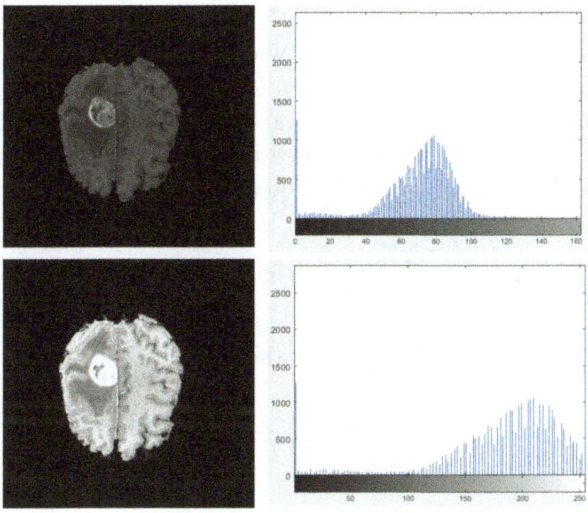

Fig. 1. The concept of the adjust intensity operation [11,12]

2.2 Multi-OTSU Method

Otsu's method is a way to automatically find an optimal threshold based on the gray levels in the image histogram. The main idea is to search for the optimal threshold to segment the image into two classes (Otsu 1979) [13].

There are L intensity levels in each gray component of a given image, which are in the range $0, 1, 2, ..., L-1$. Then, we can present the probability distribution as following:

$$p_i = \frac{h_i}{N}, \sum_{i=0}^{L-1} p_i = 1 \tag{1}$$

where i is a particular intensity level between 0 to $L-1$. N is the total number of pixels in the image. h_i denotes the number of pixels for the intensity level i. Similarly, image histogram is denoted by h_i which can be normalized and considered as the probability distribution p_i for a component of the image. The total mean, noted μ_T, can be easily calculated as follows:

$$\mu_T = \sum_{i=0}^{L-1} i\, p_i \tag{2}$$

Image thresholding (on 2 levels) can be defined as an operation that consists of dichotomizing the pixels of a given image into two classes U_1 and U_2 using a threshold level t. It can be expressed by the following equation:

$$G(x,y) = \begin{cases} 0, \ g(x,y) \leq t \\ L-1, \ g(x,y) > t \end{cases} \tag{3}$$

where the indices x and y are the pixel width and height of the image $g(x,y)$. $G(x,y)$ is the segmented image. The class U_1 groups together the pixels characterized by an intensity lower than or equal to t. However, the class U_2 regroups together all the pixels having an intensity higher than t. Hence, occurrence probabilities of classes U_1 and U_2, noted respectively w_1 and w_2, are expressed as follows:

$$w_1 = \sum_{i=0}^{t} p_i \text{ and } w_2 = \sum_{i=t+1}^{L-1} p_i \tag{4}$$

Moreover, the mean of each class can be defined as:

$$\mu_1 = \sum_{i=0}^{t} \frac{ip_i}{w_1} \text{ and } \mu_2 = \sum_{i=t+1}^{L-1} \frac{ip_i}{w_2} \tag{5}$$

Therefore, the variance between classes can be expressed by:

$$\sigma_B^2 = w_1(\mu_1 - \mu_T)^2 + w_2(\mu_2 - \mu_T)^2 \tag{6}$$

Indeed, maximizing the variance between classes is one of the most efficient approaches for calculating the optimal threshold. For that, the problem of 2-level thresholding can be expressed as an optimization problem that allows finding the threshold maximizing the objective function, commonly defined by:

$$\varphi = \max_{1<t<L-1} \sigma_B^2(t) \tag{7}$$

In the same way, the Otsu method thresholding of 2 levels can be extended to the generic thresholding of n levels (multi Otsu). In this case, $n-1$ threshold levels t_j, $j = 1, 2, ..., n-1$, are needed and the operation is carried out as follows:

$$G(x,y) = \begin{cases} 0, & g(x,y) \le t_1 \\ \frac{1}{2}(t_1 + t_2), & t_1 < g(x,y) \le t_2 \\ \cdot \\ \cdot \\ \frac{1}{2}(t_{n-2} + t_{n-1}), & t_{n-2} < g(x,y) \le t_{n-1} \\ L-1, & g(x,y) > t_{n-1} \end{cases} \tag{8}$$

where the indices x and y are the pixel width and height of the image. As mentioned above, the easiest way to determine the optimal threshold is the method that maximizes the variance between classes, which can be resumed as follows:

$$\sigma_B = \sum_{j=1}^{n} w_j \left(\mu_j - \mu_T\right)^2 \tag{9}$$

where j indicates a specific class, w_j and μ_j are occurrence probability and the mean of class j, respectively. The occurrence probabilities of w_j of classes $U_1U_2...U_n$ is calculated as follows:

$$w_j = \begin{cases} \sum_{i=0}^{t_j} p_i, & j = 1 \\ \sum_{i=t_{j-1}+1}^{t_j} p_i, & 1 < j < n \\ \sum_{i=t_{j-1}+1}^{L-1} p_i, & j = n \end{cases} \tag{10}$$

One can determine the average of each class μ_j as follows:

$$\mu_j = \begin{cases} \sum_{i=0}^{t_j} \frac{ip_i}{w_j}, & j = 1 \\ \sum_{i=t_{j-1}+1}^{t_j} \frac{ip_i}{w_j}, & 1 < j < n \\ \sum_{i=t_{j-1}+1}^{L-1} \frac{ip_i}{w_j}, & j = n \end{cases} \tag{11}$$

Hence, the problem of n-level thresholding can be reduced to an optimization problem in order to find the threshold t_j that maximizes the objective function which usually defined by:

$$\varphi = \max_{1<t_1....<t_{n-1}<L-1} \sigma_B(t_j) \tag{12}$$

2.3 Morphological Reconstruction

Morphological reconstruction (MR) is part of a group of image operators that are mostly called geodesics. MR was usually known in the binary case, where it simply extracts the connected components of an image "marked" by another image. However, reconstruction is also used for gray-scale images [14]. This process is based on using two images, called mask and marker, respectively [15]. The mask L which constrains the transformation and the marker I is the starting point of the transformation. The mask and the marker can be two binary images or intensity. Moreover, the marker must have the same size as the mask, and its elements (the structuring element SE) must be less than or equal to the elements of the mask [16]. Noting, that the processing is repeated until stability. There are two basic operations for morphological reconstruction:

1. The morphological dilation reconstruction (R^{δ}) of L from I is defined as

$$R_L^{\delta}(I) = \delta^{(n)}_{(L)}(I) \tag{13}$$

   Where,
   For one time: $\delta^{(1)}(I) = \delta(I) \wedge L$
   For n time: $\delta^{(k)}(I\ ) = \delta(\delta^{(k-1)}(I)) \wedge L$, for $2 \leq k \leq n$; k, $n \in N^{+}$
   δ is the symbol for the elementary morphological dilation operation of I by a structuring element (SE), and $\wedge$ stands for the pointwise minimum.
2. The morphological erosion reconstruction (R^{ε}) of L from I, is denoted by

$$R_L^{\varepsilon}(I) = \varepsilon^{(n)}_{(L)}(I) \tag{14}$$

   Where,
   For one time: $\varepsilon^{(1)}(I) = \varepsilon(I\ ) \vee L$
   For n time: $\varepsilon^{(k)}(I\ ) = \varepsilon(\varepsilon^{(k-1)}(I)) \vee L$, for $2 \leq k \leq n$, k, $n \in N^{+}$
   ε is the symbol for the elementary morphological erosion operation of I by a structuring element, and $\vee$ stands for the pointwise maximum.

In image processing, for more effective reconstruction and to obtain better performance, the erosion and dilation reconstruction composition were used to get morphological opening and closing reconstructions (R^{γ} and R^{φ}) of L from I as follows [17]:

$$R_L^{\gamma} = R_L^{\delta}(R_L^{\varepsilon}(I)) \tag{15}$$

$$R_L^{\varphi} = R_L^{\varepsilon}(R_L^{\delta}(I)) \tag{16}$$

The performance of morphological reconstruction (MR) in medical image processing depends mainly on some assumptions about the size and shape of the tumor. They are applied to the medical image obtained after segmentation at multiple thresholds to correctly identify the tumor. In this study, the disk-shape form is utilized because human brains possess an approximately oval shape.

3 Proposed Methodology

In this section, We explain in detail the proposed method for segmenting a brain tumor in MRI. Firstly, we start by applying a pre-processing operation on the image that we want to segment. The objective of this step is to improve the contrast of the considered image by adjusting its intensity. After that, we apply a method based on multi-threshold represented by the multi-Otsu method. It segments the image into several clusters by determining the n-1 thresholds. Then, the morphological reconstruction process was applied to the segmented image. The purpose of the last step is to detect the tumor and isolate it from the rest of the brain tissues. Figure 2 is a clear explanation of the proposed method. Based on the above described, the proposed algorithm can be summarized as follows:

Algorithm 1: Proposed method

1. **Input:** Load Image MRI.
2. **Initialize:** Number of thresholds (t=3) and structuring element (SE=9).
3. Apply an intensity adjustment on the loaded image.
4. **Multi-thresholding:** Segment the image into clusters using the multi-Otsu method.
5. **Morphological reconstruction operation:** Apply morphological closing reconstruction on the segmentation image.
6. **Output:** Tumor separation.
7. **Evaluation:** Performance indicators.

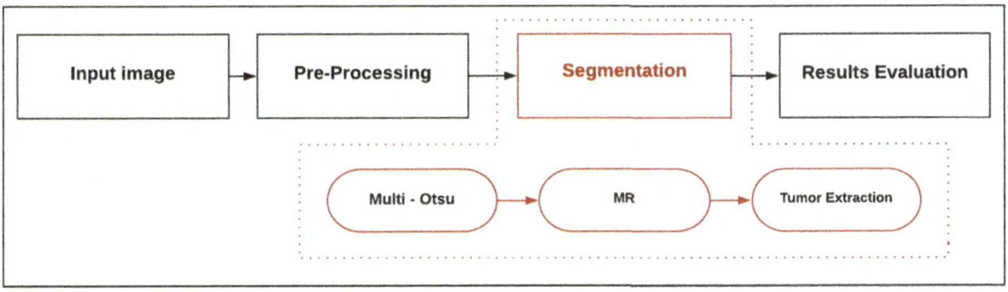

Fig. 2. The framework of the proposed algorithm.

4 Experimental

4.1 Dataset

Brain tumors are usually classified according to their source or degree of aggression. Until, now the most used grading system presented by the World Health Organization (WHO), where brain tumors are classified from grade 1 to grade 4 with increased aggression. Among these tumors are WHO grade 4 gliomas, also known as glioblastoma multiform (GBM). These are the most common primary malignant brain tumors in humans, showing very rapid growth [1].

In this paper, we use an improved T1-weighted MRI image which is selected from a set of images acquired from patients with multiple glioblastoma (GBM). The segmentation of ground truth images has been performed by expert neurosurgeons [11,12].

4.2 Validation

In this subsection, a comparative study is performed to clarify and demonstrate the effectiveness of the proposed method for image segmentation of brain tumors. Indeed, the results of the proposed method will be compared to some state-of-the-art clustering algorithms such as multi-Otsu [13], K-means (KM) [18], HKM [19], Fuzzy c-means (FCM) [20],FRFCM [21] and PSO [22]. For all methods, the number of segmented regions is four.

As regards the evaluation of medical image segmentation, we utilize some commonly-used metrics. In what follows, we define some indicators to use in this evaluation study:

- *Specificity*: It refers to the value of the properly segmented healthy (non-tumor) region and that should not be segmented.

$$Sp = \frac{TN}{TN + FP} \tag{17}$$

- *Sensitivity*: It considers the value of the region of interest (tumor) correctly segmented.

$$Se = \frac{TP}{TP + FN} \tag{18}$$

- *Precision*: It describes the purity of our positive detection from the truth on the ground.

$$Pr = \frac{TP}{TP + FP} \tag{19}$$

Where,

- TP: Number of pixels in the region of interest (ROI) detecting correctly.
- FP: Number of healthy pixels detected as the region of interest pixels.
- FN: Number of the region of interest (ROI) detected as healthy pixels.
- TN: Number of healthy pixels detected correctly.

The more the sensitivity, specificity and precision are close to one, the less there are detection errors.

In addition, to assess the validity of the proposed segmentation method, the accuracy index is measured. Accuracy of a segmentation technique refers to the degree to which the segmentation results agree with the true segmentation. It is formulated as a ratio between the number of correctly segmented pixels and the total number of pixels in an MRI slice. The accuracy is defined as:

$$Ac = \left[\frac{TP + TN}{N}\right] \times 100 \tag{20}$$

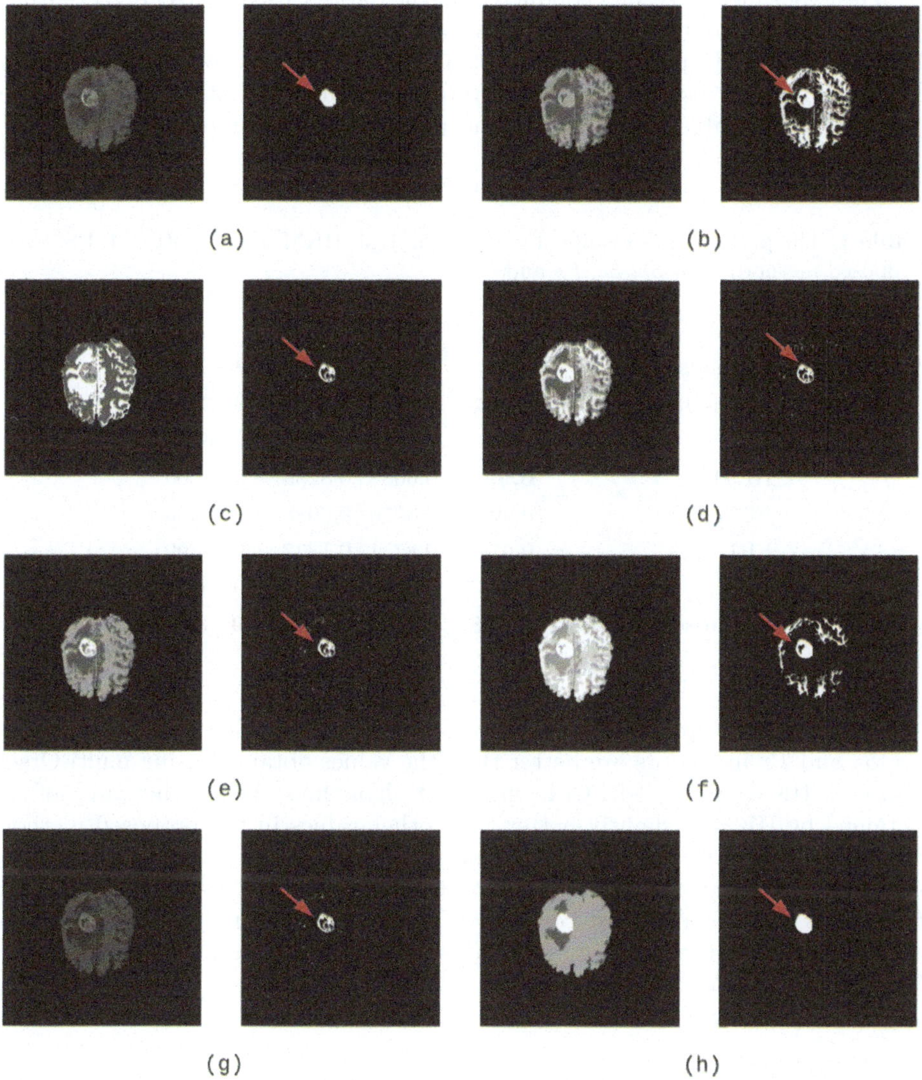

Fig. 3. Segmentation results for different methods on the MRI image (T1-weighted Axial slice, number 190). (a) Original image and ground truth. (b) Result of the multi-Otsu and tumor separation. (c) Result of the KM and tumor separation. (d) Result of the HKM and tumor separation. (e) Result of the FCM and tumor separation. (f) Result of the FRFCM and tumor separation. (g) Result of the PSO and tumor separation. (h) Result of the proposed method and tumor separation.

N = TP+TN +FP +FN (Number total of pixels) [23,24].

The experimental results conducted on the dataset used a MATLAB (2020a) software and a computer with the following characteristics: an Intel Core i3-2328 MCPU@ 2.20 GHz processor and 1.89 RAM system. The results and images obtained from our experience of the proposed method and other methods have been shown in Table 1 and Fig. 3 (The red arrow in the figures indicates the tumor region).

Table 1. The performance results of multi-Otsu, KM, HKM, FCM, FRFCM, PSO and Proposed method, with *Sp*, *Se*, *Pr* and *Ac*.

Figure 1				
Methods	*Sp*	*Se*	*Pr*	*Ac*
Multi-Otsu	0.9435	0.8922	0.0811	94.3180
KM	0.9987	0.6026	0.7286	99.6544
HKM	**0.9993**	0.6088	0.8259	99.7000
FCM	0.9985	0.6376	0.6985	99.6456
FRFCM	0.9783	0.8751	0.1837	97.7699
PSO	0.9987	0.6150	0.7226	99.6548
Proposed method	**0.9988**	**1.000**	**0.8302**	**99.8772**

As regards the proposed method, we can observe in Table 1 that the values of *Sp*, *Se* and *Pr* indicators are better than the values obtained using multi-Otsu, K-means HKM, FCM, FRFCM, and PSO algorithms. While the value of *Sp* obtained by HKM is slightly better than other values in the proposed method algorithm. Moreover, the value of *Se* and *Pr* is much better than other methods, as we seeing the value of *Se* is equal to 1, which means that the region of interest (tumor) is segmented almost correctly. Furthermore, accuracy indicator values demonstrate that the proposed method is more effective than other methods to isolate brain tumors in MRI images.

5 Conclusion

Localization and discovery of the tumor are necessary to treat the tumor. Therefore, an effective method with certain characteristics such as robustness and accuracy is necessary to segment medical images. In this paper, we used an image segmentation technique based on threshold and some morphological reconstruction operation to detect the brain tumor, contour, and calculate the tumor area. The performances of the proposed method were compared with the multi-Otsu, K-means, HKM, FCM, FRFCM, and PSO clustering algorithms using some indexes frequently utilized in the field of medical image segmentation. The results obtained clearly indicate the effectiveness of the proposed method, where the percentage of accuracy with the proposed approach is higher than in other methods.

References

1. Bauer, S., Wiest, R., Nolte, L.P., Reyes, M.: A survey of MRI-based medical image analysis for brain tumor studies. Phys. Med. Biol. **58**(13), R97–R129 (2013)
2. Dougherty, G.: Digital Image Processing for Medical Applications. Cambridge University Press, Cambridge (2009)
3. Senthilkumaran, N., Vaithegi, S.: Image segmentation by using thresholding techniques for medical images. Comput. Sci. Eng. Int. J. **6**(1), 1–13 (2016)
4. Dawngliana, M., Deb, D., Handique, M., Roy, S.: Automatic brain tumor segmentation in MRI: hybridized multilevel thresholding and level set. In: 2015 International Symposium on Advanced Computing and Communication (ISACC), Silchar, pp. 219–223. IEEE (2015)
5. Sujan, M., Alam, N., Noman, S.A., Islam, M.J.: A segmentation based automated system for brain tumor detection. Int. J. Comput. Appl. **153**(10), 0975–8887 (2016)
6. Ilhan, U., Ilhan, A.: Brain tumor segmentation based on a new threshold approach. Procedia Comput. Sci. **120**, 580–587 (2017)
7. Mehidi, I., Belkhiat, D.E.C., Jabri, D.: An improved clustering method based on K-means algorithm for MRI brain tumor segmentation. In: 6th International Conference on Image and Signal Processing and their Applications (ISPA), Mostaganem, Algeria, pp. 1–6. IEEE (2019)
8. Vala, H.J., Baxi, A.: A review on Otsu image segmentation algorithm. Int. J. Adv. Res. Comput. Eng. Technol (IJARCET) **2**(2), 387–389 (2013)
9. Kaur, N., Kaur, R.: A review on various methods of image thresholding. Int. J. Comput. Sci. Eng. **3**, 3441 (2011)
10. Kalavathi, P., Prasath, V.S.: Methods on skull stripping of MRI head scan images-a review. Journal of digital imaging **29**(3), 365–379 (2016)
11. Lindner, L., Narnhofer, D., Weber, M., Gsaxner, C., Kolodziej, M., Egger, J.: Skull-stripped MRI GBM Datasets. Figshare (2018)
12. Egger, J., Kapur, T., Fedorov, A., Pieper, S., Miller, J.V., Veeraraghavan, H., Kikinis, R.: GBM volumetry using the 3D slicer medical image computing platform. Sci Rep. **3**, 1364 (2013)
13. Otsu, N.: A threshold selection method from gray-level histograms. IEEE Trans. Syst. Man Cybern. **9**(1), 62–66 (1979)
14. Udayakumar, R., Khanaa, V., Saravanan, T., Saritha, G.: Retinal image analysis using curvelet transform and multistructure elements morphology by reconstruction. IEEE Trans. Biomed. Eng. **58**(5), 1183–1192 (2010)
15. El Allaoui, A., Nasri, M.: Medical image segmentation by marker controlled watershed and mathematical morphology. Int. J. Multimedia Appl. **4**(3), 1 (2012)
16. Vincent, L.: Morphological grayscale reconstruction in image analysis: applications and efficient algorithms. IEEE Trans. Image Processing **2**(2), 176–201 (1993)
17. Lei, T., Jia, X., Liu, T., Liu, S., Meng, H., Nandi, A.K.: Adaptive morphological reconstruction for seeded image segmentation. IEEE Trans. Image Process. **28**(11), 5510–5523 (2019)
18. Norouzi, A., Rahim, M.S.M., Altameem, A., Saba, T., Rad, A.E., Rehman, A., Uddin, M.: Medical image segmentation methods, algorithms, and applications. IETE Tech. Rev. **31**(3), 199–213 (2014)

19. Mehidi, I., Belkhiat, D. E. C., Jabri, D.: A fast K-means clustering algorithm for separation of brain tissues in MRI. In: The 2nd International Conference on Mathematics and Information Technology (ICMIT 2020), Adrar, Algeria, pp 1–6 IEEE (2020)
20. Bezdek, J.C., Ehrlich, R., Full, W.: FCM: the fuzzy c-means clustering algorithm. Comput. Geosci **10**(2–3), 191–203 (1984)
21. Lei, T., Jia, X., Zhang, Y., He, L., Meng, H., Nandi, A.K.: Significantly fast and robust fuzzy c-means clustering algorithm based on morphological reconstruction and membership filtering. IEEE Trans. Fuzzy Syst. **26**(5), 3027–3041 (2018)
22. Eberhart, R., Kennedy, J.: A new optimizer using particle swarm theory. In: MHS 1995, Proceedings of the Sixth International Symposium on Micro Machine and Human Science, pp. 39–43. IEEE (1995)
23. Taha, A.A., Hanbury, A.: Evaluation metrics for medical organ segmentation and lesion detection. In: Cloud-Based Benchmarking of Medical Image Analysis, pp. 87–105. Springer, Cham (2017)
24. Fenster, A., Chiu, B.: Evaluation of segmentation algorithms for medical imaging. In: 27th Annual Conference on IEEE Engineering in Medicine and Biology, Shanghai, pp. 7186–7189. IEEE (2005)

Epileptic Seizure Detection from EEG Signal Using the Imbalanced Data Techniques

Romaissa Debeche(✉), Mostafa El Habib Daho, and Mohammed Amine Chikh

Biomedical Engineering Laboratory, Tlemcen University, Tlemcen, Algeria
debecheroumaissa@gmail.com,
{mostafa.elhabibdaho,mohammedamine.chikh}@univ-tlemcen.dz

Abstract. In neurology, the Electroencephalogram (EEG) remains one of the most effective tools for the diagnosis and the management of patients suffering from epileptic seizures. The aim of our work is the automatic detection of this disease from the EEG signal record. The study is based on three major steps. First, a feature selection step was realized using the Discrete Wavelet Transform (DWT). The extracted data were rebalanced using two resampling techniques (Synthetic Minority Over Sampling Technique (SMOTE) and Adaptive Synthetic Sampling Approach (ADASYN)). Then, the rebalanced data were classified using the Random Forest (RF) model. RF was used to distinguish between abnormal and normal patients. The proposed algorithm is tested on five data sets. Each data set contains one hundred single-channel, four data sets were considered as normal and one as seizures. Average classification accuracy of 98.64% has been reached and the proposed method surpasses the performances of the state-of-the-art methods using the same EEG benchmark.

Keywords: Electroencephalogram (EEG) · Seizure detection · Discrete wavelet transformation · Imbalanced data · SMOTE · ADASYN

1 Introduction

Epilepsy is considered one of the most common and disabling neurological conditions [1]. Epileptic seizures are characterized by recurrent and unprovoked paroxysmal alteration of neurological function caused by the excessive, hypersynchronous discharge of neurons in the brain. It can affect all people including all ages, races, and ethnic backgrounds [2]. Despite the development of the epilepsy diagnosis tools, the EEG signal interpretation is considered the most important and effective test since it is painless, low cost, available, and has an efficient temporal resolution of long-term monitoring [2]. On the other hand, for long EEG recording, the visual interpretation task is expensive and considered as a tedious exercise that can cause errors and the result of interpreting an EEG

M. R. Senouci et al. (Eds.): CSA 2020, LNNS 199, pp. 301–312, 2021.
https://doi.org/10.1007/978-3-030-69418-0_27

record may vary from one doctor to another [2]. For that, there is a need for an automatic method for detecting seizures to help the doctors to make the right diagnosis and treatment decision. In this work we have developed a method to detect seizure from EEG signal using imbalanced EEG data sets; the study was based on three steps:

1. Extracting the features from the time domain using raw EEG and from the frequency domain after applying the DWT.
2. Rebalancing the class distribution of the data extracted using SMOTE and the ADASYN.
3. Classifying the rebalanced output by the RF classifier.

The obtained results were compared to a state-of-the-art method of epileptic seizure detection using the same benchmark EEG data-set. The rest of this paper is organized as follows: Sections 2 summarize the prior work related to feature extraction and classification methods. The details of the proposed method and the material used are described in Sect. 3. In Sect. 4 the evaluation procedure, the experimental and the comparison results are presented. Finally, we finish by adding a conclusion and perspectives in Sect. 5.

2 Related Works

Many significant works proposed different methodologies for the automatic detection of epileptic seizures using EEG signals. M. Mursalin et al. [2] have used the improved correlation-based feature selection after the feature extraction using the time-domain entropy and frequency domain using DWT. They have used the random forest classifier for the classification step. The proposed algorithm has been tested on 5 sets of the EEG signal. Each dataset consists of 100 channels. Two sets were recorded from a normal person, two others were recorded from epileptic patients out of the seizure and the last data set contain a seizure record. S. Yasmeen and M.V. Karki [3] proposed a method for seizure detection based on statistical parameters extraction and frequency domain parameters extraction using DWT. The classification was based on a neural network classifier to classify the signal as normal or epileptic. The method was tested on data set consists of two sampling rates, one 128 Hz and another with a sampling rate 1024 Hz.

Among the recent works, an automatic features extraction and classification of the epileptic seizure from the EEG method based on deep learning was proposed in [4]. The proposed algorithm used the architecture of convolutional neural networks (CNN) to classify the EEG signal.

Q. Yuan et al. [5], have proposed a technique based on imbalanced classification and wavelet packet transform. Since the period of the seizure is to short compared to the duration of continuous EEG recording, the researchers treated the EEG signal as imbalanced, they used the wavelet packet transform to analyze and extract feature from the time-frequency domain and the weighted extreme learning machine for the classification.

In [6], a new method of early detection of seizure was proposed to be implemented on an electronic tool. So, the researchers were limited by the conservation of energy and calculation speed constraints. Consequently, they opted for a simple algorithm based on selecting features from the time-frequency domain then classify them with a random forest classifier.

3 Proposed Model

In this study, we have used the DWT algorithm to analyze the EEG data sets and to extract the five physiological bands of the EEG signal (gamma (30–60 Hz), beta (15–30Hz), alpha (8–13 Hz), theta (4–8 Hz) and delta (0–4 Hz)), using five decomposition levels with fourth order Daubechies (db4). Features were extracted from the raw EEG data and wavelet decomposed EEG. The obtained data were rebalanced using the SMOTE resampling technique. The rebalanced data are used as input to the Random Forest classifier to classify the EEG signal as normal or seizure. Figure 1 represents the diagram of the proposed method.

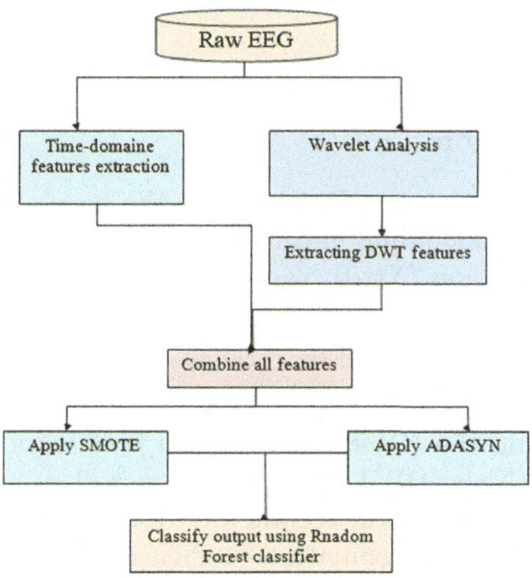

Fig. 1. The diagram of the proposed method.

3.1 EEG Dataset

The EEG database used in this study is collected by R. G. Andrzejak et al. at the Department of Epileptology, University of Bonn, Germany [7]. The database includes five EEG data sets (denoted as Set A, B, C, D, and E). the data was collected from different persons, contains mainly two classes, normal (A, B, C, D) and seizure (E), each data set contains 100 single-channels with the duration of 23.6s for each channel. The recorded data was digitized at 173.61 data points per second, the normal data-set collected from five healthy persons with 200 single -channels, the (C,D) data-sets were collected from five epileptic patients out of the seizure period with 200 single-channel, a 100 single "channels were recorded from the same five epileptic patients during the seizure (E), Fig. 2 represent an example of EEG signal from each data.

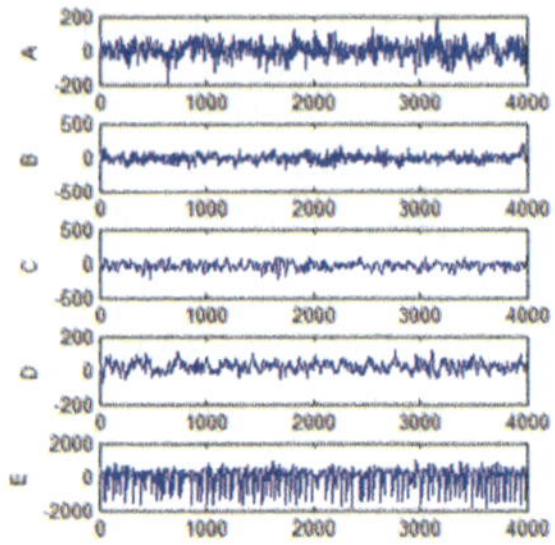

Fig. 2. Examples of EEG signal from (A-E) data sets

3.2 Feature Extraction

Time Domain Features from Raw EEG: We have extracted time domain features from raw EEG including mean (Xmean), minimum (Xmin), maximum (Xmax), standard deviation (Xstd), variance (Xvar),median (Xmed) and skewness (Xsks).

Frequency Domain Features Using DWT: Since the EEG signal is a non-stationary signal, using the DWT is one of the best ways to extract the most efficient features. With the DWT, the raw signal will pass through two filters. The first filter is a low pass and the second filter is a high pass filter. The low pass frequencies represent the approximation coefficient (A) and the high pass filter coefficient represents the detail coefficient (D). The Approximation coefficient is again decomposed to get the second level of approximation coefficient (A) and the detail coefficient (D). The procedure will be repeated until we get the desired decomposition [3]. Making a good choice of the type and the order of wavelet is a very important factor to get good decomposition results. According to [2] and [3], the Daubechies-4 wavelet is the most used for seizure detection. In this work, the EEG signal was decomposed into five levels as shown in Table 1. The frequency bands extracted were used to extract features, mean, max, min, energy; Shannon entropy and spectral entropy were extracted.

Table 1. Noise and frequency bands of EEG using 5th level decomposition.

Details and approximation	Frequency Band (Hz)	Decomposition level	Band name
D1	>60	1	noise
D2	30-60	2	gamma
D3	15-30	3	beta
D4	8-13	4	alpha
D5	4-8	5	theta
A5	0-4	5	delta

3.3 Data Re-balance

Since the user database consists of two classes with 400 normal EEG signals and 100 seizure signals. In this case, the data is considered as imbalanced. So, to treat this type of data we have to use specific methods dedicated to deal with the imbalanced data problems. In this work, we have preferred to use one of the data levels approaches. These techniques aim to rebalance the data since they add a preprocessing step to resample this later and they are unrelated to the correspondent classifier. So, with the use of data level approaches, we can work easily with an existing classification algorithm without any changes [8].

Synthetic Minority Over-Sampling. Synthetic Minority Over-sampling (SMOTE) is a method inspired by a technique that recognizes letters written by hand. It consists of adding synthetic examples of the minority class. First, it is necessary to choose a number K of the nearest neighbors and a desired percentage of over-sampling. Then, calculate the difference between a minority example and its nearest neighbors. Finally, it multiplies the difference calculated previously by a random number between 0 and 1. This last step allows creating a new synthetic instance located between the minority instance and its neighbor's [9].

Adaptive Synthetic Sampling Approach. The ADASYN algorithm is used for the two-class (binary) classification problem, the idea is to adaptively generate minority data samples according to their distributions in the data set: more synthetic data is generated for minority class samples that are harder to learn compared to those minority samples that are easier to learn. The use of the ADASYN method offers double affectivity it can at the same time can reduce the learning bias introduced by the original imbalance data distribution, and also adaptively shift the decision boundary to focus on those difficult to learn samples [11].

3.4 Classification

For the classification task, we have used the Random Forest classifier, which is a supervised learning algorithm, one of the most important ensemble algorithms in the classification field. Leo Breiman introduced the random forest classification method, it is a method based on the combination of two classification techniques: an ensemble method, especially bagging and the use of the decision tree as a classifier more precisely the CART decision tree [10].

Using bootstrapping the initial base is divided into sub-bases that contains the same number of instances of the initial base. These samples are selected by sampling with replacement, to form what are called bags. For each bag, a decision tree is trained. The induction of trees is done without pruning and according to the CART algorithm, however, at the level of each node, the selection of the best partition, based on the Gini index, is done only on a sub-set of attributes of prefixed size randomly selected from the original space of characteristics, the overall prediction of the forest is calculated by taking the majority of votes of each of its trees. The elements that do not appear on the bags(out of bags) are used as a validation set. The random forest has a random double process, at first in the random choice of the instances of each bag and, secondly, in the random choice of the subset attributes. This random double process ensures data diversity as shown in the algorithm below.

Algorithm 1: Random Forest Algorithm

Data: A training set S := $(x_1,y_1),...,(x_n,y_n)$, features F, and number of trees in forest B.
Result: RandomForest
Function: RandomForest
initialization: RandomForest(S, F) 2 H ←Φ
for $i \in 1,...,B$ **do**
  S(i) ← A bootstrap sample from S
  hi ← RandomizedTreeLearn(S(i),F)
  H ← H∪hi
Function: RandomizedTreeLearn(S, F)
At each node:
f very small subset of F
Split on best feature in f
return The learned tree

3.5 Evaluation Metrics

The evaluation was based on five metrics since we are dealing with imbalanced data metrics: Accuracy (AC), Sensitivity (SE), Specificity (SP), Geometric Mean (G-mean), and Precision.

1. Accuracy (AC): defines the true classification rate. It is calculated by the formula:

$$AC = \frac{(TP + TN)}{(TP + TN + FP + FN)} * 100 \quad (1)$$

2. Geometric mean (G-mean): Provides a simple way to assess the model's ability to correctly classify the minority and majority class by combining sensitivity and specificity in a single metric. G-mean is considered a measure of balanced accuracy and is defined as follows:

$$G - mean = \sqrt{Sensitivity * Specificity} \tag{2}$$

3. Precision: Is the fraction of the relevant instances among the retrieved instances..

$$Precision = \frac{TP}{(TP + FP)} \tag{3}$$

4. Sensibility (SE): The percentage of positive instances that are correctly classified.

$$SE = \frac{TP}{(TP + FN)} 100 \tag{4}$$

5. Specificity (SP): the percentage of negative samples that are correctly classified.

$$SP = \frac{TN}{(TN + FP)} * 100 \tag{5}$$

with:

- True positive (TP): positive cases classified as positive.
- True negative (TN): negative cases classified as negative.
- False positive (FP): positive cases classified as negative.
- False negative (FN): negative cases classified as positive.

4 Results and Discussion

In this work, DWT of level five with fourth-order Daubechies (db4) is applied to the 500 channels of EEG to extract the different sub-bands (gamma, beta, alpha, theta, and delta). Decomposition results are showing in Fig. 3, for the feature extraction raw EEG was used to extract the statistical features. The wavelet features have been extracted from D2, D3, D4, D5, and A5 while the D1 is considered as noise. After the feature extraction, the rebalanced and classification

Table 2. Classification cases.

Case	Class 1	Class 2	Imbalance rate (%)
1	AB	E	33.33
2	CD	E	33.33
3	ACD	E	25
4	BCD	E	25
5	ABCD	E	20

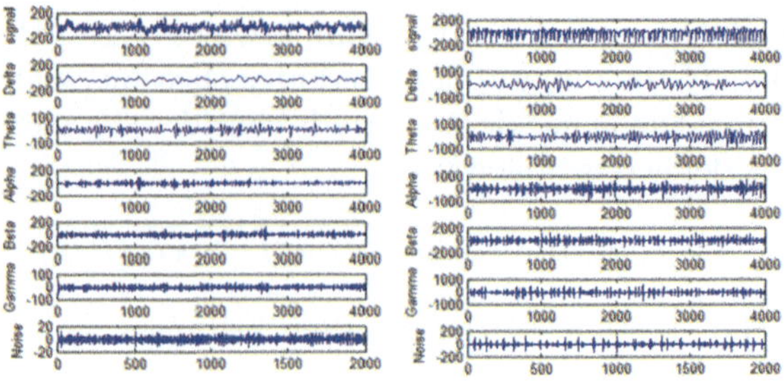

Fig. 3. EEG signal decomposition (at left normal EEG at right epileptic EEG)

tasks are applied to five different cases showed in Table 5 using SMOTE with different rates and 7 neighbors. Then, the data were divided to test and training sets using K-fold –cross-validation with 5 folds, and the ADASYN applying the Random Forest algorithm with 100 trees on the divided and rebalanced data sets, provided some very high classification performances as showed in Table 3 and Table 4. Results of the classification without resampling the data are described in Table 2. A comparison between the performances of the resampled and non-resampled data are described in Fig. 4. Also, the performances obtained with the use of the SMOTE and ADASYN outperform those of [2] in all most all the cases, as described in Table 6.

Table 3. Classification results before the resampling step.

Case	Classification Accuracy (%)	Sensitivity(%)	Specicity (%)	G-mean (%)	Precision(%)	Average CA (%)	Average G-mean(%)
1	98.00	100	93.33	96.61	100		
2	96.45	95.21	98.54	96.86	95.21		
3	96.24	98.99	88.24	93.46	98.99		
4	97.74	98.99	94.12	96.52	98.99		
5	95.78	97.67	89.19	93.34	97.67	96.84	95.36

As described in Table 2, Table 3, Table 5, and Fig. 4, the comparison between the proposed method before and after applying the SMOTE oversampling technique and the ADASYN showed that there is an improvement in the classification performances where the average accuracy jump from 96.84% with non-resampled

Table 4. Classification results after using ADASYN resampling technique.

Case	Classification Accuracy (%)	Sensitivity(%)	Specicity (%)	G-mean (%)	Precision(%)	Average CA (%)	Average G-mean(%)
1	99.03	98.57	100	99.28	98.57		
2	96.12	94.52	100	97.22	94.52		
3	98.80	98.11	100	99.05	98.11		
4	98.79	98.95	98.57	98.76	98.95		
5	99.58	99.28	100	99.64	99.28	98.46	98.79

Table 5. Classification results after using the SMOTE resampling technique.

Case	Classification Accuracy (%)	Sensitivity(%)	Specicity (%)	G-mean (%)	Precision(%)	Average CA (%)	Average G-mean(%)
1	99.18	99.28	98.99	99.13	99.28		
2	97.00	95.71	100	97.83	95.71		
3	98.97	99.31	97.97	98.64	99.31		
4	99.02	99.22	98.35	98.78	99.22		
5	99.00	99.27	97.96	98.61	99.27	98.64	98.60

data to 98,46% in the case of ADASYN and 98.64% in the case of SMOTE with and the most important is that the average geometric mean has been improved going from 95.36% to 98.79% in the case of ADASYN and 98.60%, in the case of SMOTE. As shown in Table 4 and Table 5, the obtained results from our method provide very high performances of accuracy. The results are associated with a very good geometric mean and high precision which describe the ability of our method to classify both minority and majority classes at the same time. The results of the geometric mean and the precision obtained from the

Table 6. Comparative study

Method		Cases	Accuracy
Mursalin et al [2]		ACD-E	98.5
		BCD-E	97.5
		CD-E	98.67
		ABCD-E	97.4
Our proposed method	SMOTE	ACD-E	98.97
		BCD-E	99.02
		CD-E	97.00
		ABCD-E	99.00
	ADASYN	ACD-E	98.80
		BCD-E	98.79
		CD-E	96.12
		ABCD-E	99.58

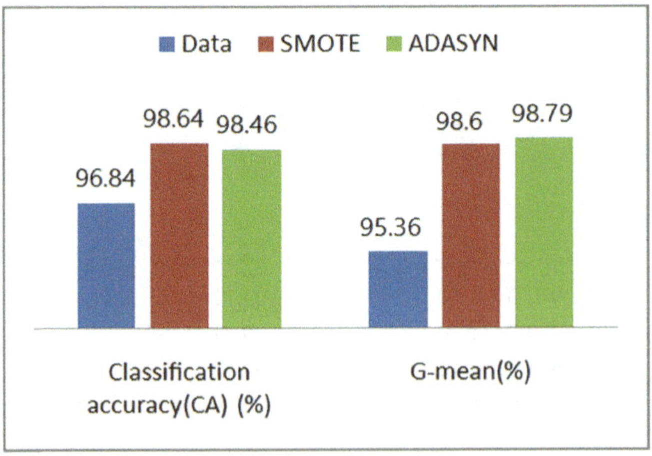

Fig. 4. Comparison between the performances of the resampled and non resmapled EEG data

ADASYN application are better than those obtained from the application of the SMOTE which makes the best compromise in this case. For the comparison between the proposed method and the method proposed by M. Mursalin et al. [2], we can observe clearly in Table 6 that the accuracy measure has increased in our proposed method for almost all the cases.

5 Conclusion

In this paper, a method of Epileptic Seizure detection from imbalanced EEG signal was proposed where an EEG database has been used. The database consists of 5 datasets. Two datasets were recorded from healthy persons, two others were recording from epileptic patients out of the seizure, and the last one contains seizure recordings. To classify the previous data sets, a feature extraction step has been established using the DWT to extract the frequency domain features. For the time domain feature, we have used the raw EEG data. We have considered that our dataset is imbalanced, so we have applied two resampling techniques (SMOTE and ADASYN) to rebalance the data class distribution and improve the classification performances. The classification task was performed using the Random Forest model. The obtained results of this work are very promising, they provide some very high performances which prove the ability of resampling methods to improve the classification performances of imbalanced data. The results also showed that the SMOTE technique improves the accuracy and the ADASYN improves the geometric mean and precision. To give more credibility to our conclusions, we have compared our proposed approach with a state-of-the-art method. the obtained results show that our proposed method outperforms the technique of Mursalin et al. [2]. For all these reasons, it can be concluded that the combination of ADASYN and RF is more effective in the seizure detection task, and the present work can be very helpful to the neurological medical stuff to make a better diagnosis and treatment decisions for epilepsy. In the future, we look to explore more features, combine those techniques, and use our model on other types of data.

References

1. Stafstrom, C.E., Carmant, L.: Seizures and epilepsy: an overview for neuroscientists. Cold Spring Harb. Perspect. Med. **5**(6), a022426 (2015). https://doi.org/10.1101/cshperspect.a022426
2. Mursalin, Md., Zhang, Y., Chen, Y., Chawla, N.V.: Automated epileptic seizure detection using improved correlation-based feature selection with random forest classifier. Neurocomputing **241**, 204–214 (2017). https://doi.org/10.1016/j.neucom.2017.02.053
3. Yasmeen, S., Karki, M.V.: Neural network classification of EEG signal for the detection of seizure. In: IEEE International Conference on Recent Trends in Electronics, Information and Communication Technology (RTEICT), Bangalore, pp. 553–558 (2017). https://doi.org/10.1109/RTEICT.2017.8256658.
4. Acharya, U.R., Oh, S.L., Hagiwara, Y., Tan, J.H., Adeli, H.: Deep convolutional neural network for the automated detection and diagnosis of seizure using EEG signals. Comput. Biol. Med. **100**, 270–278 (2018). https://doi.org/10.1016/j.compbiomed.2017.09.017
5. Yuan, Q., Zhou, W., Zhang, L., Zhang, F., Xu, F., Leng, Y., Wei, D., Chen, M.: Epileptic seizure detection based on imbalanced classification and wavelet packet transform. Seizure **50**, 99–108 (2017). https://doi.org/10.1016/j.seizure.2017.05.018

6. Donos, C., Dümpelmann, M., Schulze-Bonhage, A.: Early seizure detection algorithm based on intracranial EEG and random forest classification. Int. J. Neural Syst. **25**(05), 1550023 (2015). https://doi.org/10.1142/S0129065715500239
7. Andrzejak, R.G., Lehnertz, K., Mormann, F., Rieke, C., David, P., Elger, C.E.: Indications of nonlinear deterministic and finite-dimensional structures in time series of brain electrical activity: dependence on recording region and brain state. Phys. Rev. E **64**(6), 061907 (2001). https://doi.org/10.1103/PhysRevE.64.061907
8. Galar, M., Fernandez, A., Barrenechea, E., Bustince, H., Herrera, F.: A review on ensembles for the class imbalance problem: bagging-, boosting-, and hybrid-based approaches. IEEE Trans. Syst. Man Cybern. Part C (Appl. Rev.) **42**(4), 463-484 (2012). https://doi.org/10.1109/TSMCC.2011.2161285
9. Chawla, N.V., Bowyer, K.W., Hall, L.O., Kegelmeyer, W.P.: SMOTE: synthetic minority over-sampling technique. J. Artif. Intell. Res. **16**(1), 321–357 (2002)
10. Breiman, L.: Random forests. Mach. Learn. **45**, 5–32 (2001). https://doi.org/10.1023/A:1010933404324
11. He, H., Bai, Y., Garcia, E.A., Li, S.: ADASYN: adaptive synthetic sampling approach for imbalanced learning. In: IEEE International Joint Conference on Neural Networks (IEEE World Congress on Computational Intelligence), Hong Kong, pp. 1322-1328 (2008). https://doi.org/10.1109/IJCNN.2008.4633969

Thyroid Nodules Recognition in Ultrasound Images Based on ImageNet Top-Performing Deep Convolutional Neural Networks

Bilal Chouiha(✉) and Abdenour Amamra

Ecole Militaire Polytechnique, Bordj El -Bahri BP 17, Algiers, Algeria
bilal.chouiha@gmail.com, amamra.abdenour@gmail.com

Abstract. Deep learning algorithms, especially convolutional neural networks, have quickly become a methodology of choice for medical image analysis. In this article, we are interested in thyroid cancer, a disease that is triggered off by the malignant transformation of some of the cells that make up the thyroid gland. For this purpose, our study focuses on the use of deep learning on ultrasound images for the recognition of malignant tissue. The proposed approach is based on ImageNet top-performing deep convolutional neural networks such as VGG16, Inception and Inception-Resnet. These networks were evaluated on the public database, Digital Database Thyroid Image (DDTI). However, the latter contains only 451 thyroid images which is little to work with as sophisticated models trained on small datasets tend to overfit quickly. Thus, we carried out several data augmentation strategies on the training images prior to classification in order to enlarge the variety of training data. The obtained results showed a test accuracy of 93%, 97% and 99% for VGG16, Inception-Resnet and Inception, respectively.

Keywords: Thyroid nodules recognition · Ultrasound image analysis · Deep convolutional neural networks

1 Introduction

The thyroid gland is a butterfly-shaped organ located in the base of the neck. It releases hormones that control metabolism. The thyroid hormones regulate vital body functions, including breathing, heart rate, nervous systems, body weight, muscle strength, cholesterol levels and much more. Thyroid disorders are very common and according to the British Thyroid Foundation, about one out of 20 people has some kind of thyroid disorder, which may be temporary or permanent. There are specific kinds of thyroid disorders that include: Thyroid Cancer, Hypothyroidism, Hyperthyroidism, Goiter, and Thyroid nodules.

Thyroid nodule is a palpable swelling in a thyroid gland with an otherwise normal appearance. Thyroid nodules are common and may be caused by a variety

M. R. Senouci et al. (Eds.): CSA 2020, LNNS 199, pp. 313–322, 2021.
https://doi.org/10.1007/978-3-030-69418-0_28

of thyroid disorders. While most are benign, about 5% of all palpable nodules are malignant [1].

There are different types of thyroid nodules:

- **Adenoma**: Macrofollicular adenoma (simple colloid), Microfollicular adenoma (fetal), Embryonal adenoma (trabecular), Hürthle cell adenoma (oxyphilic, oncocytic), Atypical adenoma, Adenoma with papillae, Signet-ring adenoma.
- **Carcinoma:** Papillary, Follicular, Medullary, Anaplastic, Other Thyroid lymphoma.
- **Cyst:** Simple cyst, Cystic/solid tumors (hemorrhagic, necrotic).
- **Colloid nodule:** Dominant nodule in a multinodular goiter.
- **Other:** Inflammatory thyroid disorders and others.

In the United States, 4–7% of the adult population have a clinically identifiable thyroid nodule [2]. Although only 5% of palpable nodules are malignant, it is important to identify and treat them properly. Imaging techniques such as ultrasonography is currently used to evaluate thyroid nodules and to provide information about their structures and characteristics, which may be defined as homogeneous or heterogeneous, solid, cystic, or mixed. However, the definition of these characteristics is operator dependent and influenced by the practitioner's experience [3]. The accuracy of thyroid ultrasound diagnosis closely depends on the experience and the ability of diagnosticians. There are usually many differences in judging ultrasound images among different practitioners. Therefore, the precise ultrasound diagnosis of thyroid carcinoma is a challenging task [4].

The computer-aided diagnosis (CAD) is an approach that may deliver a second decision for radiologists and medical specialists of traditional image analysis. It has the potential to reduce subjectivity, hence improving the accuracy and reduce radiological examinations time and cost. Besides, it allows doctors in worldwide regions who have limited access to experienced radiologists to gain insight into the health of the thyroid gland.

Our approach is to adapt the ImageNet top-performing deep convolutional neural networks (DCNN), VGG16, Inception and Inception-Resnet for thyroid ultrasound image recognition through a fine-tuning approach. The training was carried out on the public database DDTI. However, this database was small and results in severe over-fitting, the reason for which we proposed a data augmentation scheme as well.

The remainder of the paper is organized as follows. In Sect. 2, we provide an analysis of some studies on the ultrasound image recognition approaches and present the respective related works. In Sect. 3, we introduce our methodology with the detailed approach that we followed. In Sect. 4, we present the experimental setup, the materials we used and experimental results are discussed, where we conduct a comparison between the selected networks. The conclusion and potential future directions are outlined in Sect. 5.

2 Related Works

Deep Learning models have shown promising results on various medical image diagnosis problems including the breast, the lung, and the thyroid nodule diagnosis. Due to space limitations, this section summarizes the most significant approaches for thyroid lesions diagnosis.

Ma et al. [5] proposed a cascaded convolutional neural network to develop and evaluate a fully automatic detection of thyroid nodules from 2D ultrasound images. The proposed model achieves an area under the curve = 98.51%. Chen et al. [6] proposed a Deep-Learning Based Ultrasound Text Classifier for Predicting Benign and Malignant Thyroid Nodules. The proposed work is a supervised classification based on a deep neural network that is trained on the labeled ultrasonic text with the benign or malignant label of pathology. Experimental results showed that this method has a highest accuracy rate of 93% and 95% both on the real medical dataset and the UCI standard dataset.

Liu et al. [7] proposed a feature extraction method for ultrasound images based on the convolution neural networks, by introducing more meaningful semantic features to the classification. Experimental results on 1037 images showed that the accuracy of their proposed method is 0.931, which outperformed other relative methods by over 10%.

Another example is the study of Chi et al. [8] which presents a computer-aided diagnosis system for classifying thyroid nodules in ultrasound images using a pre-trained GoogleNet model, which leads to superior feature extraction. The extracted features of the thyroid ultrasound images are sent to a classifier to decide whether the images are related to malignant or benign cases. The experimental results showed good performance, attaining 98.29% classification accuracy, 99.10% sensitivity and 93.90% specificity for the images in an open-access database.

In 2017, Ashnil Kumar and et al. [9] introduced a new method for classifying medical images using a new feature extractor by fine-tuning different CNNs that have been initialized on a large dataset of natural images. The method was evaluated by experiments on the Image CLEF 2016 medical image public dataset (30 modalities, 6776 training images, and 4166 test images) and it showed that the ensemble of fine-tuned CNNs achieves a higher accuracy than established CNNs and achieved a higher accuracy than alternative methods in the literature, evaluated on the same benchmark and is only overtaken by those methods that have additional training data. Based on this work Moussa et al. [10] propose a fine-tuning approach based on deep learning using a Convolutional Neural Network model named resNet-50 for the classification of thyroid nodules in ultrasound images. Experiments have been conducted on 814 ultrasound images and the results show that the proposed approach dramatically improves the accuracy of the classification of thyroid nodules and outperforms the VGG-19 model.

The main contribution of the present work is the exploitation of the ImageNet top performing fine-tuned CNN (VGG16, Inception-Resnet and Inception) in the recognition of thyroid nodules in ultrasound images. Up to our knowledge, this is a distinguished study that has been carried out through tuning Imagenet state-of-the-art deep neural architectures.

3 Methodology

In this section, we present the three network models used in this work: VGG16, Inception and Inception-Resnet. We then describe the public dataset used, namely Digital Database Thyroid Image. Then the basic idea of the proposed comparative approach as well the approach for the fine-tuning of the pre-trained DCNN.

3.1 VGG16 Model

VGG16 is a convolutional neural network model proposed by K. Simonyan and A. Zisserman [11]. The model achieved 92.7% top-5 test accuracy on ImageNet and it was one of the famous models submitted to ILSVRC-2014. It made an improvement over AlexNet by replacing large kernel-sized filters (11 and 5 in the first and second convolutional layer, respectively) with multiple 3×3 kernel-sized filters one after another. One of the main characteristics of this network is its simplicity, it has 3×3 filter convolution layers of 3×3 filter with a stride 1 using the same padding and max-pooling layer of 2×2 filter of stride 2. It follows this arrangement of convolution and max-pooling layers consistently throughout the whole architecture. Finally, it has 2 fully connected layers followed by a softmax layer that plays the role of a classifier. The 16 in VGG16 refers to the 16 layers that have trainable weights. This network is relatively sizable as it contains 138 million parameters.

3.2 Inception Model

Proposed by Christian Szegedy et al. in [12], it was an important milestone in the development of CNN classifiers. Prior to its inception, most popular CNNs just stacked convolution layers deeper and deeper, hoping to get better performance. The Inception network on the other hand, was complex and heavily engineered. The main hallmark of this network is the improved utilization of the computing resources inside the network. This was achieved by a carefully crafted design that allows for increasing the depth and width of the network while keeping the computational budget constant. Its constant evolution leads to the creation of several versions of the network. The popular versions are as follows:

- Inception V1.
- Inception V2 and Inception V3.
- Inception V4 and Inception-ResNet.

3.3 Inception-Resnet Model

Inception-ResNet is a combination of two networks, residual connections and a recent version of Inception architecture [14]. The Inception models are famous for their multi-branch architectures. They have a set of filters (1×1, 3×3, 5×5, etc.) that are merged with concatenation in each branch. The residual model is famous for training very deep architectures. The hybrid Inception-ResNet-V2 network uses residual connections with good efficiency.

3.4 Image Collection and Data Augmentation

The proposed solution was evaluated by experiences carried on thyroid ultrasound images on the open-access Digital Database of Thyroid Ultrasound Images (DDTI) [15]. The latter is an open-access resource for scientific researchers provided by the *Universidad Nacional de Colombia* and IDIME (*Instituto de Diagnostico Medico*). This dataset is composed of a set of B-mode Ultrasound images including a complete annotation and diagnostic description of suspicious thyroid lesions by expert radiologists. The diagnostic description of malignant lesions was confirmed by biopsy. The dataset contains the analysis of 451 thyroid ultrasound images, performed by two experts on 299 patients, with the size 560×360, including 376 malignant and 75 benign images, which is a rather scare quantity. Therefore, the first step was to augment these images to generate new data for modeling. To generate them, some images were randomly chosen to be horizontally flipped or mirrored into new images. Additionally, the image contrast was varied randomly from the original contrast, as well as brightness and finally, the rotation transformation of some images. The final, augmented data set therefore included 4000 images (451 original and 3549 augmented). This was split into training, validation and testing sets (2000, 1000 and 1000 images respectively).

3.5 Proposed Method

Our study was limited in only classifying the thyroid nodules into two classes of malignant and benign. In this work, we present approaches for a thyroid nodules recognition in ultrasound images using a fine tunning of the ImageNet top performing CNNs: VGG16, Inception-Resnet and Inception.

Fine tuning an existing Deep CNN saves us from resorting to a large training samples and thus allows to generate easily a domain-specific deep learning network, which can effectively extract features for thyroid ultrasound images and classify malignant and benign thyroid images with high accuracy.

As illustrated in Fig. 1, we start by pre-training the networks on the ImageNet data set that contains 1000 classes of natural images. Then, we transfer the learned features and parameters from the general domain to our specific classification problem.

The proposed solution was evaluated by experiences carried on an augmented data set included 4000 thyroid ultrasound images, but processing this important number of images means a regular, personal laptop is not sufficient. As such, the

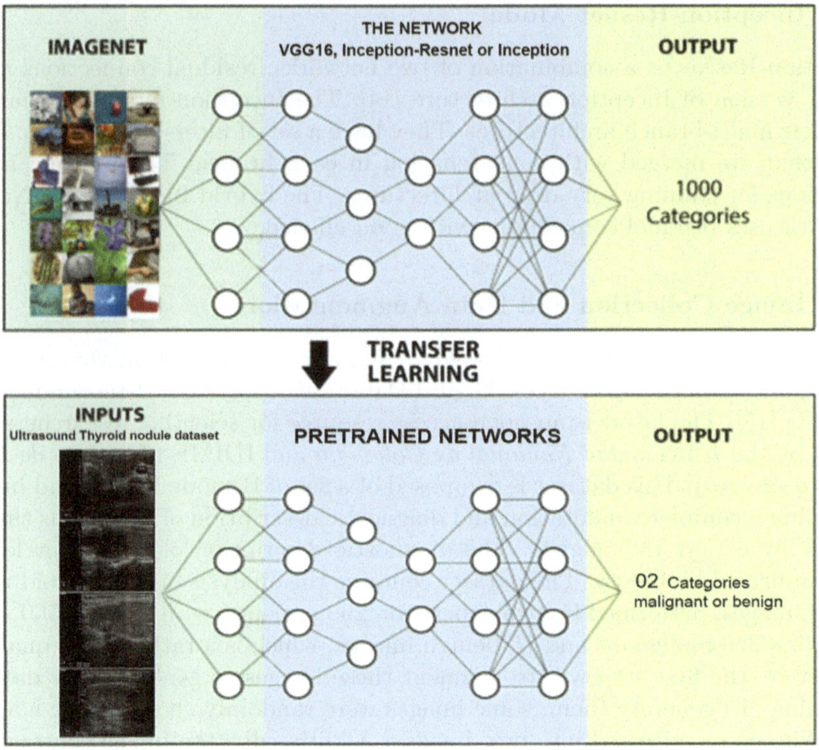

Fig. 1. Transfer learning from object recognition in ImageNet to Thyroid nodules recognition in ultrasound images

images treatment required the use of GPUs which were available for public use through the Google Colab service.

To summarize, it consists of the following basic steps:

1. Pre-train the networks : VGG16, Inception-V3 and Inception-Resnet-V2, on the ImageNet data set.
2. Assign a label to each ultrasound thyroid image of the data set.
3. Define the region of interest of thyroid nodule in each image as delineated by radiologist.
4. Rescale each thyroid nodule to 224×224 dimensions.
5. Fine-tune the pre-trained networks by modifying the weights of the last fully connected layer that contains 1000 classes with layer by a two fully connected softmax layer using the back-propagation technique.
6. Set the appropriate values of the parameters of back-propagation algorithm.

3.6 Fine-Tuning Parameter Selection

The proposed method includes fine-tuning the pre-trained DCNN: VGG16, Inception-V3 and Inception-Resnet-V2, to classify the thyroid nodules into two classes of malignant and benign. We adapted the networks parameters in the last fully-connected layer to our data set. The most influential parameters of backpropagation algorithm which affect the learning performance of the DCNN are: Learning rate, momentum rate, and activation function. However, there are no clear guidelines available about the selection of the optimal values of these parameters. The only way to check which parameters perform better is through several passes of trial and error. In our study, the weights of all fully connected layer are randomly initialized. Then, the model is fine-tuned using an optimizer with a momentum of 0.9 and learning rate of 10^{-6}. We run the backpropagation algorithm for 20 epochs, which optimizes the network parameters using stochastic gradient descent.

4 Results

4.1 Performance Measurement

Metrics: In order to measure the performances of the three trained models of recognition we used the standards metrics, classification accuracy, sensitivity, specificity, and area under receiver operating characteristic (ROC) curve (AUC). Sensitivity indicates the percentage of benign nodules correctly identified as benign. Specificity measures the percentage of malignant nodules correctly classified as malignant. Accuracy evaluates the percentage of correctly diagnosing thyroid nodules. ROC curve is a tool to measure the balance between finding true positives and avoiding false positives.

- Accuracy: Is the ratio of correct classifications to the total number made by the network, we can find it using: $Acc = (TP+TN)/(TP+TN+FP+FN)$.
- Precision: Or positive predictive value, it can be found using: $Pr = TP/(TP+FP)$.
- Sensitivity or recall: Corresponds to the probability of correctly classifying thyroid sample in our case, to find it we use: $R = TP/(TP+FN)$.
- F-score: Is a measure that combines both precision and recall as determined in the formula: $F_1 = 2 \times \frac{P \times R}{P+R}$.
- Average.
- Weighted average.

4.2 Discussion

Table 1 shows the test accuracy, precision, recall and f1-score of the VGG16 model classification. The experimental results show the proposed fine-tuned VGG16 model achieves excellent classification performance, attaining 93% classification accuracy, 93% precision and 93% sensitivity.

Table 2 shows the test accuracy, precision, recall and f1-score of the Inception-V3 model classification. The experimental results show the proposed fine-tuned Inception-V3 model achieves excellent classification performance, attain-

Table 1. Results of the classification of thyroid nodules by the VGG16 model

	Precision	Recall	F1-Score	Support
Negative	0.94	0.91	0.92	500
Positive	0.91	0.94	0.93	500
Accuracy			0.93	1000
Macro avg	0.93	0.93	0.93	1000
Weighted avg	0.93	0.93	0.93	1000

ing 99% classification accuracy (the best score obtained), 99% precision and 99% sensitivity.

Table 2. Results of the classification of thyroid nodules by the Inception-V3 model

	Precision	Recall	F1-Score	Support
Negative	0.99	0.99	0.99	500
Positive	0.99	0.99	0.99	500
Accuracy			0.99	1000
Macro avg	0.99	0.99	0.99	1000
Weighted avg	0.99	0.99	0.99	1000

Table 3 shows the test accuracy, precision, recall and f1-score of the Inception-Resnet-V2 model classification. The experimental results show the proposed fine-tuned Inception-Resnet-V2 model achieves excellent classification performance, attaining 97% classification accuracy, 97% precision and 97% sensitivity.

Table 3. Results of the classification of thyroid nodules by the Inception-Resnet-V2 model

	Precision	Recall	F1-Score	Support
Negative	0.95	1.00	0.97	500
Positive	1.00	0.95	0.97	500
Accuracy			0.97	1000
Macro avg	0.97	0.97	0.97	1000
Weighted avg	0.97	0.97	0.97	1000

These results clearly indicate that the Inception-V3 method is the one with the highest score, in terms of test accuracy, which reached 99%, which means that almost all nodules can be identifies correctly, in terms of AUC which reached

0.99, and in terms of sensitivity metric, with a value of 99%. From the curve in Fig. 2, it is clear that the area under the curve of the Inception-V3 method represents the largest area, which is the largest value of AUC. The values provided in Table 3 confirms the results of Fig. 2. The form of the curve, proves the best compromise between sensitivity and specificity.

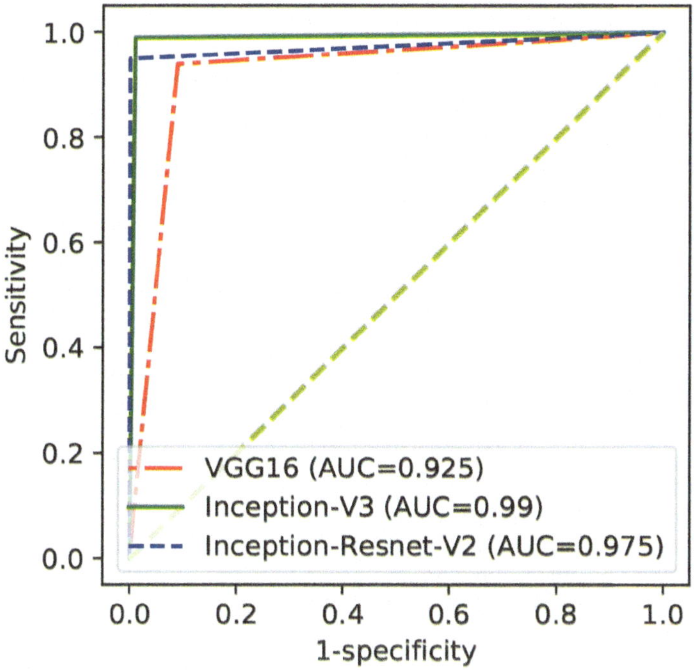

Fig. 2. ROC curves for the deep convolutional neural networks

5 Conclusions

In summary, we propose a thyroid image recognition method by fine-tuning ImageNet top-performing deep convolutional neural networks: VGG16, Inception and Inception-Resnet. Our work has shown that deep learning, especially fine-tuning an existing deep learning network, can improve the performance of thyroid nodule recognition as a computer-aided diagnosis system. The experimental results on 1000 ultrasound images showed an test accuracy of 93%, 97% and 99% for VGG16, Inception-Resnet and Inception, respectively.

In our future work, we aim to continue further tuning other DCNN in order to achieve better performance. Furthermore, we would like to enhance Thyroid ultrasound images by applying several image pre-processing techniques such as histogram equalization, contrast limited adaptive histogram equalization and Unsharp mask.

References

1. Welker, M.J., Orlov, D.: Thyroid nodules. Am. Fam. Phys. **67**, 559–566 (2003)
2. Hegedüs, L.: The thyroid nodule. N. Engl. J. Med. **351**(17), 1764–1771 (2004)
3. Sollini, M., Cozzi, L., Chiti, A., Kirienko, M.: Texture analysis and machine learning to characterize suspected thyroid nodules and differentiated thyroid cancer: where do we stand? Eur. J. Radiol. **99**, 1–8 (2018)
4. Li, H., Weng, J., Shi, Y., et al.: An improved deep learning approach for detection of thyroid papillary cancer in ultrasound images. Sci. Rep. **8**(1), 1–12 (2018)
5. Ma, J., Wu, F., Jiang, T., Zhu, J., Kong, D.: Cascade convolutional neural network for automatic detection of thyroid nodules in ultrasound images. Med. Phys. **44**, 1678–1691 (2017)
6. Chen, D., Niu, J., Pan, Q., Li, Y., Wang, M.: A deep-learning based ultrasound text classifier for predicting benign and malignant thyroid nodules. In: 2017 International Conference on Green Informatics, pp. 199-204 (2017)
7. Liu, T., Xie, S., Yu, J., Niu, L., Sun, W.: Classification of thyroid nodules in ultrasound images using deep model based transfer learning and hybrid features. In: 2017 IEEE International Conference on Acoustics, Speech and Signal Processing (2017)
8. Chi, J., Walia, E., Babyn, P., Wang, J., Groot, G., Eramian, M.: Thyroid nodule classification in ultrasound images by fine-tuning deep convolutional neural network. J. Digit. Imaging **30**(4), 477–486 (2017)
9. Kumar, A., Kim, J., Lyndon, D., Fulham, M., Feng, D.: An ensemble of fine-tuned convolutional neural networks for medical image classification. IEEE J. Biomed. Health Inform. **21**(1), 31–40 (2017)
10. Moussa, O., Khachnaoui, H., Guetari, R., Khlifa, N.: Thyroid nodules classification and diagnosis in ultrasound images using fine-tuning deep convolutional neural network. Int. J. Imaging Syst. Technol. **30**(1), 185–195 (2019)
11. Simonyan, K., Zisserman, A.: Very deep convolutional networks for large-scale image recognition. International Conference on Learning Representations (2015)
12. Szegedy, C., Liu, W., Jia, Y., Sermanet, P., Reed, S., Anguelov, D., Erhan, D., Rabinovich, A.: Going deeper with convolutions. In: 2015 IEEE Conference on Computer Vision and Pattern Recognition (CVPR) (2015)
13. He, K., Zhang, X., Ren, S., Sun, J.: Deep residual learning for image recognition. In: 2016 IEEE Conference on Computer Vision and Pattern Recognition (2016)
14. Kamble, R.M., Chan, G.C.Y., Perdomo, O., Kokare, M., Gonzalez, F.A., Muller, H., Meriaudeau, F.: Automated diabetic macular edema (DME) analysis using fine tuning with inception-resnet-v2 on OCT images. In: 2018 IEEE-EMBS Conference on Biomedical Engineering and Sciences (IECBES) (2018)
15. Pedraza, L., Vargas, C., Narváez, F., Durán, O., Muñoz, E., Romero, E.: An open access thyroid ultrasound image database. In: 10th International Symposium on Medical Information Processing and Analysis (2015)

Roadside Parking Spaces Image Classification Using Deep Learning

Khalfi Ali[1(✉)] and Guerroumi Mohamed[2]

[1] Department of Mathematics and Computer Science, University Djilali Bounaama, Khemis Miliana, Algeria
khalfiali@gmail.com

[2] Department of Electronics and Computer Science, University of Science and Technology Houari Boumediene, Bab Ezzouar, Algeria
guerroumi@gmail.com

Abstract. The shortage and imbalance of parking spaces have become serious problems in recent years on the roadside. Drivers may park incorrectly, which may take more parking space and disturb other drivers. To mitigate these problems, real-time parking monitoring with a detection system, is very indispensable. In this paper, we propose a classification system of images associated with the parking spaces for vehicles on a roadside based on the convolutional neuron networks (CNN). The image is supposed to be captured by a camera or other visual devices installed on the roadside and sent to a central computer for processing. The treatment is followed by a classification based on DL (deep learning) to determine the occupation of the parking spaces. The proposed classifier model uses fewer calculation parameters that involved a faster model by applying the principle of dilation. The proposed model has been evaluated and compared with the previous models using PKlot dataset, a set of well-known images publicly available. The evaluation results show the reliability of our proposal comparing with the previous works.

Keywords: Classification of images · Convolutional Neuron Network (CNN) · Parking lot · Pklot

1 Introduction

Recently monitoring a parking lot (also called smart parking lot or smart parking) attracted a lot of attention from scientists and community and industry [1]. In particular, the parking of roadside vehicles can be unbalanced because drivers choose illegal areas when parking. Intelligent roadside parking allows drivers to find vacant parking spaces near their destinations quickly and easily. This provides several benefits, such as reducing CO2 emissions from cars, reducing traffic congestion and finally, less stress and happier citizens. In many parking lots, floor sensors are used to determine the status of different parking spaces. This requires installation and maintenance of the sensors in each parking space, which could be expensive, especially in car parks with a high number of places

M. R. Senouci et al. (Eds.): CSA 2020, LNNS 199, pp. 323–333, 2021.
https://doi.org/10.1007/978-3-030-69418-0_29

available. To avoid these problems, researchers propose several solutions for detecting the occupancy of vehicle parking lot [2–5] using video cameras. However, despite these efforts, the detection of vacant parking spaces using only visual information remains an open problem. Most of these approaches are based on specialized visual techniques adapted to the specific scenario and the lack of generalization when applied to different car parks. In this paper, we propose an efficient and robust system for the detection of parking lot occupancy on a roadside based on convolutional neural networks (CNN). The model shows satisfactory results with a minimal number of calculation parameters. This is due to the associated parameterization for the layers of the proposed model. A standard value of dilation is used for the first convolution layer, then it increases one step, from the first to the last layer. The reporting mechanism with a progressive rate was used to have a global perceptual view and progressive on the images to be classified on the one hand and on the other hand to speed up the learning process of the model. Particularly, the main contributions of this work are as follows.

- We propose a smart monitoring system for detecting parking lot occupancy based on CNN network. We use a dilation parameter associated with each convolution layer and we use an increasing filter number from layer to another.
- We validate the performance of our solution we used PKlot [5]. The classification phase requires a few computer resources, which makes it suitable for embedded environments such as Raspberry Pi.

The rest of the article is organized as follows. Section 2 is a state-of-the-art review of research work for the classification of images in the context of detecting the occupancy of a vehicle parking lot. Section 3 presents the proposed solution characteristics. Section 4 shows the experimentation result and discussion part. The paper ends with Sect. 5 which presents a conclusion and perspectives.

2 Related Work

Deep Learning (DL) [6] is a branch of artificial intelligence that aims to develop techniques that allow computers to learn complex tasks in order to predict results with a precision close to that of humans. DL and in particular CNNs have been widely used in several fields such as the classification of images, segmentation of images, compression of images [7, 8], detection of objects, speech recognition, natural language processing, etc. Many systems are implemented using a network of sensor nodes, typically buried under the asphalt so that each node detects a parking spot. Different types of sensors have already been used. In [9] and [10], the authors use light sensors and evaluate the performance of their system using remote-controlled toy cars. In [11], the authors utilize a wireless sensor network (WSN) composed of optical sensors, while in [12] a combination of magnetic and ultrasonic sensors were used. In [12] a smart parking monitoring system based on WSN is proposed and magnetic sensors have been used to monitor the occupancy of a vehicle parking lot in [13]. Also, a particular type of sensor was used for the detection of parking spaces for Berth vehicles in Carpack [14]. In [15], a comparison of the performance of different sensors for intelligent parking is presented. IParking [16] is real-time monitoring parking. The system uses an event recorder embedded in the car. The driver can send his search query to the calculation grid through the wifi. To detect the occupation of the car park along the road, the system uses Canny Edge Detection based image recognition of the OpenCV. Another detecting parking occupancy is proposed in [17]. This system uses a smart camera and DL. It proposes a CNN scheme that is executed locally and in real-time by a smart camera. In [18] a comparative study between CTA (compress-then-Analyze) and ATC (Analysis-Then-Compress) has been studied to show the difference between the local processing of the image (at the level of the intelligent camera) and remote processing (on the central computer), the classification is done by the central computer using the SVM (Support Vector Machine). The authors in [19] propose a solution to manage the parking spaces, which consists of a wireless camera distribution system. Also, a decentralized and efficient solution is proposed in [20] for visual parking lot occupancy detection based on a deep Convolutional Neural Network (CNN) specifically designed for smart cameras. The processing of the parking request was analyzed statistically by work based on the operation of a single camera to monitor the detection of the occupancy of a parking lot [21]. With the use of a surveillance camera, the authors [22] propose a system to determine the occupation of such a vehicle parking lot. The model proposed in this paper is based on the AlexNet architecture [23] while exploiting the characteristics of the CarNet model [24] to minimize the number of calculation parameters, by reducing the number of filters associated with each convolution layer. The acceleration of the learning time was taken into account in our study by the setting of the dilation value for each convolution layer.

3 Proposed Solution

The proposed solution consists of designing a real-time monitoring system for roadside parking using a smart camera and providing an overall view of the roadside parking spaces. This system has two main steps, the first step is performed by the smart camera. It consists of extracting the associated images for each parking space from the captured image (Sect. 3.1). The second step classifies the images to determine the state of the parking spaces (free or occupied). This step is the main goal of our work which will be detailed in Sect. 3.4. The result of the classification obtained will be sent to a server for later exploitation by the users.

3.1 Extracting Captured Images

The acquisition and pre-processing of the captured image by the camera (Fig. 1) is based on [18]. The road has C parking spaces distributed along the road. Each parking space is characterized by a center Xc, a rotation angle dc, a length hc and a width wc with $c = 1 \ldots C$.

We associate for each parking space a square mask of min size (hc, wc). The mask is projected on the entire image for extraction of all the images associated with the parking space.

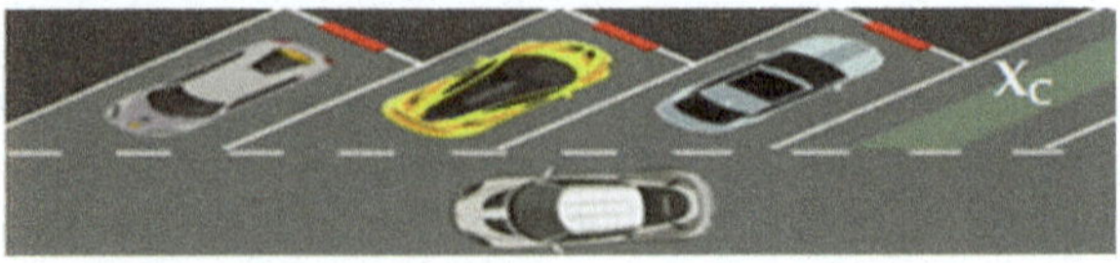

Fig. 1. A region centered by Xc for the extraction of the parking space.

3.2 Dilation Concept for Convolution

Dilated convolution is a means of exponentially increasing the receiving view (global view) by piling more and more dilated convolutional layers. The dilation rate k = 1 is a normal convolution and k = 2 means ignore one pixel per input and k = 3, ignore 2 pixels. The Fig. 2 [24] shows a dilated convolution with different parameters.

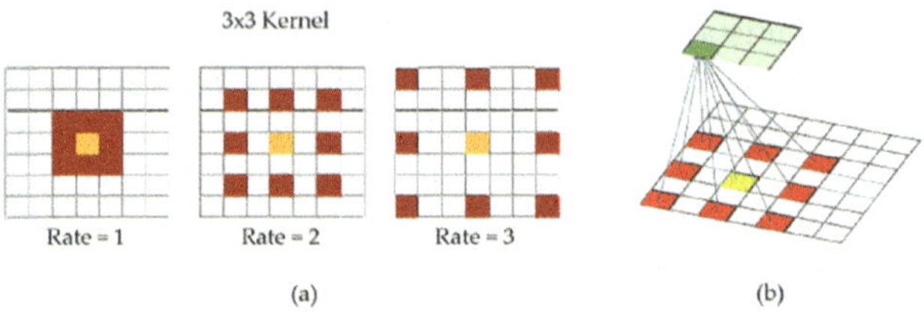

Fig. 2. (a) Examples of dilated convolutions. When the dilation rate is 1, it becomes a normal convolution. (b) Three-dimensional appearance of the dilated convolution (rate = 2).

A model using dilated convolution runs faster than a model using normal convolution. This is where the idea of using dilated convolutions comes from to reduce the learning time of the model. A good proposal of segmentation and action detection using the dilation parameter presented in [25]. The dilated convolution is also applied in areas other than vision. A good example is in [26] which presents a speech synthesis solution and in [27] the learning of text translation on time.

3.3 Structure of Our Model

The proposed model is based on AlexNet [23] where we remove the fourth convolutional layer. All convolutional layers conv1–3 with linear rectification (ReLU) followed by the overlapping max pooling. The number of conv1–3 filters is the number of neurons in fc4 and fc5 are drastically reduced to match the dimension of the problem. Our model contains the main points: (i) A dilation parameter is associated with each convolution layer with a rate that increases one step from the first layer to the third layer. (ii) The number of filters is fixed at 64 for the first layer, 128 for the second layer and finally 256 filters for the last layer. (iii) Three fully connected have been added. The details of the proposed architecture are reported in Table 1. The structure of our model consists of three convolutional layers with the same size 3 x 3 x 3 for the kernels. For each convolution layer, we apply dilation technique with a rate that varies gradually. We start with a standard dilation equal to 1 for the first layer, and then it increases by one step for the rest of the layers.

Table 1. Architecture of our model

Net	Conv1	Conv2	Conv3	Fc4	Fc5	Fc6
Our Model	64 x 54 x 32 Pool 2 x 2 ReLU Dilatation = 1	128 x 27 x 16 Pool 2 x 2 ReLU Dilatation = 2	256 x 13 x 8 Pool 2 x 2 ReLU Dilatation = 3	4096 ReLU	4096 ReLU	1 Segmoid

This is the first indicator that indicates the difference between the present work and [24]. Three fully connected layers have been added. The number of units for the two fully connected is 4096. For the last layer that represents the output of the model, it has two units, since our goal is a binary classification. The three convolutional layers were provided with a linear rectification (ReLU) with a maximum pooling process. The number of filters is reduced that indicates the second indicator for comparing with [24]. It is fixed at 64 for the first layer and it will be increased layer by layer. The number of filters for the second layer is 128 and finally 256 filters for the last layer. To avoid overfitting of the model, Dropout regularization was used for a fully connected layer and standard batch moralization has been used for each convolution layer. The proposed model accepts input images in 54×32 RGB format, like has been retained by the work [24]. To expand the size of our training dataset, we used the Keras deep learning library [27] that provides the ability to use data augmentation automatically when training a model. The data augmentation techniques can create variations of the images that can improve the ability of the fit model to generalize what they have learned to new images. The training of our model on more data results in more skillful models. Learning the model takes about 8 h for 75 iterations on a machine I3 (2.4 GHZ) of 4G RAM using the CPU mode. The batch size is 100 and the number of epochs is 50. A Stochastic Gradient Descent algorithm is used with learning rate = 0.0001, weight decay = 0.0004 and momentum = 0.99.

3.4 CNN Execution Time

A CNN model is designed of one or more convolutional layers whose execution time is based on the execution of each convolutional layer. The generation of feature maps is the result of the execution of a convolutional layer. For this purpose, our comparison method is based on the generation time of the feature map.

a. Let F the number of filter associate to a convolution layer.
b. Let FM the number of feature map witch equal to the number of filter F
c. Let Tg the generation time for each FM.

The Tgc represent the generation time apply to one of convolution layer, it is given by the Eq. (1):

$$\text{Tgc} = \sum\nolimits_{i=1}^{FM} Tg \tag{1}$$

with the Eq. (1), the time generation Tgm for a model (CNN) witch designed by N convolution layer is presented by the Eq. (2):

$$\text{Tgm} = \sum_{j=1}^{N} \sum_{i=1}^{FM} Tg \quad (2)$$

The evaluation of the Eq. (2) makes it possible to calculate the execution time of all the convolution operations of a CNN since it represents the heart of a CNN.

4 Experimental and Discussion

4.1 Dataset

To validate and test the performance of the proposed model, we use PKLot dataset [5]. PKLot dataset is an important resource for researchers and practitioners whose goal is to create parking space vacancy detection systems. This dataset contains images captured in sunny, rainy and cloudy weather. PKLot dataset contains 12,417 images of parking areas that can be segmented into 695,899 images of parking spaces. All images were captured on the parking lots of the Federal University of Parana (UFPR), and the Pontifical Catholic University of Parana (PUCPR), both located in Curitiba, Brazil. The images taken from the cameras were treated to generate parking space images. Each image is associated with a single parking space that can be occupied or free. The resulting images were organized into three subsets by UFPR04, UFPR05 and PUCPR. The subset UFPR04 is more complex than UFPR05 and PUCPR because it contains images with various obstacles and terrain patterns. It contains 3791 images captured in the aforementioned weather conditions. The three subsets, PUCPR, UFPR04 and UFPR05, were manually classified into cloud subfolders, sunny, rainy or cloudy depending on the observed weather conditions.

4.2 Experimental Results

In Fig. 3, we evaluate the accuracy and loss of our proposed model. These metrics are evaluated using the three PKLot subsets, PUCPR (Fig. 3 (a)), UFPR04 (Fig. 3 (b)) and UFPR05 (Fig. 3 (c)). We use a part of each subset for training and we utilize the rest part for validation. We divided each subset of the dataset into five parts and 80% is used for training while 20% for validation. In addition, shuffling was performed after each epoch. We observe that training and validation accuracy increases with the increasing of number of epoch. Epoch represents the number of part on which the data is divided. When data is too big, we divide it into smaller sizes and give it to our computer one by one and update the weights of the neural networks at the end of every step to fit it to the data given. We observe that the PUCPR subset gives highest performance (high accuracy and low loss). We conducted nine types of experiments. we used 50 training epochs in our experiments. The highest results were obtained for the PUCPR subset, which implies that the PUCPR set is the least difficult. It is clear that the subsets UFPR04 and UFPR05 are getting closer, which means that the two subsets are getting closer in terms of complexity. In the nine experiments shown in Table 2, our model presents some interesting results in terms

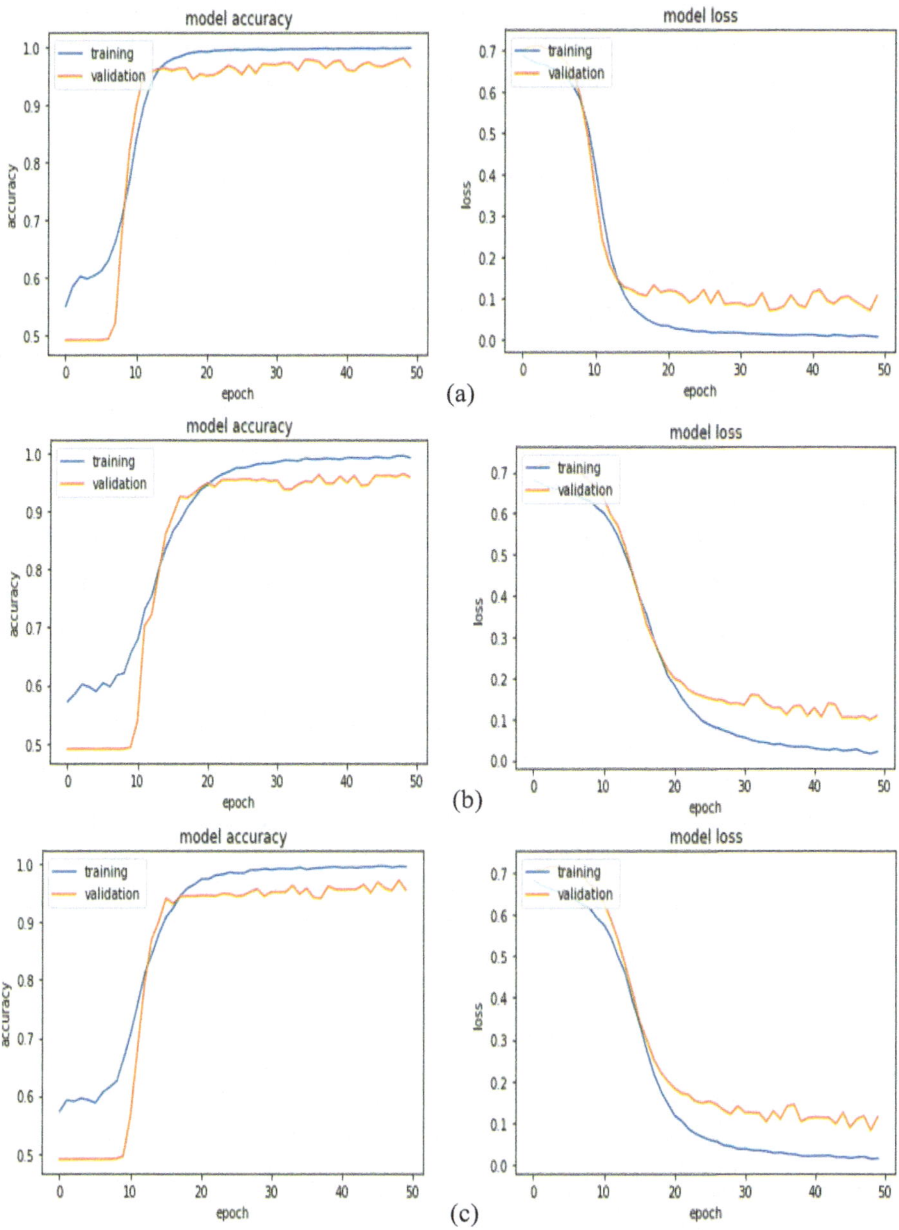

Fig.3. Different subsets, (a) PUCPR, (b) UFPR04, (c) UFPR05. Training and validation. x-direction: number of epochs (50), on the left y-direction: accuracy; on the right y-direction: loss

of accuracy, 98.94%, 98.27 and 92.89 for PUCPR, UFPR04 and UFPR05 respectively. In particular, with the dilation technique, our model is faster in time of learning than a model which does not use the dilation. Reducing the number of filter for the three convolution layers of our model lower than that of the CarNet model [24] produces a model with fewer parameters, which gives a very adaptable model to be embedded in a camera intelligent. Moreover, UFPR04 subset shows the lowest achievements.

Table 2. Accuracy values

Dataset	Training	Testing	Training accuracy (%)	Validation accuracy (%)	Accuracy %
PUCPR	PUCPR	PUCPR	99.91	98.51	98.94
	PUCPR	UFPR04	99.88	96.66	96.59
	PUCPR	UFPR05	99.81	99.96	98.99
UFPR04	UFPR04	UFPR05	99.34	99.85	98.27
	UFPR04	PUCPR	99.45	97.89	94.74
	UFPR04	UFPR04	99.23	98.09	94.92
UFPR05	UFPR05	UFPR05	99.00	98.00	92.89
	UFPR05	PUCPR	99.36	96.08	95.98
	UFPR05	UFPR04	99.36	96.85	96.92

This subset contains the most difficult images, which contain more obstacles, shadows, trees, etc. For more precision and comparison, Table 3 displays the score values of this experiment. We compared the results obtained by our model with the AlexNet [23] and CarNet model [24]. This comparison was made on the PKLot dataset. The accuracy and number of calculation parameters for each model are shown in Table 3.

Table 3. Comparison AlexNet, CarNet and our model.

Model name	Training sebset	Training accuracy (%)			Mean (%)	Number of parameters
		PUCPR	UFPR04	UFPR05		
AlexNet [23]	PUCPR	98.60	88.80	83.40	91.06	62.378.344
	UFPR04	89.50	98.20	87.60		
	UFPR05	88.20	87.30	98.00		
CarNet [24]	PUCPR	98.80	94.40	97.60	97.04	61.523.721
	UFPR04	98.30	95.60	97.60		
	UFPR05	98.40	95.20	97.50		
Our model	PUCPR	98.94	96.59	98.99	96.64	33.933.441
	UFPR04	94.75	98.27	94.92		
	UFPR05	95.98	96.92	92.89		

Comparing AlexNet and CarNet accuracy results with our model, we find that for some subsets, the CarNet model is more efficient for our model and conversely for others.

Moderately, the CarNet model is more efficient for our model. The miniaturization and embedding factor always requires optimized programs for good consumption of material resources. Our model clearly presents this need because it has fewer computational parameters than AlexNet and CarNet. Our model uses a progressive dilation value greater than that of CarNet. This makes our model faster in learning time. The number of parameters for our model clearly indicates the smaller value in comparing with AlexNet and CarNet, this implies our model is executing faster than Alexnet and CarNet.

4.3 Evaluation a CNN Execution Time

To evaluate the execution time of our model, we use the Eq. (2) to elaborate the comparison Table 4.

Table 4. Comparison of the time generation AlexNet [23], CarNet [24] and our model

Model name	Order of convolution layer	FM-Number of feature map (unite)	Time generation for convolution layer (Tg)	Tgm- Time generation to the model (Tg)
AlexNet [23]	1	96	96	1276
	2	256	256	
	3	384	384	
	4	384	384	
	5	256	256	
CarNet [24]	1	96	96	672
	2	192	192	
	3	384	384	
Our model	1	64	64	448
	2	128	128	
	3	256	256	

Comparing AlexNet, CarNet time generation with our model, we find clearly that needed time execution for executing the convolution layers for our model is lower than the execution time of the other model. Witch shows the effectiveness of our model in term of execution time.

5 Conclusion

In this paper, we have proposed a robust CNN-architecture that can detect vehicle parking spaces on a roadside. The model was parameterized by a dilated convolution layer with a gradual increase of one step from the first convolution layer to the third layer. The reduction of the number of filters, allowed us to produce an architecture with fewer training parameters which makes it possible to present a more optimized model comparing to other existing works in the literature and consequently to adapt it to be executed faster on a smart camera such as Raspberry PI. Using our model in the context of a smart city helps drivers to find a parking place on a roadside that resulting in less traffic and less CO2 emissions. From perspective, we plan to test our model on another dataset with other configuration parameters and real-world experience to detect roadside parking spaces and efficiently test the model. An installation of our model is highly appreciated for a smart city structure to fulfill the expected aspects of the smart city.

Acknowledgment. This research is part of project entitled "Green and Smart Communication in Smart Cities". The project is registered in the PRFU projects (Research Formation -University Projects) and approved by the Algerian Ministry of teaching and Scientific Research with the reference: C00L07UN160420180020.

References

1. Sanchez, L., Munoz, L., Galacche, J.A., et al.: SmartSantander : IoT experimentation over a smart city testbed. Comput. Netw. **61**, 217–238 (2014). 2. N. DAN. Parking management system and method, US Patent App. 10/066,215, January 32. 2002
2. WU, Q., Huang, C., Wang, S.-Y., et al.: Robust parking space detection considering inter-space correlation. In : 2007 IEEE International Conference on Multimedia and Expo, pp. 659–662. IEEE (2007)
3. del Postigo, C.G., Torres, J., Menéndez, J.M., et al.: Vacant parking area estimation through background subtraction and transience map analysis. IET Intell. Transp. Syst. **9**(9), 835–841 (2015)
4. de Almeida, P.R., Oliveira, L., Luiz, S., Britto, J.R., Alceu, S., et al.: PKLot–A robust dataset for parking lot classification. Expert Syst. Appl. **42**(11), 4937–4949 (2015)
5. Bengio, Y., et al.: Learning deep architectures for AI. Found. Trends® Mach. Learn. **2**(1), 127 (2009)
6. Giuseppe, V., Andrei, P., Hulusic, V., et al.: Quality Assessment of Deep-Learning-Based Image Compression (2018)
7. Liu, Z., Liu, T., Wen, W., et al.: DeepN-JPEG: a deep neural network favorable JPEG-based image compression framework. In: Proceedings of the 55th Annual Design Automation Conference, p. 18. ACM (2018)
8. Tang, V.W.S., Zheng, Y., Cao, J., et al.: An intelligent car park management system based on wireless sensor networks. In: 2006 First International Symposium on Pervasive Computing and Applications, pp. 65–70. IEEE(2006)
9. Srikanth, S.V., Pramod, P.J., Dileep, K.P., et al.: Design and implementation of a prototype smart parking (spark) system using wireless sensor networks. In: 2009 International Conference on Advanced Information Networking and Applications Workshops, pp. 401–406. IEEE (2009)
10. Chinrungrueng, J., Sunantachaikul, U., Triamlumlerd, S., et al.: Smart parking: an application of optical wireless sensor network. In: 2007 International Symposium on Applications and the Internet Workshops, pp. 66–66. IEEE(2007)
11. Lee, S., Yoon, D., Ghosh, A., et al.: Intelligent parking lot application using wireless sensor networks. In: CTS, pp. 48–57 (2008)
12. Wolff, J., Heuer, T., Gao, H., et al.: Parking monitor system based on magnetic field senso. In: 2006 IEEE Intelligent Transportation Systems Conference, pp. 1275–1279. IEEE (2006)
13. Chunhe, Y., Jilin, L., et al.: A type of sensor to detect occupancy of vehicle berth in carpark. In: Proceedings 7th International Conference on Signal Processing, 2004. Proceedings. ICSP 2004, pp. 2708–2711. IEEE (2004)

14. Kumar, R., Chilamkurti, N.K., Soh, B.: A comparative study of different sensors for smart car park management. In: The 2007 International Conference on Intelligent Pervasive Computing (IPC 2007), pp. 499–502. IEEE (2007)
15. Yang, C.-F., Ju, Y.-H., Hsieh, C.-Y., et al.: iParking–a real-time parking space monitoring and guiding system. Veh. Commun. **9**, 301–305 (2017)
16. Amato, G., Carrara, F., Falchi, F., et al.: Car parking occupancy detection using smart camera networks and deep learning. In: 2016 IEEE Symposium on Computers and Communication (ISCC), pp. 1212–1217. IEEE (2016)
17. Baroffio, L., Bondi, L., Cesana, M., et al.: A visual sensor network for parking lot occupancy detection in smart cities. In: 2015 IEEE 2nd World Forum on Internet of Things (WF-IoT), pp. 745–750. IEEE (2015)
18. Vítek, S., Melničuk, P.: A distributed wireless camera system for the management of parking spaces. Sensors **18**(1), 69 (2018)
19. Amato, G., Carrara, F., Falchi, F., et al.: Deep learning for decentralized parking lot occupancy detection. Expert Syst. Appl. **72**, 327–334 (2017)
20. Jermsurawong, J., Ahsan, U., Haidar, A., et al.: One-day long statistical analysis of parking demand by using single-camera vacancy detection. J. Transp. Syst. Eng. Inf. Technol. **14**(2), 33–44 (2014)
21. Delibaltov, D., Wu, W., Loce, R.P., et al.: Parking lot occupancy determination from lamp-post camera images. In: 16th International IEEE Conference on Intelligent Transportation Systems (ITSC 2013), pp. 2387–2392. IEEE (2013)
22. Krizhevsky, A., Sutskever, I., Hinton, G.E.: ImageNet classification with deep convolutional neural networks. In: Advances in Neural Information Processing Systems, pp. 1097–1105 (2012)
23. Nurullayev, S., Lee, S.-W.: Generalized Parking Occupancy Analysis Based on Dilated Convolutional Neural Network. Sensors **19**(2), 277 (2019)
24. Lea, C., Flynn, M.D., Vidal, R., et al.: Temporal convolutional networks for action segmentation and detection. In: Proceedings of the IEEE Conference on Computer Vision and Pattern Recognition, pp. 156–165 (2017)
25. van den Oord, A., Dieleman, S., Zen, H., et al.: WaveNet: a generative model for raw audio. arXiv preprint arXiv:1609.03499 (2016)
26. Kalchbrenner, N., Espeholt, L., Simonyan, K., et al.: Neural machine translation in linear time. arXiv preprint arXiv:1610.10099 (2016)
27. https://keras.io/search.html?q=data+augmentation+. Accessed 20 Nov 2019

Fast Segmentation of Markov Random Fields Corrupted by Correlated Noise

Ahmed Habbouchi[1(✉)], Mohamed El Yazid Boudaren[1], Amar Aïssani[2], and Wojciech Pieczynski[3]

[1] Ecole Militaire Polytechnique, PO Box 17, Bordj El Bahri, 16111 Algiers, Algeria
ahabbouchi@gmail.com, boudaren@gmail.com

[2] Université des Sciences et de la Technologie Houari Boumediene, Algiers, Algeria
aaissani@usthb.dz

[3] Télécom SudParis, 9 rue Charles Fourier, 91011 Evry Cedex, France
wojciech.pieczynski@telecom-sudparis.eu

Abstract. Markov Random fields (MRF) represent a powerful mathematical model and they are used in several areas, but it is almost impossible to perform exact analytical calculations when using MRF and we must use approximations and iterative methods that are greedy in terms of time and computing resources. In the literature, proposed MRF methods for Bayesian and parameters estimation are complicated for implementation and represent many disadvantages in practice.

We propose in this work, and in order to remedy the problems mentioned above, a very simple MAP-MRF framework based mainly on local conditional probabilities, contrary to the existing solutions in literature where we rely on the energy function model.

Two powerful models based on the proposed framework are then presented. They will be compared with two recent works to show how they are more efficient with respect to classical models when it comes to the unsupervised segmentation of corrupted data with correlated noise.

Keywords: Markov Random Field (MRF, CRF) · Segmentation · Parameters estimation · Correlated noise · MPM · ICM · ICE

1 Introduction

Let $X = (X_1, X_2, \ldots, X_n)$ and $Y = (Y_1, Y_2, \ldots, Y_n)$ be two random variables, with $X_i \in \Omega = \{\omega_1, \ldots, \omega_K\}$ and $Y_i \in \mathbb{R}^d$. X is a Markov random field (MRF) if and only if:

$$\forall x \in \Omega^n,\ P(X = x) > 0 \tag{1a}$$

$$P(X_i = x_i / \{x_t\ /t \in \{1..n\} \wedge t \neq i\}) = P(X_i = x_i / \mathcal{N}(i)) \tag{1b}$$

$\mathcal{N}(i)$ in Eq. 1b represents the set of the neighbors of x_i. That means that each variable x_i depends only on its neighborhood and not on all the field X. For a given realization $Y = y$, the variable X/Y is a MRF if and only if:

$$\forall x \in \Omega^n,\ P(X = x/Y = y) > 0 \tag{2a}$$

M. R. Senouci et al. (Eds.): CSA 2020, LNNS 199, pp. 334–343, 2021.
https://doi.org/10.1007/978-3-030-69418-0_30

$$P(X_i = x_i/\{x_t\ /t \in \{1..n\} \wedge t \neq i\}, Y = y) = P(X_i = x_i/\mathcal{N}(i), Y = y) \quad \text{(2b)}$$

The three main problems treated with in the area of MRF's are:

1. Bayesian estimation: Where the probability distribution in Eq. 1b (resp. 2b) is known and we need to estimate $\hat{X}$ (resp. $\hat{X}/y$) that maximizes it.
2. Parameters estimation with complete data: Where we have a set of X realizations $X^{(1)}, X^{(2)}, \ldots, X^{(m)}$ (resp. X/Y realizations $(X^{(1)}, Y^{(1)})$, $(X^{(2)}, Y^{(2)}), \ldots, (X^{(m)}, Y^{(m)})$) and we need to estimate the probability distribution in Eq. 1b (resp. 2b).
3. Parameters estimation with incomplete data: Where we have a set of Y realizations $Y^{(1)}, Y^{(2)}, \ldots, Y^{(m)}$ and we need to estimate the probability distribution in Eq. 2b.

In literature, the three problems are treated basing on the MRF-Gibbs equivalence. So, MRF can be represented by a function called the energy and the tree problems are converted to optimization problem (minimization-maximization of the energy function). More details are found on next sections where we show difficulties and problems that can be encountered.

We introduce in this work a new representation of MRF's that relies on conditional probabilities to avoid problems due to energy function representation. Then, we retreat the tree problems using the proposed representation.

We propose two new MRF models, based on the proposed representation, to highlight the power of our framework and to treat the problem of unsupervised segmentation with correlated noise. Results are compared with two recent works to show the real importance of the new proposed models.

In the next section, reader can find basis of MRF and how the three problems are treated in literature, while in the third section the new MRF representation is treated. The fourth section is devoted to unsupervised segmentation with the new two proposed models. Finally, we end the paper with conclusions and perspectives.

2 Markov Random Field

Thanks to the equivalence between Markov field and Gibbs distribution we can calculate the probability of a Markov random field realization x:

$$P(X = x) = \frac{1}{Z_U} \exp\left(-U(x)\right) \quad (3)$$

with $Z_U = \sum_{x \in \Omega^n} \exp\left(-U(x)\right)$ is a normalizing constant called partition function. The energy function U depends on the neighborhood system $\mathcal{N}$ and it can be chosen according to needs. Usually we use the Auto-Models where:

$$U(x) = \sum_{s=1}^{n} V(x_s) + \sum_{s=1}^{n} \sum_{t \in \mathcal{N}(s)} W(x_s, x_t) \quad (4)$$

Recall that the evaluation of Z_U is intractable, and we can't evaluate in practice Eq. 3, but we are still able to calculate local conditional probabilities:

$$P(X_i = x_i/\mathcal{N}(i)) = \frac{\exp -(U(x_i, \mathcal{N}(i)))}{\sum_{v \in \Omega} \exp -(U(v, \mathcal{N}(i)))} \tag{5}$$

Thanks to the ability to evaluate Eq. 5 we can sample MRF, using iterative algorithms such as Gibbs sampler [7] and Metropolis [11].

Let X and Y be the two variables defined in the introduction. Supposing that X is a Markov random field, we get:

$$P(X = x/Y = y) = \frac{P(x) \times P(y/x)}{P(y)} = \frac{\exp(-(U(x) - log(P(y/x))))}{Z \times P(y)} \tag{6}$$

This result can justify the choice of, both, X and X/Y as Markov random fields:

$$P(X = x/Y = y) = \frac{1}{\psi} \exp(-(U(x) - log(P(Y = y/x)))) \tag{7}$$

The energy function of the MRF X/Y, also known as conditional random field (CRF), is $\Phi_{(U,y)}(x) = U(x) - log(P(Y = y/x))$.

2.1 Bayesian Estimation

It's interesting to establish the most probable realization $\hat{x} = (\hat{x}_1, \hat{x}_2, \ldots, \hat{x}_n)$ for a given markov random field (known energy function U). The two most used estimators are the maximum a posteriori estimator (MAP) where $\hat{x} = \underset{x}{argmax}(P(X = x))$ and marginals a posteriori mode estimator (MPM) where $\forall i \in \{1 \ldots n\}$, $\hat{x}_i = \underset{x_i}{argmax}(P(X_i = x_i/\mathcal{N}(i)))$. And as a local approximation of the estimator MAP, the Iterated Conditional Mode estimator (ICM) is proposed in [4]. Steps of ICM are illustrated in Algorithms 1.

Algorithm 1. Iterated Conditional Mode Estimator

```
procedure ICM( U, x̂)                    ▷ In: U. Out: x̂ = argmax_x(P(X = x))
    Initialize randomly x̂.
    for t = 1 to Max_Iteration do
        for i ∈ {1...n} do                ▷ Sequential or random scan of x
```

$$\hat{x}_i = \underset{x_i}{argmax}(P(X_i = x_i/\mathcal{N}(i))) \tag{8}$$

```
        end for
    end for
end procedure
```

2.2 Parameters Estimation with Complete Data

The second treated problem in random Marko field context, is how to estimate parameters (the energy function U) from a given realization x. The usual estimator is the maximum likelihood one:

$$\hat{U} = \underset{u}{argmax}(P(x)) = \underset{u}{argmax}(-U(x) - \log(Z_U)) \tag{9}$$

Resolving Eq. 9 is computationally intractable due to the famous Z_U. Several other estimators have been proposed in the literature.

Pseudo-Likelihood and Coding Method. Conditional local probabilities are used in the Pseudo-Likelihood (PL) estimator [3]:

$$\hat{U}_{PL} = \underset{u}{argmax}(\prod_{i=1}^{n} P(X_i = x_i/\mathcal{N}(i))) \tag{10}$$

The consistence of the pseudo-likelihood estimator, when n is important, have been proved [8]. Z_U doesn't figure in Eq. 10 and we are able to find, with difficulties, a solution by resolving a non-linear equations system.

The coding method [2] relies on the same previous principle, but by independent region. if the set $\{1 \dots n\}$ is subdivided on disjoint sets noted $S^{l \in \{1 \dots L\}}$ representing the independent x_i, then the likelihood of each sub-set is:

$$P^l = \prod_{i \in S^l} P(X_i = x_i/\mathcal{N}(i)) \tag{11}$$

And the coding method estimator of the energy function is:

$$\hat{U}_{Coding} = \underset{u}{argmax}(\frac{1}{L}\sum_{l=1}^{L} \underset{u}{argmax}(P^l)) \tag{12}$$

This method suffers from several problems well explained in [6].

Mean Field Approximation and Stochastic Gradient. The previous estimators avoid explicitly the partition function Z_U contrary to the Mean Field Approximation [12] and the Stochastic Gradient [9] estimators.

In the first one, we replace (approximate) the partition function by the mean of the field X (more details can be found in [10]). While we use, in the second estimator, the stochastic gradient method to resolve Eq. 9. This method treats only the case where the objective function accepts an unique solution, so the stochastic gradient method can get close to the optimum.

Least Squares Fit. The least squares fit is a statistical estimator, based on the relation between the conditional local probabilities and the energy function:

$$U(x_t = i, \mathcal{N}(x_t)) - U(x_t = j, \mathcal{N}(x_t)) = \log(\frac{P(X_t = j/\mathcal{N}(x_t))}{P(X_t = i/\mathcal{N}(x_t))}) \tag{13}$$

The left side terms in Eq. 13 are composed from U to be estimated, while the right part can be easily estimated from a realization x using the histogram method ($P(x_i, \mathcal{N}(i))$ is substituted by the apparition frequency of $(x_i, \mathcal{N}(i))$ in x). Then, we can develop from Eq. 13 a linear equations system, and we resolve it using least squares method to estimate U. [10] contains more details with advantages and disadvantages of this method.

2.3 Parameters Estimation with Incomplete Data

Let X/Y be a CRF defined by Eq. 7. We need to estimate $\Phi_{(U,y)}$ from a realization $Y = y$. Using methods cited in previous Sect. 2.2, we need realizations x and y to estimate $\Phi_{(U,y)}$, but we have only y. And to estimate x, we need $\Phi_{(U,y)}$. To break this vicious circle, we can use iterative labeling-estimation (ILE) algorithm [5,14]. We estimate initially x from y using another simple scheme like k-means, then we estimate $\Phi_{(U,y)}$ using y and estimated x. We, alternatively, estimate x from estimated $\Phi_{(U,y)}$ and $\Phi_{(U,y)}$ from x, getting better estimation each iteration. ILE steps are illustrated in Algorithm 2.

Algorithm 2. Iterative Labeling-Estimation (ILE)

procedure ILE($y, \hat{\Phi}_y$) ▷ In: y. Out: $\hat{\Phi}_y$
  Initialize $\hat{x}^{(0)}$ using y ▷ ex: k-means
  for $t = 1$ **to** $Max\_Iteration$ **do**
    Estimate $p^{(t)} = P(y/\hat{x}^{(t-1)})$ ▷ ex: normal distribution
    Estimate $u^{(t)} = U$ from $\hat{x}^{(t-1)}$ ▷ see section 2.2
    $\Phi_y^{(t)}(x) = u^{(t)}(x) - log(p^{(t)}(Y = y/x))$
    Estimate $x^{(t)}$ from $\Phi_y^{(t)}$ ▷ MAP or MPM, see section 2.1
  end for
  $\hat{\Phi}_y = \Phi_y^{(t)}$
end procedure

3 Energy-Model Free MRF

The implementation of MRF frameworks is encircled by many problems, such as conception difficulties, existence of a lot of exceptions and high time execution. It is due, principally, to the evaluation and the estimation of the energy function that is characteristic of MRF. But we can notice that the three treated problems in MRF context (Bayesian estimation, parameters estimation with complete and incomplete data) can be hold using directly conditional local probabilities (Eq. 5) and avoiding the explicit energy function. Therefor, we propose a new representation of MRF, based on conditional local probabilities and called Energy-Model Free MRF (**EFMRF**).

3.1 Conditional Matrix

In our framework, MRF will be characterized by a matrix holding its conditional local probabilities. If X, as defined in the introduction, is a MRF, then with respect to Eq. 5 we have $K^{(1+|\mathcal{N}|)}$ conditional local probabilities:

$$M(x_i, \mathcal{N}(i)) = P(X_i = x_i / \mathcal{N}(i)) \tag{14}$$

Example 1. Consider 4-neighborhood $\mathcal{N}(i) = \{x_{up(i)}, x_{right(i)}, x_{down(i)}, x_{left(i)}\}$. The conditional matrix M has $1 + |\mathcal{N}| = 5$ dimensions, and its elements are $M(a, b, c, d, e) = P(x_i = a / \{x_{up(i)} = b, x_{right(i)} = c, x_{down(i)} = d, x_{left(i)} = e\})$.

We notice that the size of the conditional matrix M is exponential to the cardinal $|\mathcal{N}|$. But memory at the actual machines disposal do make the size of conditional matrix negligible, even with relatively big classes count (K).

Example 2. If $K = 10$, we use 4 neighborhood system and a real number is represented by 8 bytes, then we can save the conditional matrix within 800 kb.

This representation considers only discrete MRF. But we can represent CRF described by the probability 7 if we consider the noise distribution $P(Y/X)$ as product $P(Y/X) = \prod_{i=1}^{n} f(Y/\{x_i, \mathcal{N}(i)\})$. We can characterize X and X/Y by the same conditional matrix:

$$P(X_i = x_i / \ \mathcal{N}(i), Y = y) = \frac{M(x_i, \mathcal{N}(i)) \times f(Y/\{x_i, \mathcal{N}(i)\})}{\sum\limits_{v \in \Omega} [M(v, \mathcal{N}(i)) \times f(Y/\{v, \mathcal{N}(i)\})]} \tag{15}$$

3.2 Bayesian Estimation

Our proposed EFMRF framework represents a real advantage in front of bayesian estimation problem (see Sect. 2.1), especially in term of execution time. In the bayesian estimation algorithms, we evaluate in intensive way the conditional local probabilities (line 5 of Algorithm 1). These evaluations consume a considerable time in classical MRF frameworks. In EFMRF, they are substituted with direct memory access to their values, already stocked in the conditional matrix (Eq. 14). In the case of CRF, they are replaced by Eq. 15, but still easily and more rapidly evaluated thanks to conditional matrix.

3.3 Parameters Estimation with Complete Data

The main problem in implementing methods of MRF parameters estimation with complete data (see Sect. 2.2) is the need to use heuristic (convergence is not guaranteed) and highly time consuming methods, such as non-linear systems resolvers and stochastic gradient.

In our framework, the MRF parameters are the conditional matrix. Their estimation from a realization x can be done simply with histogram method:

$$\hat{M}(x_i, \mathcal{N}(i)) = \begin{cases} \frac{Frq_x(\{x_i, \mathcal{N}(i)\})}{Frq_x(\mathcal{N}(i))} & \text{if } Frq_x(\mathcal{N}(i)) \neq 0 \\ \frac{1}{K} & \text{otherwise} \end{cases} \tag{16}$$

With $Frq_x(B)$ is apparition frequency of the set B in the realization x.

3.4 Parameters Estimation with Incomplete Data

The same used paradigm in classical MRF framework for parameters estimation with incomplete data is reused in our energy-free model framework (see Sect. 2.3). The alternative ILE (Algorithm 2) used in our framework is called Iterative Labeling-Conditional Matrix Estimation (ILCME) and is listed in Algorithm 3.

Algorithm 3. Iterative Labeling-Conditional Matrix Estimation (ILCME)

```
procedure ILCME( y, (M̂, p̂_y))                          ▷ In: y. Out: (M̂, p̂_y)
    Initialize x̂^(0) using y                              ▷ ex: k-means
    for t = 1 to Max_Iteration do
        Estimate p^(t) = P(y/x̂^(t−1))                     ▷ ex: normal distribution
        Estimate M^(t) = M from x̂^(t−1)                   ▷ see section 3.3
        Estimate x^(t) from M^(t) and p^(t)               ▷ MAP or MPM, see 3.2
    end for
    (M̂, p̂_y) = (M^(t), p^(t))
end procedure
```

4 Image Unsupervised Segmentation

4.1 Independent and Pseudo-correlated Noise EFMRF

We propose two models based on the energy-model free MRF framework. The first one is simple and proposed, principally, to prove the efficiency of EFMRF comparing to classical MRF frameworks, in terms of execution time and of segmentation error rate. It's called Independent Noise EFMRF (EFMRF-IN) and it's defined by the energy function denoted in Eq. 17.

$$P(x/y) = \frac{1}{\psi_i} \exp(-(\sum_{i=1}^{n}(U(x_i, \mathcal{N}(i)) - log(\frac{1}{\delta_{x_i}\sqrt{2\pi}} \exp\frac{-(y_i - \mu_{x_i})^2}{2\delta_{x_i}})) \quad (17)$$

In this model, the noise distribution ($P(Y/X)$) is simplified by considering the Y_i variables independent conditionally to X, respecting gaussian distribution ($P(Y/X) = \prod_{i=1}^{n} P(Y_i/X_i)$ and $P(Y_i/X_i) \sim \mathcal{N}(\mu(X_i), \sigma(X_i))$).

The EFMRF-IN model is represented by the conditional matrix M of X, by K means and by K variances. We can easily evaluate the Eq. 15 by substituting f with Eq. 18.

$$f(Y/\{x_i, \mathcal{N}(i)\}) = \frac{1}{\delta_{x_i}\sqrt{2\pi}} \exp\frac{-(y_i - \mu_{x_i})^2}{2\delta_{x_i}} \quad (18)$$

The second model is called Pseudo-Correlated Noise EFMRF (EFMRF-CN). The noise in this model is supposed approximately correlated, where each bloc $\{Y_i, \mathcal{N}_y(i)\}$ is considered as independent from other blocs conditionally to X.

It depends only on X_i ($P(Y/X) = \prod_{i=1}^{n} P(\{Y_i, \mathcal{N}_y(i)\}/X_i)$) and it's multivariate gaussian distribution ($P(\{Y_i, \mathcal{N}_y(i)\}/X_i) \sim \mathcal{N}(\mu(X_i), \Sigma(X_i))$). The energy function of EFMRF-CN model is represented by Eq. 19.

$$P(x/y) = \frac{1}{\psi_m} \exp(-(\sum_{i=1}^{n}(U(x_i, \mathcal{N}(i)) - log(\frac{1}{\sqrt{|\Sigma_{x_i}|(2\pi)^{(1+|\mathcal{N}|)}}} \exp(-\frac{1}{2}((y_i, \mathcal{N}_y(i)) - \mu_{x_i})^T \Sigma_{x_i}^{-1}((y_i, \mathcal{N}_y(i)) - \mu_{x_i}))))) \quad (19)$$

The EFMRF-CN model is also represented by the conditional matrix M of X, by K means and by K variances. And we can easily evaluate the Eq. 15 by substituting f with Eq. 20.

$$f(Y/\{x_i, \mathcal{N}(i)\}) = \frac{1}{\sqrt{|\Sigma_{x_i}|(2\pi)^{(1+|\mathcal{N}|)}}} \exp(-\frac{1}{2}((y_i, \mathcal{N}_y(i)) - \mu_{x_i})^T \Sigma_{x_i}^{-1}((y_i, \mathcal{N}_y(i)) - \mu_{x_i})) \quad (20)$$

4.2 Experimentations

As a proof of concept, we will reproduce a 2D image segmentation example treated in a recent work [1]. The authors compare their models with a recent method called SVA-B [13] and we will compare our results with their best model called CCNGN. In this example, two synthetic images are noised with independent gaussian and correlated noises, as illustrated in Fig. 1.

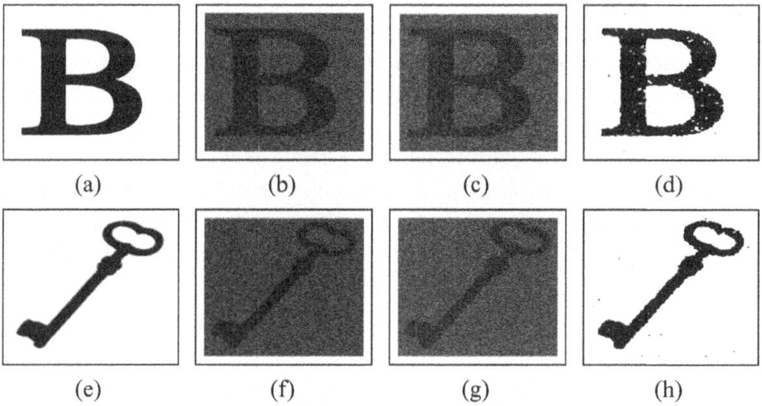

Fig. 1. Synthetic Images Unsupervised Restoration. (a)(e): Original class-images; (b)(f) Images corrupted by independent noise; (c)(g) Images corrupted by correlated noise; (d)(h) Restored images from (b)(g) using $EFMRF - CN$.

We apply the ILCME algorithm to segment the four corrupted images, using the both models EFMRF-IN and EFMRF-CN. Results are detailed in Table 1.

Table 1. Segmentation accuracy and Kappa.

Image	Model			
	SVA-B	CCNGN	EFMRF-IN	EFMRF-CN
Overall accuracy(%)				
B-IN	94.92	95.13	96.19	**96.32**
B-CN	89.36	92.88	93.03	**94.82**
Key-IN	97.67	98.01	98.45	**98.72**
Key-CN	90.27	97.17	97.80	**97.95**
Kappa				
B-IN	0.887	0.886	0.911	**0.913**
B-CN	0.775	0.846	0.840	**0.879**
Key-IN	0.904	0.909	0.932	**0.944**
Key-CN	0.674	0.870	0.906	**0.909**

The results show the performances of our models EFMRF-IN and EFMRF-CN in terms of segmentation error rate and Kappa measure. Our simple model EFMRF-IN gives better results comparing to the others both methods thanks to the estimation accuracy and rapidity (more iterations) of our algorithm ILCME. The EFMRF-CN model, and because it handles the correlated noise, performs the best results. As another advantage of our framework, we can mention the time consumption where the segmentation using our algorithm can be performed in really negligible time.

The use of conditional matrix instead of the energy function makes a huge difference between execution times and performances of our framework and classical MRF frameworks, because the conditional matrix estimation process (line 5 in Algorithm 3) is: (*i*) very fast comparing to the energy function estimation methods, allowing us to go for more iterations. (*ii*) more accurate estimation then results of heuristics methods used for the energy function estimation. At each iteration of Algorithm 3, comparing to classical MRF frameworks, we have more accurate estimation of parameters and we can perform more iterations in negligible time. This justifies the performance of our proposed EFMRF representation.

5 Conclusions

MRF methods based on energy function suffer from several problems and they are complicated to be implemented in practice. We propose in this work a new MRF's representation, based on local conditional probabilities, that allows more accurate parameters estimation and fast segmentation. A set of appropriate algorithms is presented to highlight the power of this new MRF representation.

We propose two new MRF models called EFMRF-IN and EFMRF-CN, based on the new proposed presentation, and we illustrate their performances compared with two recent works when confronted to unsupervised segmentation of data corrupted with correlated noise.

More complicated MRF models can be proposed using our MRF's representation. We will, soon, present a triplet MRF model that can handle the non-stationarity of data corrupted with correlated and non-gaussian noise.

References

1. An, L., Li, M., Boudaren, M.E.Y., Pieczynski, W.: Unsupervised segmentation of hidden Markov fields corrupted by correlated non-gaussian noise. Int. J. Approximate Reasoning **102**, 41–59 (2018). https://doi.org/10.1016/j.ijar.2018.08.001
2. Besag, J.: Spatial interaction and the statistical analysis of lattice systems. J. Roy. Stat. Soc. Ser. B (Methodological) **36**(2), 192–225 (1974). https://doi.org/10.1111/j.2517-6161.1974.tb00999.x
3. Besag, J.: Efficiency of pseudolikelihood estimation for simple Gaussian fields. Biometrika **64**(3), 616–618 (1977). https://doi.org/10.1093/biomet/64.3.616
4. Besag, J.: On the statistical analysis of dirty pictures. J. Roy. Stat. Soc. Ser. B (Methodological) **48**(3), 259–279 (1986). https://doi.org/10.1111/j.2517-6161.1986.tb01412.x
5. Bouman, C.A., Shapiro, M.: A multiscale random field model for Bayesian image segmentation. IEEE Trans. Image Process. **3**(2), 162–177 (1994). https://doi.org/10.1109/83.277898
6. Derin, H., Elliott, H.: Modeling and segmentation of noisy and textured images using gibbs random fields. IEEE Trans. Pattern Anal. Mach. Intell. **PAMI–9**(1), 39–55 (1987). https://doi.org/10.1109/TPAMI.1987.4767871
7. Geman, S., Geman, D.: Stochastic relaxation, Gibbs distributions, and the Bayesian restoration of images. IEEE Trans. Pattern Anal. Mach. Intell. **PAMI–6**(6), 721–741 (1984). https://doi.org/10.1109/TPAMI.1984.4767596
8. Geman, S., Graffigne, C.: Markov random field image models and their applications to computer vision. In: Gleason, A.M. (ed.) ICM 1986, Proceedings of the International Congress of Mathematicians, vol. 1, pp. 1496–1517. American Mathematical Society, Berkeley (1986)
9. Habbouchi, A., Boudaren, M.E.Y., Aissani, A., Pieczynski, W.: Unsupervised segmentation of Markov random fields corrupted by nonstationary noise. IEEE Signal Process. Lett. **23**(11), 1607–1611 (2016). https://doi.org/10.1109/lsp.2016.2609887
10. Li, S.Z.: Markov Random Field Modeling in Image Analysis. Springer, Tokyo (2001). https://doi.org/10.1007/978-4-431-67044-5
11. Metropolis, N., Rosenbluth, A.W., Rosenbluth, M.N., Teller, A.H., Teller, E.: Equation of state calculations by fast computing machines. Tech. rep. (1953). https://doi.org/10.2172/4390578
12. Parisi, G., Machta, J.: Statistical field theory. Am. J. Phys. **57**(3), 286–287 (1989). https://doi.org/10.1119/1.16061
13. Pereyra, M., McLaughlin, S.: Fast unsupervised Bayesian image segmentation with adaptive spatial regularisation. IEEE Trans. Image Process. **26**(6), 2577–2587 (2017). https://doi.org/10.1109/tip.2017.2675165
14. Qian, W., Titterington, D.: Stochastic relaxations and EM algorithms for Markov random fields. J. Stat. Comput. Simul. **40**(1–2), 55–69 (1992). https://doi.org/10.1080/00949659208811365

Analysis and Comparison of Audiovisual Quality Assessment Datasets

Fatima Boudjerida[1(✉)], Atidel Lahoulou[1], and Zahid Akhtar[2]

[1] LAOTI laboratory, University Mohammed Seddik Ben Yahia, Jijel, Algeria
fatimaboudjrida@gmail.com

[2] State University of New York Polytechnic Institute, Utica, USA

Abstract. Estimating perceived audiovisual quality by user is a vital component in many multimedia networks and applications. The need of ground truth data for quality assessment leads to elaboration of subjective quality assessment databases. Over the years, several datasets dedicated to audiovisual quality have been released in the public domain that are built upon extensive psychophysical experiences. In this paper, we study four audiovisual quality datasets: PLYM, TUM1080p50, VQEG MM2, and INRS. We also present detailed description of each dataset with the intention to facilitate their use for the researchers' community. Moreover, different summarization tables have been reported in order to accomplish clear and fair comparison between the audiovisual quality databases. It is hoped that the present work will help researchers in selecting the appropriate databases for their experimentations on objective audiovisual quality metrics.

Keywords: Audiovisual dataset · Subjective quality assessment · PLYM · TUM 1080p50 · VQEG MM2 · INRS

1 Introduction

To resolve the problem of multimedia quality assessment, we need to tackle it from two sides; we need to understand both the objective and the subjective models. On the one hand, the subjective quality evaluation approach is the obvious and the ultimate gauge of multimedia quality since it allows to reliably determining the perceptual quality of the visual content. In fact, it is the best representation of the perceived quality feedback of the human visual/hearing systems. Despite the predominant importance of the subjective quality evaluation techniques, they are time consuming and consequently cannot be incorporated in real-time application systems.

On the other hand, the objective approach of quality measurement is developed to design computational models that can automatically estimate the perceptual quality of multimedia content for real-time applications. There exists a limited array of audiovisual quality metrics in the literature such as in [3,12]. Nevertheless, these models should be consistent with the Human Visual/hearing

M. R. Senouci et al. (Eds.): CSA 2020, LNNS 199, pp. 344–353, 2021.
https://doi.org/10.1007/978-3-030-69418-0_31

System in a manner that agrees with subjective ratings. In this regard, it is always necessary to supplement objective measurements with subjective rating tests. These latter allow the benchmarking of the predictive performances of the developed objective automatic multimedia quality models in the end-user applications.

In this paper, we are interested in the subjective assessment methodologies used to build the audiovisual quality datasets. Some of the databases presented in the present work have some limitations in terms of real-time communication use cases, mainly because they were generated with a different scope in mind. The PLYM and INRS datasets are of interest for real-time communications since they have explored coding configurations and network degradations often observed in contemporary videophone applications. However, in PLYM the video frame rates and the selected video resolution no longer reflect contemporary applications.

In [4], Qualinet white paper provides a list of publicly available quality audio visual databases. All of them have been built upon extensive psycho-visual experiences carried out within specific framework but under different conditions. Due to the lack of review papers that allow thought-process for deep understanding of the existing subjective approaches, our aim is to provide a comprehensive analysis of the audiovisual quality datasets methodologies as well as a comparison between them. Hence, we study four audiovisual available quality datasets, where:

- Audiovisual sequences were coded with AVC codec.
- Audiovisual stimuli cover a wide range of perceived quality (very low to very good).
- Different subjective rating methods have been applied namely ACR (Absolute Category Rating) and SSMM (Single Stimulus Multimedia).
- Mean opinion scores (MOS) are discrete.

The remainder of this paper is organized as follows: In Sect. 2, the experimental setup as recommended by the International Telecommunications Union is presented. Next, four public quality audiovisual datasets are presented, described and analyzed. In Sect. 4, we give a comparison between the studied datasets. Finally, in Sect. 5, the paper ends with our conclusions and remarks.

2 Experimental Setup

To reliably gauge the perceived quality of audiovisual content, psycho-visual experiments are conducted under standardized conditions. Volunteer observers provide their opinion scores of the quality of the visual content they see and/or hear. The visual dataset is generated by applying some common distortions to the reference data. This assessment framework is known as subjective quality assessment. It results in attributing, to each digital content, a value which represents the average perceived quality scores [1].

The recommendations commonly used for audiovisual quality testing are those of the International Telecommunication Union as ITU-T P.920 [8], ITU-T

P.1401 [9] and ITU-T P.913 [11]. These recommendations are in fact a guide to the test methods, the equipment to be used as well as the test environment. The subjective evaluation of the perceived quality of audiovisuals can be classified according to the evaluation method in two categories:

- **double stimulus methodology**, subject is presented with the source as reference and test samples to evaluate their qualities.
- **single-stimulus methodology**, the subject is presented with the test data only without the source as reference to evaluate quality.

The single-stimulus methodology is more useful in realistic test environments, since it takes much less time and does not require the availability of the pristine data. The scale for rating can be either numerical or categorical, and either continuous or discrete. The rating can be obtained after or during stimulus presentation to acquire overall quality or temporal quality variations, respectively [1].

The two most commonly used scales for Absolute Category Rating (ACR) are the 5-point and the 11-point scales. The 5-point category scale is most widely used in telecommunications. The labels "excellent", "good", "acceptable", "mediocre", and "bad" refer to values 5, 4, 3, 2 and 1 when calculating the MOS [10]. The final quality score is obtained by averaging the rating values registered by multiple subjects, which is referred to as mean opinion score (MOS).

The Single Stimulus MultiMedia (SSMM) method is a variation of the standard Single Stimulus Impairment Scale (SSIS). Instead of a degradation scale given by the SSIS test method, the SSMM one uses a discrete quality scale [6], a discrete quality scale is used. In order to avoid context effects, each coding sequence and condition was repeated once in a different context, i.e., a different predecessor sequence and a different coding condition.

3 Publicly Available Audiovisual Datasets

If we look at the existing databases for audiovisual quality assessment, we can conclude that there are different approaches to build audiovisual quality models [3]. We have categorized the models according to the type of data as follows:

- Audiovisual quality can be estimated directly via a function regardless of how the impairments individually affect the audio and video quality.
- The audio and intermediate video quality are individually predicted, then they are integrated into a global audiovisual quality as the Parametric quality models predict the impact of network encoding configurations and impairments on multimedia quality. These models typically use information extracted from packet headers and do not have access to the payload data of packets. They are well appropriate to cases where the payload data is encrypted [3].

The audiovisual datasets described in this paper attempt to analyze the impact of environmental factors (different display devices, wireless and mo bile environments) or contextual changes (network compression and distortion parameters) on audiovisual quality predictions. The following datasets are the publicly available audiovisual databases in the audiovisual quality assessment:

3.1 University of Plymouth Dataset (PLYM)

PLYM Audiovisual Dataset was published in 2010 by Goudarzi *et al.* [5] at the EU FP7 ADAMANTIUM project of the University of Plymouth, UK. PLYM was created to study audiovisual quality predictions for video calling via wireless applications, and the relationship between the audio and video modalities in wireless and mobile environments. It also allows to predict the effects of network/application parameters on the overall audiovisual quality in video calls. Subjective tests for 60 audio samples, 60 video and 60 audiovisual samples have been conducted. The videos were coded with the H.263 and G.711 μlaw codecs where each video had a 176×144 resolution and 7 to 14 s duration,

The distortions applied to the reference sequence: to generate the 60 test sequences, Goudarzi *et al.* [5] relied on Parametric quality models. They applied two types of attack to the six reference sequences for each subjective test session video, audio, and audiovisual.

- **parameter frame rate (FR)**: PLYM authors used two video frame rates (8 and 15 fps) to simulate the degradation of the resolution during the coding
- **network parameter packet error rate (PER)**: Packet Error Rate (PER) is used to test the performance of an access terminal's receiver [3]. Five values of packet loss rates (0,01, 0,05, 0,1, 0,15 and 0,20) have been used to simulate packet loss after transmitting multimedia content.

The subjective tests have been carried out to assess audio, video and audiovisual quality with the help of 16 observers in each part. It consisted of 3 sessions of approximately one hour. The test sessions were performed for 3 consecutive days. Samples were presented to subjects in groups of 15 and a 3 to 5 min break was observed after each group of samples. There was a total of 4 groups in each session. ACR with a discrete 9-level quality scale for low bit-rate evaluations was followed during the experiments. The dataset description is summarized in Table 1.

3.2 TUM 1080p50 Dataset (TUM)

The objective of creating this dataset was, on the one hand, to compare the visual quality perceived on different devices, from a reference display to a home theater setup, including audio. On the other hand, the objective was to get data for 1080p50 material. The tests were carried out in the video quality evaluation laboratory of the Data Processing Institute of the Technical University of Munich, Germany in a room compliant with Recommendation ITU-R BT.500

Table 1. Summary of PLYM database description.

Encoding scenarios				
Content of the reference scenes	Scene	Video	Audio	Duration
	Reading	Male speaker, head and shoulder, light background	Normal male voice	12,90 s
	Jobinterview	Female speaker, head and shoulder, light background	High-pitched female voice	7,50 s
	Cbsnews	News speaker, colored background	Normal female voice	12,50 s
	Newspart	TV presenter, hand movements, colored background	Fast speaking male voice	3.74 s
	Gold	Lecturer behind a desk, reading from a note	Normal male voice	13.32 s
	Conversation	A male and a female in front of camera, male speaking, hand movement	Normal male voice	13.84 s
Encoding		Audiovisual containers .avi	Codec Video H.263 AVC Codec audio G.711 u-law	
Distortions		Packet Error Rate (PER): 0.01, 0.05, 0.1, 0.15, 0.20	Frame Rate(FR): 8, 15	
Total number of test scenes			60 sequences	
Test methodology				
Subjects	48 (26 men and 22 women) Staff members and students from Plymouth University		16 for audio test only (1st day) 16 for video test only (2nd day) 16 for audiovisual test (3rd day)	
Method	ACR (Absolute Category Rating)			
Subjective evaluations	Quality scale	discrete at 9 levels: 1: bad 2:/3: poor 4:/5: fair 6:/7: good 8:/9: excellent		
	Final score	MOS		
	Scale of scores	1..9		

[2]. Keimel *et al.* [6] have created subjective quality datasets in 1080p50 formats. This data is intended for the evaluation of the audiovisual quality of high definition videos. The video sequences were encoded with the H.264/AVC video codec; they were selected in the SVT High Definition multi-format test set, as shown in Table 2.

The Distortions Applied to the Reference Sequence: To generate the 20 test clips, Keimel *et al.* [6] exploited parametric quality models. Then, each of these reference videos was encoded at the four selected bitrates (RP1, RP2, RP3, and RP4). The result is a quality range from "not acceptable" to "very good" (in the range of 2 Mbit/s to 40 Mbit/s). The artifacts introduced into the videos by this coding scheme include induce a panel of impairments including, blocking, blurring, ringing artifacts, flicker and similar effects on visible elements.

The subjective test was carried out a projector and a surround sound system following the Single Stimulus MultiMedia (SSMM) methodology. To enable test subjects to distinguish between relatively low quality differences, a discrete voting scale with eleven notes ranging from 0 to 10 was used. Before each test, a short training was offered to the subjects. It consisted of ten video sequences with quality levels and coding artifacts similar to the video sequences tested [6].

Twenty-one observers participated in the study and rated the perceived quality for Twenty different data points using a discrete 11-level quality scale from 0 to 10.

Table 2. Summary of TUM 1080p50 database description.

Encoding scenarios						
Content of the reference scenes	Scene	Difficulty of coding	Distortions Bit rate [Mbit / s]			
	FlagShoot	Easy (supposed)	2	3	6	10
	CrowdRun	Difficult	8	20	30	40
	TreeTilt	Way	2	3	6	10
	PrincessRun	Difficult	8	20	30	40
	DanceKiss	Easy	2	3	6	10
Encoding	Audiovisual containers .avi		codec Video H.264 AVC			
			Binary format YUV 4:2:0 8			
Total number of test scenes			20 sequences			
Test methodology						
Subjects	21 (19 men and 2 women) students aged 16 to 27					
Method	Single Stimulus MultiMedia (SSMM)					
Display devices	Reference screen (Sony BVM-L230, 23")					
	LCoS projector (JVC DLA-HD750, 2.8m)					
	Consumer LCD Display(Sony KDL-55X4500, 56")					
	LCoS projector and surround sound (JVC DLA-HD750, 2.8 m)					
Subjective evaluations	Quality scale	Discrete at 11 levels: 0,1,2: Bad / 3,4: Poor /5,6: Satisfactory / 7,8: Good / 9,10: Excellent				
	Final score	Mos				
	Scale of scores	0..10				
Objective evaluations	PSNR					

3.3 VQEG MM2 Dataset (VQEG)

The dataset VQEG MM2 [7] contains 10 audiovisual subsets produced by six different international laboratories in 2013 as part of a study on the impact of environmental or contextual factors on Subjective Scores. As detailed in Table 3, the attacks applied to the reference sequences (Stimuli and scale) have been kept constant through experiments and laboratories; only the subjective test environment was varied. Some subjective tests were conducted in controlled environments and others in public environments (cafeteria, terrace or corridor). Audiovisual stimuli covered a wide range of quality since audiovisual sequences have been coded to achieve three coding qualities: high, medium and low. In particular, the H.264/AVC video codec and advanced audio coding (AAC) at 6 and 3 rate levels, respectively, were used for coding.

The results show that these subjective audiovisual tests were highly reproducible from one laboratory and from one environment to another. The number of subjects was the most important factor. Based on this experience, 24 or more

Table 3. Summary of VQEG MM2 database description.

Encoding scenarios				
Content of the reference scenes	N° Dataset	Lab	Language used	Notes on the environment
	1	NTIA	English	Laboratory: acoustic insulation chamber, gray walls, adjustable lighting
	2	Intel	English	Public cafeteria with a variable number of people who speak. Indirect solar radiation and fluorescent lighting
	3	Intel	English	Laboratory: semi-anechoic chamber, gray walls
	4	IRCCyN	French	Laboratory: furniture and gray walls, background illumination (of the screen)
	5	IRCCyN	French	Public cafeteria with a variable number of people who speak. Indirect sun
	6	Technicolor	French	Laboratory: black walls, 20 lux background lighting, very quiet
	7	Technicolor	French	Patio with a variable number of people speaking. Dazzling sunlight
	8	AGH	Polish	Laboratory
	9	AGH	Polish	Corridor with some people passing by. Indirect sun
	10	Opticom	German	Well-lit home office. Quiet environment
Encoding	Windows Media Video (WMV)			
	Format compression ratio ranging from 5 Mb/s to 15 Mb/s			
	Codec audio AAC codec video H.264 AVC			
Distortions	Bit rates for video (kbps)			Bit rates for audio (kbps)
	100, 192, 250, 448, 500, 1000			8, 32, 64
Total	60 sequences			
Test methodology				
Subjects	**Lab 1**:28 paid temporary workers – **Lab2**: 9 –**lab3**: 34			
	Lab4 + lab5: 25 (Half of these subjects evaluated dataset4, then 5 while the other half evaluated dataset 5, then 4			
	Lab6 + lab7: 24 unpaid volunteers –**lab8**: 14			
	Lab9: 15 paid temporary workers –**lab10**: 15 unpaid volunteers			
Method	ACR (Absolute Category Rating)			
Subjective evaluations	Quality scale		Discrete at 9 levels: (1: bad 2:/3:poor 4:/5:fair 6:/7: good 8:/9: excellent)	
	Final score		MOS	

subjects are recommended for absolute category testing (ACR). In the public sector, 35 subjects were required to obtain the same sensitivity to the subjects test. The second most important variable was individual differences between subjects. Other environmental factors had minimal impact, such as language, country, lighting, background noise, wall color and monitor calibration. Analyzes indicate that mean opinion scores (MOS) are relative rather than absolute.

3.4 Institut National de la Recherche Scientifique Dataset (INRS)

In 2016, Demirbilek created a subjective quality database named INRS Audiovisual Dataset [3]. It consists of 160 unique configurations for audiovisual content. The INRS audiovisual quality dataset was designed to cover the main factors influencing network compression and distortion, these parameter range values are selected to simulate when using real-time communications. These factors

Table 4. Summary of INRS database description.

Encoding scenarios			
Content of reference	Scene		Duration
	ntia_HeadShouldersFemale15		42 s
Encoding	Audiovisual containers .3gp	Codec Video H.264 AVC	
	Resolution: 720p	Codec audio AMR-WB	
Distortions	Type of distortion	Video	Audio
	Frame rate	0,15,20,25	Monochrome channel, 16 kHz sampling frequency, 24 kbps bit rate
	Quantization parameter	23,27,31,35	
	Noise reduction	0, 999	
	Packet loss rate (%)	0,1-0,5-1-5	0,1-0,5-1-5
Total	160 sequences		
Test methodology			
Subjects	30 (are from the higher level INRS who mastered The evaluation of multimedia quality, from various backgrounds and fluent in English)		Age: 20–37 years
Method	ACR (Absolute Category Rating)		
Subjective evaluations	Quality scale	Discrete at 5 levels: 1: bad,2: poor, 3: fair, 4: good, 5: excellent	
	Final score	MOS	
	Scale of scores	1..5	

are typically the video frame rate, quantization, filters, and network packet loss rate. Following collaboration with the industrial partner Summit-Tech Multimedia Communications Inc., they chose the range of network parameters for H.264 video coding as follows: (0, 0.1, 0.5, 1 and 5%) for the packet loss rate of the video and audio streams, (10, 15, 20 and 25 fps) for the video frame rate, (23, 27, 31 and 35) for the quantization parameter, and (0 and 999) for the noise reduction filter [3]. The dataset description is summarized in Table 4.

4 Comparison Between the Audiovisual Quality Datasets

A comparison between the audiovisual quality datasets that we described in the previous section is established in Table 5. As can be seen, the data are different from one database to another in terms of duration, resolution and audio encoding algorithm. Different levels of bitrates (BR) and frame rates (FR) have been experienced on the reference data of the four datasets while the packet error rate (PER) has only applied to reference content of the INRS and PLYM databases. This latter is distinguished by the fact that both audio and video quality were rated separately by observers as well as the overall audio-visual quality, leading to two supplementary raw vectors of subjective scores, namely MOS-audio and MOS-video. This is very important for research works that aim at designing objective metrics relying on the modality fusion approach.

We can also note that some of the quality databases have some limitations in terms of real-time communication use cases, mainly because they were generated with a different scope in mind. The PLYM and INRS datasets are of interest for

Table 5. Comparison between the audiovisual quality datasets

Dataset		PLYM	TUM	VQEG	INRS
Characteristics	Reference scenes	6	5	5	1
	Duration (s)	6	7–14	10	42
	Resolution	144p	1080p	480p	720p
	Codec video	H.263	H.264	H.264	H.264
	Codec audio	G.711	PCM	AAC	AMR-WB
	Motion	Low	Low-high	Low	Low-high
Distortions	Bit Rate	1x	4x	3x	4x
	Frame Rate	8,15	50	30	10–30
	Packet Error Rate	0,01–0,2	/	/	0–5
Total sequences		60	20	60	160
Subjective assessment		MOS-audio MOS-Video MOS-Audiovisual	MOS-Audiovisual	MOS-Audiovisual	MOS-Audiovisual

real-time communications since they have explored coding configurations and network degradations often observed in contemporary videophone applications. However in PLYM, the video frame rates and the selected video resolution no longer reflect contemporary applications.

For the other datasets TUM and VQEG, they are of interest for the various devices and display environments for audiovisual sequences. However, the TUM 1080p50 dataset aimed to compare the visual quality perceived on different devices (from a reference display to a home cinema configuration) while in the VQEG dataset, the audiovisual quality was highly reproducible from one laboratory/environment to another.

5 Conclusion

As we have explained in the beginning of this paper, there are two approaches to resolve the problem of multimedia evaluation: the objective and the subjective measurement models. In this paper, we have been interested in the subjective assessment methodologies. We have provided a comprehensive analysis and comparison of four public audiovisual quality databases. We have presented the standard framework recommended by of the International Telecommunications Union, the test methodologies as well as the types of distortions the multimedia can undergo. We have also given a description of each dataset with the intention to facilitate their use for the researchers' community. Summarizing tables have also been set to accomplish clear and fair comparison between the audiovisual quality databases.

It is known that subjective ratings are usually characterized by large variations between subjects probably due to multiple factors that affect human decision making, such as general conditions (viewing/hearing distance, digital media player), vision/hearing ability, subject's mood and background experience. Indeed, evaluating subjective multimedia quality is a complex decision process that is suggested to evolve with experience. Hence, a visual content that has been evaluated to be of fair quality some years ago can become judged to be of poor or even very poor quality in a few years by the same observers' panel. In the same way, a visual data that has been very well appreciated can become of medium quality and so forth. Though this phenomenon raises the question about the evolution of the use of the quality databases in the future, subjective ratings still provide the ground-truth data for the objective models.

References

1. Akhtar, Z., Falk, T.H.: Audio-visual multimedia quality assessment: a comprehensive survey. IEEE Access **5**, 21090–21117 (2017)
2. BT, R.I.R.: Methodology for the subjective assessment of the quality of television pictures. Int. Telecommun. Union (2002)
3. Demirbilek, E.: Modèles d'apprentissage automatique d'estimation de qualité perçue dans les communications en temps réel/Machine learning based perceived quality estimation models in realtime communications. Ph.D. thesis, Université du Québec, Institut national de la recherche scientifique (2017)
4. Fliegel, K.: QUALINET multimedia databases v5. 0. Czech Technical University in Prague, Faculty of Electrical Engineering, Multimedia Technology Group (2014)
5. Goudarzi, M., Sun, L., Ifeachor, E.: Audiovisual quality estimation for video calls in wireless applications. In: 2010 IEEE Global Telecommunications Conference GLOBECOM 2010, pp. 1–5. IEEE (2010)
6. Keimel, C., Redl, A., Diepold, K.: The TUM high definition video datasets. In: 2012 Fourth International Workshop on Quality of Multimedia Experience, pp. 97–102. IEEE (2012)
7. Pinson, M.H., et al.: The influence of subjects and environment on audiovisual subjective tests: an international study. IEEE J. Sel. Top. Sign. Proces. **6**(6), 640–651 (2012)
8. Rec, I.: P. 920: Interactive test methods for audiovisual communications. Int. Telecommun. Union (1996)
9. Rec, I.: P. 1401, methods, metrics and procedures for statistical evaluation, qualification and comparison of objective quality prediction models. International Telecommunication Union, Geneva, Switzerland (2012)
10. Recommendation, I.: P. 910, subjective video quality assessment methods for multimedia applications. Technical report, International Telecommunication Union (2008)
11. Recommendation, I.: P. 913, methods for the subjective assessment of video quality, audio quality and audiovisual quality of internet video and distribution quality television in any environment. Int. Telecommun. Union (2016)
12. Saidi, I.: Analyse et modélisation de la qualité perçue des applications de visiophonie. Ph.D. thesis, INSA de Rennes (2018). https://tel.archives-ouvertes.fr/tel-01977199

Experimental Investigation into the Influence of the Distance Between Microphones for 2D Real-Time Sound Source Localization Using GCC-PHAT Technique

Belgacem Douaer[1(✉)], Farid Ykhlef[1], and Fayçal Ykhlef[2]

[1] Signal and Image Processing Laboratory, University of Blida 1, Blida, Algeria
douaer.belkacem@gmail.com, ykhlefarid@gmail.com
[2] Division Architecture des Systèmes et Multimédia, Centre de Développement des Technologies Avancées, Algiers, Algeria
fykhlef@cdta.dz

Abstract. In this paper, we conduct an experimental investigation into the effect of the distance between two microphones for real-time sound localization using the generalized cross-correlation phase transform technique (GCC-PHAT). We have used the microphone array incorporated in the Microsoft Kinect device to acquire sound waves in real-time. The microphone array is composed of four audio sensors (microphones) which have been used in pairs to mount the experimental setup for helicopter sound localization. We have considered four pairs of microphones in our experiment. For each single pair, the distance between the couple of microphones, which composes that pair, was defined beforehand. Later, the angle of the source has been varied. The impact of the distance between the sound source and the pair of audio microphones has been investigated by varying gradually the distance between the source and the microphones. Experimental results show that the localization performance is improved as we increase the distance between the pair of microphones provided that the distance between the couple of microphones of an array is less than the distance between the source and the acquisition microphones.

Keyword: GCC-PHAT · Time delay · Sound source localization

1 Introduction

The goal of sound source localization system is to estimate the exact position of sound sources [1]. This task is achieved by processing the set of acoustic waveforms captured by a microphone array [2]. The technology of source localization has been widely used in several applications; one can mention: automatic camera steering in a conference hall, audio surveillance, traffic control, hearing aid devices and radars. In most of these applications, it is essential to accurately estimate the sound-source location in real-time.

One of the most important systems for acoustic source localization is the Boomerang [3]. It allows to precisely detect and locate the position of a shooter in indoor and

M. R. Senouci et al. (Eds.): CSA 2020, LNNS 199, pp. 354–362, 2021.
https://doi.org/10.1007/978-3-030-69418-0_32

outdoor environments. Extensive research has been carried out to propose sophisticated source localization techniques. Most of the proposed schemes are relying on time-delay and explore the Time Difference of Arrival (TDOA) between two spatially separated microphones to locate the exact position of the sound source. The localization of the sound must be conducted using relatively short segments to achieve high processing rates. Unfortunately, short segments are severely affected by reverberation and background noises which reduce the performance of the pairwise Time-Delay Estimation (TDE) based techniques.

The generalized cross-correlation (GCC) is the most common method used to estimate the Angle of Arrival (AOA) of sound sources due to its accuracy and moderate computational complexity [3]. However, the incorporation of weighting functions is necessary to improve the performance of GCC based techniques in noisy and reverberant environments [4, 5]. Several weighting functions have been proposed in the literature, one can mention: Smoothed Coherence Transform (SCOT) [6], ROTH processor [7] and Phase Transform (PHAT) [8, 9]. Bestagini et al. [5] have mentioned that the PHAT is the most appropriate weighting function in noisy conditions.

In this paper, we have conducted an experimental investigation into the influence of the distance between two microphones for real-time sound source localization using the GCC-PHAT technique. We have used the microphone array incorporated in the Microsoft Kinect device to acquire sound waves in real-time [10, 11]. The microphone array is composed of four microphones that have been used in pairs to mount the experimental setup for the localization of the helicopter sound. We have considered four pairs of microphones in our experiment (P_1, P_2, P_3 and P_4). The angle of the sound source has been estimated in real-time using the GCC-PHAT technique. To implement such schematic, we have disassembled the Kinect device to modify the distance H_{P_i} between the couple of microphones (A and B) that composes a single pair (P_i) along a straight line. For each pair of microphones (P_i), the distance H_{P_i} between the couple of microphones was defined beforehand. The global configuration is given as follows: $H_{P1} = 0.3$ m, $H_{P2} = 0.5$ m, $H_{P3} = 0.7$ m, and HP4 = 1.5 m. The angle of the source sound has been varied from 10° to 80°. The impact of the distance (Radius) between the rotating sound source and microphone-pairs has been investigated by varying the distance between the source and the microphones. Three scenarios have been established: (a) Radius = 1 m, 2 m and 3 m.

It is worth noticing that the experimental setup we have mounted is noncomplex and allows us to conduct a single experiment to estimate the angle of the source sound in real-time using four different configurations of microphone pairs. One configuration corresponds to a specific distance between the couple of microphones which form a single pair. Therefore, the results regarding the influence of the distance between microphones for real-time sound localization are obtained exactly for the same acoustic conditions (level of background noise, reverberation and the impulsive response of the room). Laboratory demonstrations in sound processing is a very important step towards implementing real-time localization systems [12, 13].

The remaining of the paper is organized as follows: A brief description of the GCC technique is described in Sect. 2. The experimental setup we have mounted is described in Sect. 3. The main results and discussions are provided in Sect. 4. Finally, Sect. 5 concludes our investigation.

2 GCC-PHAT

The GCC-PHAT method is composed of two-steps: (i) the estimation of the TDOA between a pair of signals, and (ii) the prediction of the incidence angle of the sound source. A pair of microphones is used to localize the sound source. Let's suppose a signal $s(n)$ that propagates in noisy environment (2D space). The signals $x_1(n)$ and $x_2(n)$ which are acquired by the aforementioned pair of microphones can be represented as follows:

$$x_1(n) = s(n) + n_1(n) \tag{1}$$

$$x_2(n) = s(n-\tau) + n_2(n) \tag{2}$$

Where $n_1(n)$ and $n_2(n)$ are additive white Gaussian noises.

The GCC of $x_1(n)$ and $x_2(n)$ is the cross-correlation of filtered versions of these signals. $w_1(\omega)$ and $w_2(\omega)$ represent respectively the Fourier Transforms of the filters which are convoluted with $x_1(n)$ and $x_2(n)$.

The weighting function $\psi_{12}(\omega)$ which encompasses the spectral representation of these filters is defined as follows:

$$\psi_{12}(\omega) = w_1(\omega).w_2^*(\omega) \tag{3}$$

Finally, the generalized cross correlation $R_{x_1x_2}(\tau)$ can be expressed in the following form:

$$R_{x_1x_2}(\tau) = \frac{1}{2\pi}\int_{-\infty}^{+\infty} \psi_{12}(\omega)X_1(\omega)X_2^*(\omega)e^{j\omega\tau}d\omega \tag{4}$$

Where $X_1(\omega)$ and $X_2(\omega)$ are respectively the Fourier Transforms of $x_1(n)$ and $x_2(n)$. [.]* represents the conjugate transpose operator. ω is the angular frequency.

Weighting functions are applied to improve the accuracy of the time delay estimation in noisy conditions.

Phase Transform is one of the most important weighting functions. Its frequency response can be expressed as follows:

$$\psi_{PHAT}(\omega) = \frac{1}{|X_1(\omega)X_2^*(\omega)|} \tag{5}$$

If the phase transform is incorporated in the GCC function (see Fig. 1), we can substitute $\psi_{12}(\omega)$ by $\psi_{PHAT}(\omega)$ in Eq. 4.

The TDOA between $x_1(n)$ and $x_2(n)$, which denotes the time delay that maximizes Eq. (4), is given as follows:

$$\tau_{x_1,x_2} = \text{Arg max}(Rx_1, x_2(\tau)) \tag{6}$$

$\tau_{x_1x_2}$ is the time instant which corresponds to the highest peak of $R_{x_1x_2}(\tau)$.

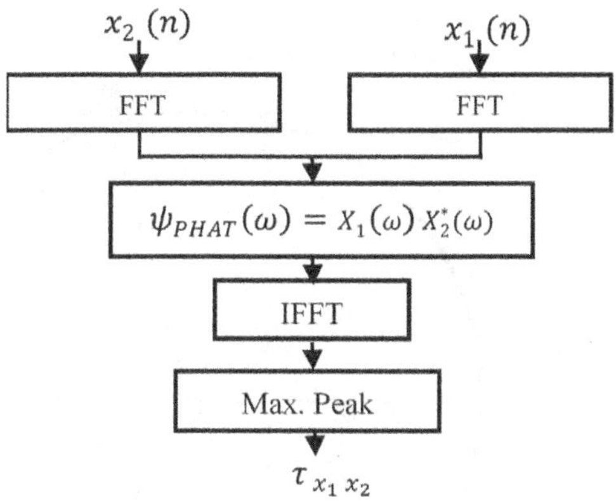

Fig. 1. Block diagram of GCC-PHAT.

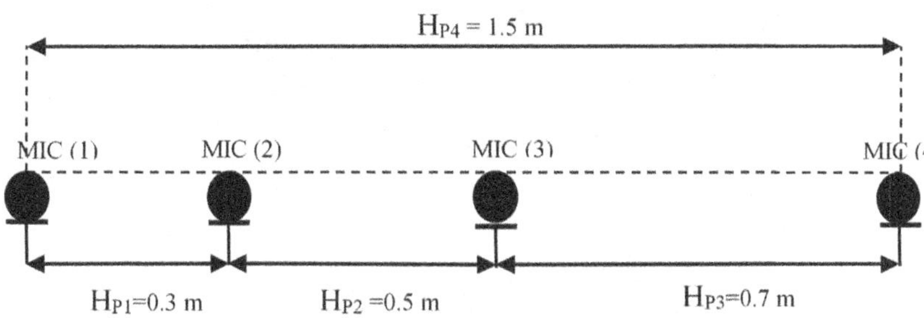

Fig. 2. Positioning of microphones.

3 Experimental Setup

The Microsoft XBOX 360 Kinect Sensor has been identified as an effective multimedia big data signal receiver in numerous practical applications [14]. It includes the following type of sensors: (i) depth, (ii) camera, and (iii) microphone array. The microphone array is composed of four small-scale linear microphones. The Kinect microphone array was originally designed for beamforming [14, 15]. In our experiment, we have disassembled the Kinect device so that we could modify the distances between the four microphones along a straight line (see Fig. 2).

In the first step of our investigation, we have considered four different pairs of microphones. For each pair (Pi), the distance H_{Pi} between microphones-pair is given as follows:

P_1: {MIC (1), MIC (2)} → $H_{p1} = 0.3$ m.
P_2: {MIC (2), MIC (3)} → $H_{p2} = 0.5$ m.
P_3: {MIC (3), MIC (4)} → $H_{p3} = 0.7$ m.
P_4: {MIC (1), MIC (4)} → $H_{p4} = 1.5$ m.

Let's consider a pair of microphones which is composed of microphones (A) and (B) (see Fig. 3). Microphone A is placed at (L, 0) coordinates while microphone B is placed at (−L, 0). The distance Hp (or H_{Pi} if we are considering the first pair) between

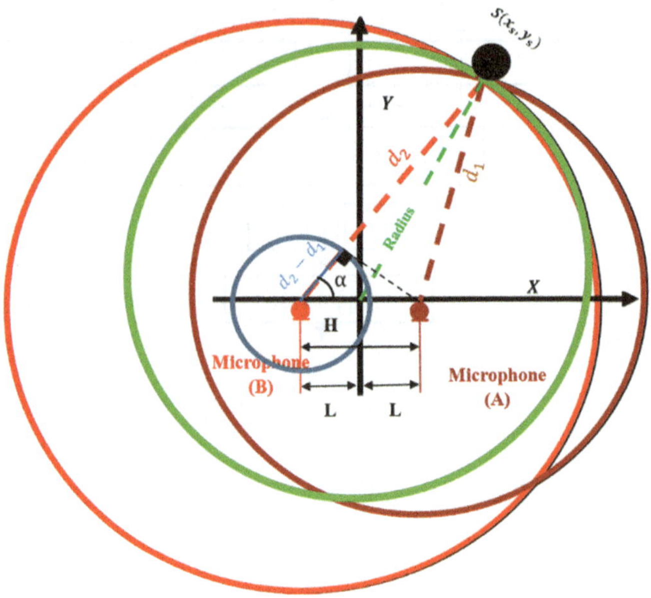

Fig. 3. Sound source acquisition using two microphones.

both microphones is related to L as follows:

$$L = \frac{Hp}{2} \tag{7}$$

The x-y coordinates of microphones A and B in 2D plan are respectively (L,0) and (−L,0). If we suppose that the sound wave is emitted at (x_s,y_s) coordinates, then, based on the general form of circle equation, the following expressions can be established:

$$\begin{cases} d_1 = \sqrt{(x_s - L)^2 + {y_s}^2} \\ d_2 = \sqrt{(x_s + L)^2 + {y_s}^2} \end{cases} \tag{8}$$

d_1 is the distance between the sound source and microphone A whereas d_2 is the distance between the sound source and microphone B.

Based on motion equations, we can connect the displacement to the time delay using the following form:

$$\Delta d = c^*\tau \tag{9}$$

Where $\Delta d = d_2 - d_1$ is the difference of the distance between microphone A and microphone B, *c is* the sound propagation velocity in air, and $\tau = \tau_{x_1 x_2}$ is the TDOA between $x_1(n)$ and $x_2(n)$. This implies that

$$\tau_{x_1 x_2} = \tau_{\text{micAmicB}} = \frac{\sqrt{(x_s + L)^2 + {y_s}^2} - \sqrt{(x_s - L)^2 + {y_s}^2}}{C} \tag{10}$$

$\cos(\alpha)$ can be estimated using the following equation:

$$\cos(\alpha) = \frac{d_2 - d_1}{|H|} \tag{11}$$

Using Eq. (9), the value of α can be expressed as follows:

$$\alpha = \cos^{-1}\left(\frac{c * \tau_{x_1 x_2}}{|Hpi|}\right) \tag{12}$$

We have used the relative error (RE) to assess the performance of sound localization:

$$RE = \frac{|\alpha_m - \alpha_R|}{\alpha_R} \tag{13}$$

α_R is the real angle of the sound source,
α_m is the angle measured (estimated) using GCC-PHAT,

The Microsoft Kinect device is connected to a laptop via USB port. Sound signals are acquired in real-time and transmitted to MATLAB software. We have used in our experiment the sound of Helicopter as an audio source [16]. It was played using a smartphone. We have implemented the GCC-PHAT technique using MATLAB software. The following parameters have been initialized:

- Sound duration = 10 min.
- Sample quantization = 16 bits.
- Sampling rate (Fs) = 16 kHz.
- Frame length = 25 ms (for short term processing).

Information about our laboratory (room) are given as follows:

- Level of noise: ~ 40 dB.
- Dimension: 4 × 4 × 2.8 m^3
- Objects: a small bed, a table, a chair, and a wardrobe.

The angle α of the sound source has been varied from 10° to 80°. We have investigated the impact of the distance between the moving sound source and microphone-pairs by changing the radius of the source (as depicted in Fig. 3). Three different scenarios have been considered: (a) scenario 1: radius = 1 m, (b) scenario 2: radius = 2 m, (c) scenario 3: radius = 3 m. For each scenario, the angle has been varied as described above. The main results are reported in the following section.

4 Results and Discussion

The relative error, as a function of the sound source angle which rotates in clockwise opposite direction of a circle, is depicted in Fig. 4. Three different values of the circle radius, which represent the distance between the sound source and the microphone array, have been experimented: radius = 1 m, radius = 2 m, and radius = 3 m. Regardless

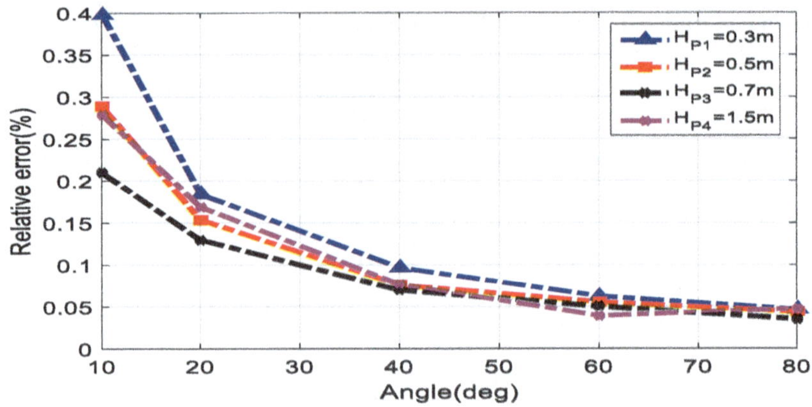

(a)

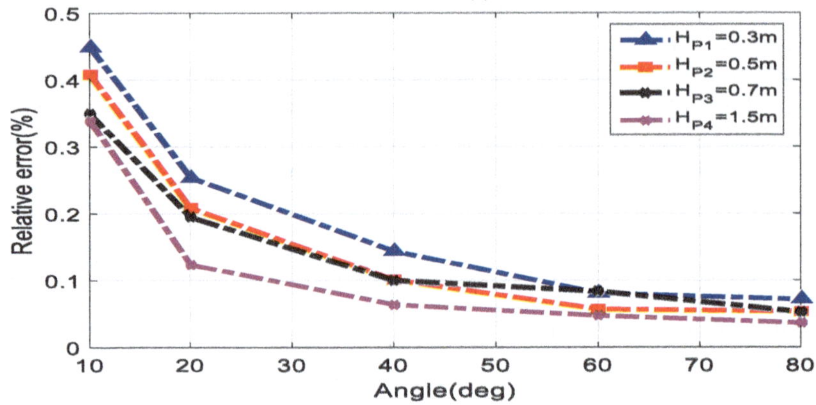

(b)

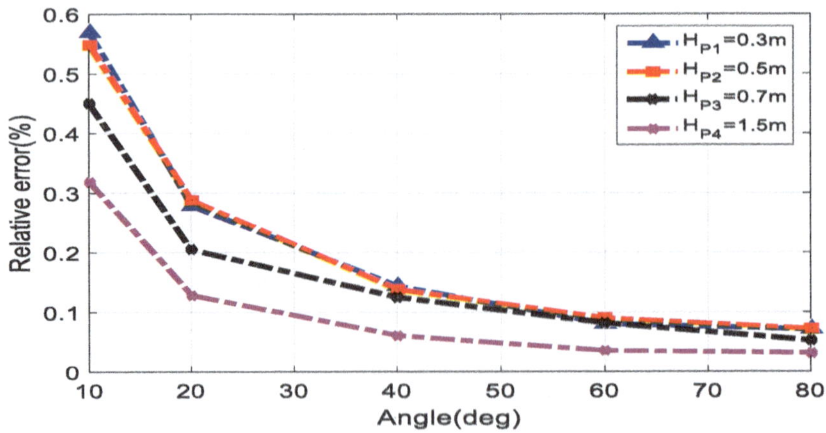

(c)

Fig. 4. Relative error of the rotating sound source localization technique: (a) Radius = 1 m, (b) Radius = 2 m, and (c) Radius = 3 m.

of the distance between microphones A and B, the RE is decreasing as the angle of the source sound increases.

However, the precise variation of the RE depends on the distance between the sound source and the microphone array.

Table 1. RE range of the sound source localization.

Hpi (m)	Radius(m)		
	1	2	3
0.3	3.38	5.15	5.25
0.5	3.25	4.45	5.1
0.7	2.6	4.8	4.6
1.5	3.15	3.45	3.2

We have studied the effect of the distance (Hpi) between these two microphones on the accuracy of localizing a rotating source, which is distant from the pair of microphones (microphone array). The RE range is reported on Table 1. Radius corresponds to the distance between the rotating sound and the microphone array (see Fig. 3).

Based on our experiments, we have noticed that for a sound source which is located at a far distance from the microphone array (Radius = 3 m), the error range is decreased as we increase the distance between microphones A and B (Table 1, column 4). This result is observed for all the values of α as shown in Fig. 4(c).

However, when the sound source is closer to the microphone array (Radius = 1 m) (Fig. 4(a)), the raise of the distance between microphones does not have a big impact on localization performance (Table1, column 1). This impact is becoming quite clear when the source is placed at a distance of 2 m (Fig. 4(b) and Table 1, column 2).

5 Conclusion

We carried out an experimental investigation into the effect of the distance between microphones to localize a sound source direction in real-time using GCC-PHAT technique. We have used the Microsoft Kinect device to mount the experimental setup.

The angle α of a rotating sound has been varied from 10° to 80°. Three different values of the circle radius have been experimented: radius = 1 m, radius = 2 m, and radius = 3 m. Based on the relative error, the preliminary results are reported in this study. The localization performance using GCC-PHAT technique is improved as we increase the distance between the pair of microphones provided that the distance between the couple of microphones of an array is less than the distance between the sound source and the acquisition microphones. In order to obtain more representative conclusions about the order of magnitude of these distances, further experiments in noisy and reverberant conditions need to be conducted.

References

1. Strumillo, P.: Advances In Sound Localization. In: IntechOpen (2011)
2. Chen, L., et al.: Analysis on the TOA tracking with DVB-T signals for positioning. IEEE Trans. Broadcas. **62**(4), 957–961 (2016)
3. Lamelas, F., Swaminathan, S.: Laboratory demonstration of acoustic source localization in two dimensions. Am. J. Phys. **87**(1), 24–27 (2019)
4. Park, J.H., Seung, H.C.: Interaural Time Difference Estimation Using Generalized Cross-correlation with Maximum Likelihood Weighting in Reverberant Environments. Int. J. Multimed. Ubiquitous Eng. **9**(4), 43–50 (2014)
5. Bestagini, P., et al.: TDOA-based acoustic source localization in the space–range reference frame. Multidimension. Syst. Sign. Proces. **25**(2), 337–359 (2014)
6. May, T., van de Par, S., Kohlrausch, A.: Binaural localization and detection of speakers in complex acoustic scenes. In: The technology of binaural listening. Springer, Heidelberg, 397–425 (2013). https://doi.org/10.1007/978-3-642-37762-4_15
7. Park, C.-S., Jeon, J.-H., Kim, Y.-H.: Localization of a sound source in a noisy environment by hyperbolic curves in quefrency domain. J. Sound Vib. **333**(21), 5630–5640 (2014)
8. Zhang, C., Florêncio, D., Zhang, Z.: Why does PHAT work well in lownoise, reverberative environments?. In: 2008 IEEE International Conference on Acoustics, Speech and Signal Processing. IEEE (2008)
9. Kwon, B., Park, Y., Park, Y.-S.: Analysis of the GCC-PHAT technique for multiple sources. In: ICCAS 2010. IEEE (2010)
10. Jia, M., Sun, J., Bao, C.: Real-time multiple sound source localization and counting using a soundfield microphone. J. Ambient Intell. Humanized Comput. **8**(6), 829–844 (2017)
11. Webb, J., Ashley, J.: Beginning Kinect Programming with the Microsoft Kinect SDK. Apress, New York (2012)
12. Douaer, B., Ykhlef, F., Ykhlef, F.: Tracking a sound source in two dimensions using microsoft kinect sensors. In: 2019 International Conference on Advanced Electrical Engineering (ICAEE). IEEE (2019)
13. Pourmohammad, A., Ahadi, S.M.: N-dimensional N-microphone sound source localization. EURASIP J. Audio Speech Music Proces. **2013**(1), 27 (2013). https://doi.org/10.1186/1687-4722-2013-27
14. Jiang, W., et al.: Two-stage Localisation Scheme Using a Small-scale Linear Microphone Array for Indoor Environments. J. Navig. **68**(5), 915–936 (2015)
15. Chiariotti, P., et al.: Average acoustic beamforming in car cabins: an automatic system for acoustic mapping over 3D surfaces. Appl. Acoust. **129**, 47–63 (2018)
16. Sound source: www.soundsnap.com. Accessed 21 Dec 2019

Author Index

M. R. Senouci et al. (Eds.): CSA 2020, LNNS 199, pp. 363–364, 2021.
https://doi.org/10.1007/978-3-030-69418-0

The manufacturer's authorised representative in the EU is Springer Nature Customer Service Centre GmbH, Europaplatz 3, 69115 Heidelberg, Germany. If you have any concerns regarding our products, please contact ProductSafety@springernature.com

Printed and bound by CPI Group (UK) Ltd, Croydon, CR0 4YY
30/09/2026
02225292-0006